Understanding Mathematics
Through Problem Solving

Problem Solving in Mathematics and Beyond

Print ISSN: 2591-7234
Online ISSN: 2591-7242

Series Editor: Dr. Alfred S. Posamentier
Distinguished Lecturer
New York City College of Technology - City University of New York

There are countless applications that would be considered problem solving in mathematics and beyond. One could even argue that most of mathematics in one way or another involves solving problems. However, this series is intended to be of interest to the general audience with the sole purpose of demonstrating the power and beauty of mathematics through clever problem-solving experiences.

Each of the books will be aimed at the general audience, which implies that the writing level will be such that it will not engulfed in technical language — rather the language will be simple everyday language so that the focus can remain on the content and not be distracted by unnecessarily sophiscated language. Again, the primary purpose of this series is to approach the topic of mathematics problem-solving in a most appealing and attractive way in order to win more of the general public to appreciate his most important subject rather than to fear it. At the same time we expect that professionals in the scientific community will also find these books attractive, as they will provide many entertaining surprises for the unsuspecting reader.

Published

For the complete list of volumes in this series, please visit www.worldscientific.com/series/psmb

**Problem Solving in
Mathematics and Beyond** Volume **02**

Understanding Mathematics
Through Problem Solving

Alfred S. Posamentier
The City University of New York, USA

Peter Poole
Mercy College New York, USA

 **World Scientific**

NEW JERSEY · LONDON · SINGAPORE · BEIJING · SHANGHAI · HONG KONG · TAIPEI · CHENNAI · TOKYO

Published by

World Scientific Publishing Co. Pte. Ltd.
5 Toh Tuck Link, Singapore 596224
USA office: 27 Warren Street, Suite 401-402, Hackensack, NJ 07601
UK office: 57 Shelton Street, Covent Garden, London WC2H 9HE

Library of Congress Control Number: 2020932884

British Library Cataloguing-in-Publication Data
A catalogue record for this book is available from the British Library.

Problem Solving in Mathematics and Beyond — Vol. 2
UNDERSTANDING MATHEMATICS THROUGH PROBLEM SOLVING

ISBN 978-981-4663-67-0 (hardcover)
ISBN 978-981-4663-25-0 (paperback)
ISBN 978-981-4663-68-7 (ebook for institutions)
ISBN 978-981-4663-69-4 (ebook for individuals)

For any available supplementary material, please visit
https://www.worldscientific.com/worldscibooks/10.1142/9532#t=suppl

Desk Editors: V. Vishnu Mohan/Tan Rok Ting

Typeset by Stallion Press
Email: enquiries@stallionpress.com

Printed in Singapore

Preface

Typically, books with "problem solving" in the title focus primarily on a variety of problems in mathematics that are challenging and at the same time, hopefully, entertaining. In this book, we take a slight deviation from the normal, offering a deeper insight into mathematics by analyzing problem situations and then buttressing a newfound understanding through a variety of mathematical discussions that should be entertaining and supportive to the mathematics previously discussed.

This shouldn't be confused with a typical school textbook where a topic is presented and then exercises offered to ensure proper understanding of the newly presented topic. On the contrary, here, we delve into mathematical situations that are not necessarily topic oriented, but rather concept oriented, and then apply this stronger understanding of mathematical situations through some entertaining problems, along with detailed solutions.

We hope to lead the reader to a new vantage point high above their own thoughts, to help them see that the mathematical thinking they are used to does not exist in isolation. Rather, every type of problem solving belongs to a vast landscape of interconnected strategies for thinking about mathematical problems. By familiarizing the reader with some of our techniques for walking this landscape, we believe they will be able to develop mastery over their own mathematical thinking, and in time, to see how mathematics extends into their daily lives.

The book is intended to provide readers with an entertaining journey through mathematical concepts that they heretofore may not have thoroughly understood. Now, through this relaxed presentation, we hope to reveal new insights and motivate the reader to pursue topics and concepts further. Above all, the power and beauty of mathematics should come through clearly throughout the book.

About the Authors

 **Alfred S. Posamentier** is currently Distinguished Lecturer at New York City College of Technology of the City University of New York. Previously, he was the Executive Director for Internationalization and Sponsored Programs at Long Island University, New York. This was preceded by a five-year period where he was Dean of the School of Education and tenured Professor of Mathematics Education at Mercy College, New York. He is now also Professor Emeritus of Mathematics Education at The City College of the City University of New York, and former Dean of the School of Education, where he was tenured for 40 years. He is the author and co-author of more than 70 mathematics books for teachers, secondary and elementary school students, and the general readership. Dr. Posamentier is also a frequent commentator in newspapers and journals on topics relating to education. After completing his B.A. degree in mathematics at Hunter College of the City University of New York in 1964, he took a position as a teacher of mathematics at Theodore Roosevelt High School (Bronx, NY), where he focused his attention on improving the students' problem-solving skills and at the same time enriching their instruction far beyond what the traditional textbooks offered. During

his six-year tenure there, he also developed the school's first mathematics teams (both at the junior and senior level). He is still involved in working with mathematics teachers and supervisors, nationally and internationally, to help them maximize their effectiveness. During this time, he earned an M.A. degree at the City College of the City University of New York in 1966.

Immediately upon joining the faculty of the City College in 1970, he began to develop in-service courses for secondary school mathematics teachers, including such special areas as recreational mathematics and problem-solving in mathematics. As Dean of the City College School of Education for 10 years, his scope of interest in educational issues covered the full gamut. During his tenure as Dean he took the School of Education from the bottom of the New York State rankings to the top with a perfect NCATE accreditation assessment in 2009. He achieved the same success in 2014 at Mercy College, which received both NCATE and CAEP accreditation during his leadership as dean of the School of Education.

In 1973, Dr. Posamentier received his Ph.D. from Fordham University (New York) in mathematics education and has since extended his reputation in mathematics education to Europe. He has been visiting professor at several European universities in Austria, England, Germany, Czech Republic, and Poland, while at the University of Vienna he was Fulbright Professor (1990). In 1989, he was awarded as an *Honorary Fellow* at the South Bank University (London, England). In recognition of his outstanding teaching, the City College Alumni Association named him *Educator of the Year* in 1994, and in 2009. New York City had the *day*, May 1, 1994, named in his honor by the President of the New York City Council. In 1994, he was also awarded the *Grand Medal of Honor* from the Republic of Austria, and in 1999, upon approval of Parliament, the President of the Republic of Austria awarded him the title of *University Professor of Austria*. In 2003, he was awarded the title of *Ehrenbürger* (Honorary Fellow) of the Vienna University of Technology, and in 2004 he was awarded the *Austrian Cross of Honor for Arts and Science*, First Class from the President of the Republic of Austria. In 2005, he was inducted into the *Hunter*

College Alumni Hall of Fame, and in 2006 he was awarded the prestigious Townsend Harris Medal by the City College Alumni Association. He was inducted into the New York State *Mathematics Educator's Hall of Fame* in 2009, in 2010 he was awarded the coveted *Christian-Peter-Beuth Prize* in Berlin, and in 2017 he received the *Summa Cum Laude nemine discrepante* award from Fundacion Sebastian, A.C. in Mexico City.

He has taken on numerous important leadership positions in mathematics education locally. He was a member of the New York State Education Commissioner's Blue-Ribbon Panel on the Math-A Regents Exams, and the Commissioner's Mathematics Standards Committee, which redefined the mathematics standards for New York State, and he also served on the New York City schools' Chancellor's Math Advisory Panel. Dr. Posamentier is a leading commentator on educational issues and continues his long-time passion of seeking ways to make mathematics interesting to both teachers, students and the general public — as can be seen from some of his more recent books:

1. *The Joy of Geometry* (Prometheus, 2020).
2. *Math Makers: The Lives and Works of 50 Famous Mathematicians* (Prometheus, 2020).
3. *Solving Problems in our Spatial World* (World Scientific, 2019).
4. *Tools to Help Your Children Learn Math: Strategies, Curiosities, and Stories to Make Math Fun for Parents and Children* (World Scientific, 2018).
5. *The Mathematics Coach Handbook* (World Scientific, 2018).
6. *The Mathematics of Everyday Life* (Prometheus, 2018).
7. *The Joy of Mathematics: Marvels, Novelties, And Neglected Gems That Are Rarely Taught in Math Class* (Prometheus, 2017).
8. *Strategy Games to Enhance Problem-Solving Ability in Mathematics* (World Scientific, 2017).
9. *The Circle: A Mathematical Exploration Beyond the Line* (Prometheus, 2016).
10. *Problem-Solving Strategies in Mathematics: From Common Approaches to Exemplary Strategies* (World Scientific, 2015).

11. *Effective Techniques to Motivate Mathematics Instruction* (Routledge, 2016).
12. *Numbers: Their Tales, Types and Treasures* (Prometheus, 2015).
13. *Teaching Secondary Mathematics: Techniques and Enrichment Units*, 9th Edition (Pearson, 2015).
14. *Mathematical Curiosities: A Treasure Trove of Unexpected Entertainments* (Prometheus, 2014).
15. *Geometry: Its Elements and Structure* (Dover, 2014).
16. *Magnificent Mistakes in Mathematics* (Prometheus Books, 2013).
17. *100 Commonly Asked Questions in Math Class: Answers that Promote Mathematical Understanding, Grades 6–12* (Corwin, 2013).
18. *What Successful Math Teachers Do: Grades 6–12* (Corwin 2006, 2013).
19. *The Secrets of Triangles: A Mathematical Journey* (Prometheus Books, 2012).
20. *The Glorious Golden Ratio* (Prometheus Books, 2012).
21. *The Art of Motivating Students for Mathematics Instruction* (McGraw-Hill, 2011).
22. *The Pythagorean Theorem: Its Power and Glory* (Prometheus, 2010).
23. *Mathematical Amazements and Surprises: Fascinating Figures and Noteworthy Numbers* (Prometheus, 2009).
24. *Problem Solving in Mathematics: Grades 3–6: Powerful Strategies to Deepen Understanding* (Corwin, 2009).
25. *Problem-Solving Strategies for Efficient and Elegant Solutions, Grades 6–12* (Corwin, 2008).
26. *The Fabulous Fibonacci Numbers* (Prometheus Books, 2007).
27. *Progress in Mathematics K-9 Textbook Series* (Sadlier-Oxford, 2006–2009).
28. *What successful Math Teacher Do: Grades K-5* (Corwin, 2007).
29. *Exemplary Practices for Secondary Math Teachers* (ASCD, 2007).
30. *101+ Great Ideas to Introduce Key Concepts in Mathematics* (Corwin, 2006).
31. *π, A Biography of the World's Most Mysterious Number* (Prometheus Books, 2004).

32. *Math Wonders: To Inspire Teachers and Students* (ASCD, 2003).
33. *Math Charmers: Tantalizing Tidbits for the Mind* (Prometheus Books, 2003).

Peter Poole knew from a young age that he wanted to pursue mathematics. In high school, he earned the maximum score on every standardized test of math, including the ACT, SAT, and SAT II. Having run out of mathematics classes to take in the United States, he spent a year overseas, where he earned the French Baccalauréat with Speciality in Mathematics. He studied mathematics at the City College of New York, where he earned the *Belden Silver Medal* for achievement in Advanced Calculus, and he won second place in the highly prestigious City University of New York Mathematics Competition. He received special notice at that competition for the depth of his solutions and the clarity of his explanations. He has worked as a Mathematics Teacher; tutored students in mathematics, physics, chemistry, and computer science; served in the U.S. Army as a Tank Crew member from 2003–2007 (with two deployments to the Middle East); and is currently employed as a Software Engineer in New York City. He has also contributed to a number of other projects with Dr. Posamentier since 2007, most frequently by proofreading and by suggesting alternate solutions to problems.

He resides with his wife in Westchester County, NY.

Contents

Introduction

You are preparing to move to a foreign country. You know a little about the country's history and language from your school days, but you're nervous because this move will put everything you learned to test, laying bare any shortcomings of your half-remembered education. How can you prepare? What steps can you take to make the most of your time abroad?

Some authors point out that mathematics is a language, but that's only half the story. Just like the start of a move to a foreign country, the process of learning mathematics effectively demands that you absorb cultural details which may not make sense in your mother tongue. Memorizing the sentence "where is the bus stop?" might seem useful at first, until you step off the plane and realize that in this country most people get around via trains instead of buses.

Think of this book as a travel guide. It does not go in-depth on any one topic, like a book on history or the geography of a region might. Rather, it focuses on highlights, introducing the reader to topics that may merit more interest, or provide context that was missing on an earlier visit. There are plenty of books on individual topics in mathematics, but, for many readers, something else is missing: an explanation of how the residents of this country think, and why the culture is the way it is. This is our attempt at providing some of that context.

Mathematics is taught in familiar terms around the world. In many ways, this makes it different from other academic topics.

The most glaring contrast can be seen when comparing mathematics to language instruction: different countries teach reading and writing in their own languages, and even different classes in the same school may use different works of the literature to teach the rules and principles of effective communication.

Meanwhile, the language of mathematics differs very little from place to place, from a makeshift classroom under a highway overpass in a developing nation to the most expensive private schools. Separate mathematics problems from their cultural context, and a good student from one school can easily succeed anywhere in the world.

It is because of this common yardstick that countries like to use mathematics instruction as a means to compare the quality of education. In fairness, this is part of the beauty and appeal of mathematics: it is a language that transcends time, place, and culture. However, this rarified presentation of abstract mathematics does a disservice to most learners.

Effective problem solving in mathematics demands the mastery of a wide array of skills. That some students develop these skills in response to mathematics classes is too often a side-effect of instruction rather than a stated goal with predictable results: too few students develop the rich understanding of mathematics as it relates to real problems and the world outside the classroom.

The aim of this book is to bring the act of problem solving back to earth, to connect it to more grounded motivations and intuitive ways of thinking.

Who Should Read This Book?

This book assumes exposure to the most well-trodden areas of K-12 mathematics: arithmetic, algebra, and geometry. However, as we are going to deviate from typical approaches and demonstrate problem solving from angles that demand much less formal competence, it should be approachable by readers from many backgrounds.

More specifically, this book is intended for readers who know a little mathematics, but feel like they've been missing the larger picture. Whether currently in school, considering going back, looking to supplement your problem-solving skills through independent study, or wondering what you may have missed when studying mathematics in school, this book can help you.

Above all, this book is for readers with an open mind, as will become clear in the pages to come.

How to Read This Book

Reading a mathematics book demands a different kind of attention than other types of books. For some readers, this is a foreign sensation: where a page in a novel may only demand two or three minutes, reading a page of the same length from a mathematics book can rightfully take ten times as long.

This book is written in a way that the sections loosely build on one another. At times, we will make callbacks to previous chapters. However, don't be fooled: the chapters can be read independently of one another, and by no means is our chosen order the only way to learn about these topics!

Try to work out problems for yourself. This cannot be emphasized enough: when we suggest that you take some time to work on something before revealing our solution, this is your chance to strengthen your problem solving skills! Some books are like soup — you can drink (or read) as fast as you're able to. A good mathematics book is like steak: each bite requires cutting off a small piece and chewing for yourself.

Sometimes, when faced with a problem, we get a "gut feeling" that something isn't right, without understanding what that means or how to fix it. For instance, a headache can be a sign of many different issues. Some people can say when a movie is bad, or a song is out of tune, or the color on a screen is wrong, without being able to articulate the underlying reason. The act of learning is no exception.

Recognizing and harnessing these feelings is a critical step in developing useful intuition. If you get the feeling that you're missing the point of a section, here are three options. First, trying skipping ahead to see where we're heading. Mathematics problems aren't like jokes; you can't ruin the punchline. However, after seeing a solution, it's still important to work through the problem yourself. We suggest following step-by-step alongside our solution, as necessary, until you understand the principles at hand, and then repeating on your own a minute, a day, or a week later. Working through a problem mechanically strengthens pathways in your brain, building mathematical skill.

Second, try skimming material to get a general idea of what topics are being covered. Sometimes, when your brain is tired, or you are confused by a particular approach, getting a general idea of the lay of the land will help you understand the path we're taking. Stubborn readers will throw themselves at a problem over and over until they understand it. Resist this urge. Just as there is nothing wrong with putting a book down for a little while to let your brain relax, so too, there is nothing wrong with reading a chapter or a section more than once, at different levels of attention, to understand it better.

Finally, don't feel constrained by the approaches we demonstrate! The point of problem solving is, in the end, about gaining the confidence and ability to get a correct answer and know that it's correct. If a different approach works better for you, use it — and recognize that what you are doing is exactly the point of this book, *Understanding Mathematics Through Problem Solving!*

How This Book Is Organized

In Chapter 1, we will spend some time identifying in further detail who you are and the kinds of skills and knowledge that will come into play throughout your journey through this book. This chapter is necessarily text-heavy because we will be preparing your mind to make deeper connections and free you of some common mental blocks that may have built up in your brain.

Chapters 2 and 3 are about developing mental flexibility. Chapter 2 explores an assortment of mathematics problems which can be solved by experimenting with physical objects, separate from the notational complexity that scares many students away. These problems broadly fit into a theme of "party tricks," both to provide entertainments to yourself while you solve them and to encourage you to discuss them with others. In fact, discussing problems (or presenting tricks) with friends and family can stimulate kinds of learning that you can't replicate on your own! Above all, this chapter encourages you to identify relevant information and learn to question facts in ways that will become incredibly productive.

Chapter 2 ends with a set of problems that will increasingly challenge your intuition, providing a natural bridge to Chapter 3. This chapter is filled with puzzles that demand a fundamental shift in perspective. These problems are by necessity simple because this chapter is about changing how you think, examining the mental baggage you bring to problem solving, and building your instinct for how to think laterally.

Chapter 4 begins a transition to answering the question, "why do we need mathematics?" The word "important" suggests there was a need that had to be fulfilled; we will examine a number of these needs, how mathematicians approached them, and how they subsequently improved on them. We will also contrast historical mathematician's successes and failures, proving a powerful lesson: how mathematics builds on itself, like interest on a bank account. Finally, we will look at a couple of modern applications of mathematics that have powered innovation in our daily lives.

Chapter 5 will give us another chance to look back at history while demonstrating the application of a powerful principle. Whereas in Chapters 2 and 3, we challenge your intuition and demonstrate how to extract meaningful information from a mathematical context. Chapter 5 focuses on how to fluently translate that information into forms that make problem solving easier. In so doing, we will witness the application of a powerful principle: mathematical truths are true no matter how you look at them!

Chapter 6 will take you back to your school days. After a long lead-up, here, you will begin to see what you probably think of as "normal mathematics." While each of the chapters in this book will have some crossover,[1] this chapter will bring together the ideas of the previous chapters to help you more fully understand and apply the lessons you learned in school to everyday life.

---

[1] For instance, there are certain principles we can't easily discuss without the use of algebra.

Chapter 1

Problem Solving in Mathematics

Didn't We Learn That in School?

We all have experiences with classroom mathematics. Think back to your most recent lesson, whether it was a day or a decade ago. What do you remember?

Let's envision a normal day in a normal US high school. It is a late fall day, and the leaves are changing colors in the crisp autumn air. Inside a low brick building, there is a rectangular cinderblock room painted beige, with about 30 desks oriented toward the front of the classroom. The walls here are covered with blackboards (or whiteboards), and slightly in front of them sits a projector on top of a table. As the day begins, students file in and find their assigned seats. They remove backpacks, taking out their tools of the day: folders filled with notes and assignments, blank paper, pens, calculators and laptop computers.

A harried teacher rushes in just before the bell rings. She pulls out a massive teacher's edition of the textbook and, flipping through pages that are marked with dog-eared corners, flags, highlights and underlined passages, she finds a relevant page. "Good morning, class," she opens, prompting the room full of teens for attention, "first of all, are there any questions on the homework?"

Papers shuffle as the students pull out homework, signifying a shift into the classroom mindset. A particularly well-prepared student pulls out a model assignment: three sheets, neatly stapled through the corner, written in clean, orderly lines, the answers highlighted via colored boxes. Other students extricate crumpled messes, finished (barely!) during the bus ride to school.

One student raises her hand. "Yes?" the teacher prompts. "I had a question about problem 32." Students and teacher alike flip pages in their books, locating the relevant item. "Ah, yes," the teacher begins, "that was a tricky one. Does anyone think they got an answer for that?" Another student raises her hand. "Yes? Could you write it on the board?"

So begins the class day. For the next 40 minutes, both students and the teacher write strings of symbols on the board, ask questions and scribble down answers, and work their way through a portion of a section in a chapter of a mathematics book. Finally, the teacher assigns another set of homework problems, reminds the students of the date of the next quiz or test, and the students pack up in preparation for their next class.

Does this vignette sound familiar to you? The particulars of your schooling may differ, especially if your school was outside the United States. Some classrooms emphasize a strict hierarchy: the students listen and the instructor speaks. Other classrooms feature desks arranged in clusters, emphasizing the role of student discussions in learning. Regardless of the precise configuration and methodology of the classroom, there are certain similarities which transcend time and place.

If you were to open a mathematics book from 100 years ago, or from another country today, what do you think you would find? It may surprise you to hear that despite differences in language, the mathematics would be recognizable: here a formula, there a diagram of a triangle, or a table of values. Depending on your personal feelings as a reader, this realization might be comforting or alienating: in a broad sense, the topics covered in mathematics classes are the same everywhere. You might then suspect that what happens in mathematics classrooms wouldn't much differ from place to place, and indeed with minor variations, this is often true.

Upon reading the previous story you, as a former student, may have remembered how you felt at that time. If you didn't, pause for a moment and consider: what did you feel? Were you ever anxious, bored, or confused? On the other hand, perhaps you remembered feeling confident and engaged. Each reader's reactions are a unique product of their own experiences, but we can make certain assertions about mathematics education in general.

Almost every student feels "math anxiety" at some point in their mathematical education. For many, these feelings are strong enough to prevent them from ever deliberately studying mathematics after high school. It is only fair, though, that students should feel that this is a viable strategy: when mathematics is presented as little more than abstract formulas and problems separated from any part of reality, and these problems are challenging enough that the student's grades (and sanity!) suffer for it, a rational response might be to stop the suffering.

Imagine this situation: pretend this is a school where chess is a mandatory subject which all students must study for 12 years. Students are told that chess is incredibly important, and that some of the most successful people in the world are chess masters. Meanwhile, a great many of these students struggle to understand, much less win, any games they play; and when speaking to adults they often hear comments like "Chess? Oh, I was never very good in school; I stopped playing once I got to college. Never really did see the point."

With apologies to anyone in our audience who enjoys chess, you can understand how "chess anxiety" might be something students — and even adults — would wrestle with in this world! Yet, if we tolerate (or at least, are unsurprised by) this kind of reaction to mathematics, it points to the existence of a gap: students' experiences are not connecting with the goals, or importance, of mathematical instruction.

It is that gap that we wish to bridge with this book.

What is a "normal" math book?

Without diving into all possible rationales for existing books, we can make some generalizations about typical mathematics books. It is by far easier to write on a narrow than on a broad topic, making certain assumptions about the reader: that they have some external motivation for learning (whether to pass a course, or intrinsic curiosity), and that they, in turn, have pressure pushing them (or desire pulling them) to learn the material, however it is explained.

As we mentioned above, the language of mathematics is almost universal. At this point, half the authors' work is done: they need not re-establish prerequisite material, such as basic algebra; furthermore, the language that surrounds mathematics ("As we can see… thus, for every *X, Y*… therefore *Z*…") tends to follow a predictable cadence. The remainder of the challenge at that point is to identify problems and exercises which, between them, sufficiently cover the book's topic, and connect these problems via prose. The writers' goal, whether it be to entertain or educate, is enabled by the assumption that the reader will be able to take this mathematical language as it is presented and internalize it, furthering their understanding. Problems can then be arranged in some logical order, while the overarching picture is left somewhat implicit, for readers to sort out on their own.

There are positives and negatives to this approach. A reader who understands the flow of mathematical language need not be bogged down by extraneous explanations, nor will the writer feel encumbered by the need to write such lengthy texts. Indeed, for readers with a certain kind of mathematical maturity, extra concise "outlines" books, heavy on exercises and light on prose, are all that's needed to gain mechanical proficiency in whole new fields.

We do not mean to trivialize the act of writing, nor disparage those authors who take that approach! Our purpose here is to highlight a deep truth: *Everything is a skill.* Writing concise mathematics books requires very specific aptitudes, as does reading these books. Playing chess well is a skill; reading deeply is a skill. Moreover, each genre of text — fictional stories, newspapers, philosophical essays, or mathematical proofs, to name a few — demands a different kind of skill, which are only superficially similar. There are general fluencies which apply across fields, but too often books will take these fluencies as unstated or understated assumptions, deserving no more than a couple of lines in an oft-skipped introductory chapter.

There are other explanations for gaps between writers and audiences. The minimal qualification for writing a book — any book, not just on mathematics — is understanding (or appearing to understand) the chosen topic. But levels of understanding lie in a spectrum. The level of understanding necessary to apply knowledge is less than the level necessary to write about it at all; the level of understanding necessary to teach a topic is less than the level needed to teach it *well.* Indeed, teaching (or explaining) a topic in an approachable manner is often a skill that is orthogonal (that is, unrelated) to the topic itself. Again, everything is a skill, even explaining complex things!

Finally, some truly brilliant people never quite realize how different their thinking and understanding are from that of the everyday person. The challenges they faced, and the methods they developed to overcome them, may be so obvious after decades of experience that they struggle to remember what life was like when they weren't so brilliant, or teach in a way that helps students become as skilled as they are. Many college students have been driven to complain about their professors: "Well, he may be world-renowned, but he's not very good at teaching."

There is a place in the world for many kinds of books for readers of all ability levels. This book strives for a specific place among them.

What is this book?

Bridging the gap

This book seeks to enumerate and make explicit a portion of the skills and meta-skills essential to understanding mathematics. The remainder of this chapter will be devoted to unpacking this sentence and describing what it means to you, the reader.

"Meta" as a prefix means applying something to itself. At the end of the previous section, we wrote about the process of writing; that could be considered meta-writing. Thinking about your own thought processes is called *meta-cognition*, and is the broadest and most widely applicable meta-skill we will consider. But there are many facets of meta-cognition, and because it is itself such a broad topic, we cannot exhaustively list every kind there is. What is most important is that you come away with a greater awareness of your internal thoughts, and that you are more and more able to take a step back from your emotions, frustrations, and intellectual dead-ends to consider how you might better approach problem solving.

What do we mean by "skills," compared to "meta-skills?" An example of a mathematical skill might be algebraically rationalizing denominators,[1] and is a topic that can be explicitly measured by an exam. In contrast, some meta-skills we can easily identify are as follows:

- How do you deal with a term you don't understand?
- When you find the definition, how do you make it "real?"
- How do you identify what is, and is not, part of the definition you've read?
- What approach can you take to surmount a tricky obstacle when the direct route isn't working?

This is not an exhaustive list, again, but rather a few examples meant to illustrate a principle. That principle is, whether you have realized it or not before now, there are skills you have used related to *the process* of mathematical problem solving that are not part of problem solving *per se*.

The first version of this chapter started by trying to identify the reason you are reading this book. Here you are (self-evidently) having read several pages into the first chapter; you must have had a motivation to pick it up in the first place! What will keep you reading? What motivation can we provide, beyond the motivation that is getting you to the end of this sentence?

If you desire to learn the mathematics necessary to become a better carpenter or accountant, a book that focuses on specific types of real-world problems may suffice.

If, rather, you wish to gain a greater understanding of what you're doing when you're solving problems, and you wish to see mathematics everywhere in the world, this is the book you should continue reading. This book is about the *how* and the *what* of mathematical

---

[1] Don't worry if you do not precisely recall what those words mean.

problem solving: *how* do you approach a novel problem, *how* do you synthesize information into a whole that makes the answer lucid; *what* skills can you apply, *what* types of problems can succumb to a mathematical approach?

"How to solve mathematical problems" and "what problems to solve" are comparatively easy questions to address. "Why," as in "why solve mathematical problems" is and will always be a much more difficult question to answer. On the one hand, asking that question too deeply can result in an existential crisis, leaving you bewildered in bed, staring at the ceiling, wondering "why do we do anything at all?"

On the other hand, the sketch of an answer presents itself by reading between the lines of the previous paragraphs.

Why do we care about mathematical problem solving?

The simple answer to this question risks being unsatisfying. We solve mathematical problems because we wish to know an answer; we are faced with some situation (involving numbers!) where mathematics can show us a way forward. These types of problems are often concrete: we wish to budget money, or to build a bridge.

A deeper answer will require some elaboration. The skills we develop and apply in effective mathematical problem solving are, quite possibly, the most universally applicable (and useful) skills in human experience.

"What a lofty claim!" you might be saying to yourself, and rightfully so. Consider this, though: a well-crafted mathematical exercise (or problem)[2] demands that you know nothing more than what is presented in the text. There are well-established rules for manipulating symbols and expressions to produce an explicit result. Mathematics skills build on each other, like compound interest in a bank account: the more you learn (principal), the more you can learn (interest). If mathematical problem solving is black and white, then the real world is gray: to solve a problem in business, or history, or philosophy, requires many of the skills we develop while learning mathematics,

---

[2] We will address the contrast between exercises and problems later in this chapter.

but also entail much more ambiguity. The simplicity and triviality[3] of school-level mathematics exercises provide learners with an environment where they can develop rich personal skills.

Indeed, the most versatile benefit of a mathematics education are the skills that learners develop on their own to surmount the obstacles they face. Have you ever found yourself reading a book, only to realize several paragraphs (or pages) in that you can't recall a single thing you read? While you might be able to check that reading off your to-do list without having absorbed anything, a mathematics assignment is never so forgiving. If your assignment on a certain day is to solve 10 mathematics problems, your grade depends on you paying attention to each and every one: when you get stuck on a problem, there is no way to coast through; it is your duty to overcome each challenge.

How someone learns these skills, and indeed whether they learn them at all, can have a significant impact on the quality of their education.

Overcoming obstacles: A case study

You're working on a mathematics assignment one evening, and you find yourself faced with a problem you can't solve. What do you do? Whether you're a current or former student, or a parent whose children need help completing their homework, the strategies are similar.

First of all, keep in mind that the exercises in a given section of the book are almost always based on examples presented earlier in that section. By pattern-matching to an existing example, you can often fill in the blanks, completing some exercises almost by rote. There is a lot of value to this approach: you can sustain your motivation by successfully finding an answer, even when you don't understand every step. But beware! The part of your brain that is thinking about your thinking should recognize that you don't yet fully understand the lesson.

---

[3] "Simplicity and triviality" refers in this case to a characteristic we will explore later on: namely, the mathematical problems students face in school are not important *per se*; rather, these problems provide a basis upon which further learning can be built.

In many books, certain exercises have solutions at the back of the book.[4] You should not shy away from using these answers, but if your ultimate goal is deep knowledge, you should be aware that you'll need to take extra steps to internalize the chapter's lessons. If your assignment is to solve the odd exercises and you get stuck, one approach might be to look at the answer to see if you're on the right track, solve the exercise, and then step away from the assignment for a few minutes or hours. When you come back, attempt the problem again. If you can solve it fresh, without "cheating," you'll be sure you've learned something!

Often, a given even-numbered exercise is a conceptual partner to the preceding odd-numbered exercise. So if your assignment consists of solving even-numbered exercises, and you get stuck on one of them, one approach you can take is to examine the previous odd-numbered exercise and look for similarities. Since you can see the solution to the latter exercise in the back of the book, you'll have the opportunity to try an approach you can understand in a more challenging situation. While you might not get confirmation that your answer is correct until the assignments come back graded, you can have some confidence in your approach.

While solving additional, unassigned exercises might sound like extra work, remember that we mentioned mathematics builds on itself like compound interest. If you understand a concept deeply today, then that understanding will save time tomorrow. Further, addressing frustration by tackling a simpler problem can save time directly: solving a problem correctly will prove that you're on the right path, and give you a sense of progress. It is more productive to solve a problem that wasn't asked for in order to make progress than to stare at a difficult problem without comprehension for 15 minutes!

Consider all these approaches and extend them. What in this scenario applies to other parts of life? For instance, in the first strategy, we suggested solving a problem by comparing it to a known example. Have you ever done this in a situation that didn't involve mathematics? Did you realize that this was a skill you had developed?

---

[4] In the United States, these are customarily the odd-numbered problems.

There are countless examples of this: when learning a language, you might wish to construct a sentence whose exact translation isn't in your dictionary. By swapping out a single noun or verb, though, you can say what you want. Or perhaps you wanted to make a meal but weren't able to find the perfect recipe. If you can find one that's close (perhaps using beef instead of chicken or tofu) and substituting your chosen ingredients, you can have a fantastic dish!

The previous paragraphs illustrated, in a long-winded way, the expansion of a collection of meta-skills that mathematics can help you develop as a learner. Indeed, there are countless meta-skills that you will develop throughout your encounters with mathematics; as you can no doubt tell we could fill endless pages with these kinds of explanations and revelations and barely scratch the surface. You can understand, then, why many books don't spend time on this!

Like a tour guide, we will often call your attention to these kinds of meta-skills as they become relevant. The most important one of all of these, though, is one we have made reference to again and again thus far in this chapter: the inner voice of meta-cognition. This inner voice is that part of your brain that watches your thinking, helping you take a step back to see how the current task fits into the larger context, identifies when the direct approach is no longer productive, and helps you apply alternate strategies in order to make progress.

While we won't bother spelling out every possible skill you might develop during your study of mathematics, there is one more topic that merits particular focus: how we can study topics that may seem trivial without becoming disillusioned.

Learning without worrying about learning

In fiction, suspension of disbelief refers to the willingness to immerse yourself in a fantastic world for the sake of enjoyment. For example, when watching a movie with laser guns and spaceships that can travel instantly to distant stars, believing in the logic of that world for two hours (instead of scoffing at everything on screen as illogical) can be very entertaining.

What does this have to do with mathematical problem solving? When learning the fundamentals of mathematics, as in any subject, students are not handed the great problems of our time. Rather, they are tasked with solving "easy" exercises to help them become comfortable with the fundamentals of the field.

Some students tackle these assignments with gusto, never looking beyond the day's lesson, motivated by great teachers, or good grades on daily assignments, weekly quizzes, and monthly tests. Other students are plagued by doubt. They ask questions: "What is the point of all this stuff we're learning? *When will we ever use this?*" Most students fall between these two extremes, their motivation highly dependent on the current situation.

If the goal is a mere two hours' entertainment, maintaining suspension of disbelief is easy. Across an educational career than can last 12 years or more, students might never consciously realize that they're learning the basics that will allow them to begin solving important problems. Their suspension of belief falls apart as doubts creep in.

Although this book can be read in a matter of days, you may find that some topic seems irrelevant to the overarching theme. If this ever happens, there are two things you can keep in mind: first, we may be building to a particular payoff. In this case, see if you can suspend your disbelief, and treat the problems we address as important, regardless of whether you can see the bigger picture at the moment. Second, we cannot possibly explain every possible connection between the problems we will present and the universe where those solutions may be useful. In these cases, try to suspend your disbelief enough to understand the problem as presented, and then take some time to consider where the skills you're learning may be useful. You may be surprised what you can discover.

Finally, if you are still in school or considering returning to pursue another degree, remember: every lesson you are taught was designed by experts in the field. While you may spend most of your time immersed so deeply in the current topic that it's hard to see the bigger picture, rest assured that there is always a connection to be made to real problems. In the meantime, if you can master the lessons one at a time, your mind will change in profound ways.

Suspension of disbelief is a useful tool for learning from day-to-day or chapter-by-chapter. However, it is not adequate on its own to gain you mastery of mathematics. For that, we will need something more powerful.

Systematic understanding

Mathematics is in many ways the art of abstraction. Children are not born knowing how to count, yet with time and effort every child learns that the numbers 1, 2, 3, ... can be used to count anything, from people to pieces of candy.[5] Counting is only the first step humans take in their mathematical career, but it is by far the most common and unquestioned mathematical abstraction we know.

The idea of converting objects to numbers points us toward surprisingly deep insight: concrete things can be turned into abstract concepts, which can be manipulated in useful ways. In time, each of us learned that there is no difference between adding apples to apples or cows to cows; two plus two always equals four. Climbing the ladder of abstraction further, we learn about subtraction (and negative numbers, how odd!), multiplication, and division; fractions and decimals; algebra, etc. and it's at this point that students begin to stumble. The great villain of algebra is that monster, x, which represents *things*. It's not always clear what x is, because in a typical curriculum algebra is the students' first exposure to the truly abstract: a black box, which is to be moved around, solved for, charted, and graphed — but never to be opened.

At this stage, students often begin trying to memorize their way around the subject. Formulas — such as the quadratic formula — are seen as a piece of solid ground in a world that is constantly shifting. They say to themselves (whether or not they realize it) "I can write this down; I can memorize this; when someone asks me on a test how to use the quadratic formula, I can put numbers in and get numbers out!"

---

[5] Pause for a moment to marvel at this: children struggle to learn to count, yet you (who were once a child) have no difficulty with this at all. Perhaps in time, you'll feel the same way about mathematics in general!

For students simply trying to keep their heads above water, this may seem to be a prudent course of action; however, the failure of this approach is that it leads to fragile (or non-existent) understanding.

Human memory is built on connections. A forgotten memory may spring to mind after the merest whiff of a scent from childhood. A scent, or scene, or souvenir from a time long past triggers journeys down mental pathways heavy with cobwebs.

Similarly, mathematical memory benefits from cross-linkages and entryways. An entryway can consist of a mnemonic,[6] the visual memory of a page, or the recollection of what the teacher looked like during a particular lesson. These approaches can be astonishingly effective, however, unpredictable.

A different way of "memorizing" mathematics is to learn the base principles firmly enough that you can, at a year's or a decade's remove from learning, reconstruct the knowledge from principles. In order to begin being able to perform such feats, though, you must learn to be as strong in the fundamentals of algebra as you currently are in counting and arithmetic.[7] Then, when asked to learn, remember, and apply some named formula (for instance, the quadratic formula), you might be able to do so 5 or 10 years in the future!

A major goal of this book is to introduce you to topics in a way that removes much of the discomfort of symbolic representation, instead presenting abstract concepts in concrete ways. Only later will we build toward the kind of symbolic manipulations you might now call "real mathematics," of the kind you saw in school. But have no fear: we will

---

[6] There are entire books devoted to memory techniques, none of which we will be addressing in the main body of this text. One such technique consists of associating the positions of cards with certain phrases and images, which makes it possible for a well-trained magician to memorize the order of cards in an entire deck almost instantly!

[7] It's a common joke among students of higher mathematics (such as advanced mathematics at the undergraduate or graduate level) that the more mathematics they learn, the more mathematics they forget. We will for the moment ignore the fact that some otherwise brilliant mathematicians squirm when asked to calculate the tip on a restaurant bill.

guide you, and regularly point backwards to where we came from, to offer you the opportunity to make connections in your own mind.

What this book is not

There is no silver bullet when it comes to learning mathematics. As we mentioned in the introduction, there is no substitute for work and thinking for yourself! This book will not think for you, nor can it put knowledge directly into your brain. When we ask a question or suggest that you put the book aside to try a problem yourself: do so!

We will be your guides, but it is your duty to think, and to try new techniques, and to learn what works for you and what doesn't. We can only provide suggestions and hints; we cannot make you understand.

We said earlier that everything is a skill. Consider the beginning of your journey through this book as an opportunity to practice the skill of reading mathematics! One approach would be to find a comfortable seat in front of a table in a room with good lighting, set aside a stack of paper and a pencil or two, shut off any distractions, and work through a single problem step-by-step.

Finally, if you are reading this book merely for entertainment, by all means skip around and skim the parts that don't capture your interest — think of this approach as akin to wandering through a store, seeing what kinds of things are for sale — but we hope that you will return to read more in depth! We encourage multiple readings, as necessary, and for you to read as slowly as you need to in order to understand the problems we present.

Why do we write?

We have spoken about what this book means to address, and how, but one open question is why have we written this book? There is some overlap here with the earlier question of why people learn mathematics at all; clearly, we wish to share those virtues with you.

Separate from those reasons, mathematics is an intrinsically fascinating subject. Understanding mathematics opens whole new worlds of entertainments and delights; even if it weren't so often useful, mathematics can be incredibly fun and satisfying. There are branches of mathematics for which no clear application exists, but simply exploring problems and rules and asking again and again, "what if," is quite possibly the most challenging game there is. And that is what mathematics, separate from pressing questions, amounts to: a game that can be played alone or in groups, where more of the game opens before you the more you learn.

Each moment of enlightenment brings a satisfaction that is incomparable to anything else in life. Who can fault us for wanting to share this with you, dear reader?

Finally, even if you are never convinced that mathematics is fun by itself, learning to think mathematically about practical matters will make you a better, more engaged citizen. The techniques we will help you develop will allow you to ask critical questions about statements made by politicians, journalists, and scientists, and begin to discern truth from fiction.

Our approach

We have mentioned that we intend for this book to be different from other mathematics books. One facet of this difference is breaking you out of any bad habits you may have developed in the past.

You personally may not have any particular mathematical bad habit; however, each example or counterexample we provide has been chosen with an archetypical student in mind. Due to the breadth of our target audience, we will necessarily fail to make some examples relevant to all readers some of the time. Please be patient; while you might not see the point of a particular approach, aside, or analogy, for some readers, those portions of this book will be a breath of fresh air.

There is no one way to do mathematics. Again, if you find an approach that works better for you, use it! But do take time to try and understand the approach we present; there may be some previously concealed trick that you can add to your mathematical toolbox, or a subtle insight that reveals a before-unknown facet of the target problem.

You may have the impression from your school days that mathematics is a rigid and stodgy subject. On the contrary, though: practical mathematical problem solving is all about creating, testing, disposing of, and refining new tools and notations to open up new pathways, simplify complicated ideas, and produce correct results. The reason that algebra is universal is that it makes it easier for people in different times and places to discuss ideas without needing to negotiate notation. While progressing through this book, we give you leave to invent whatever tricks and notations you need to claim ownership of each new concept. You will often discover that your inventions have already been discovered by someone else. Take heart, then, because it's much more difficult to learn a new idea than a new syntax; by learning, by *understanding,* new ideas by any means necessary, you'll be in a much better position to learn the existing syntax than the other way around.

Problem, Solving, in Mathematics...?

In this section, we will dissect the title of this book to give you a clear idea of where we are headed. There are two purposes to this. The first is straightforward: we want to set down the ground rules for what you'll be reading in the chapters to come. The second is more abstract. By defining terms and analyzing our assumptions, we have an opportunity to see a key skill in formal mathematics, without yet bringing mathematics *per se* into our discussion.

Exercises, puzzles, and problems

The use and creation of terms

> *"When I use a word," Humpty Dumpty said, in rather a scornful tone, "it means just what I choose it to mean — neither more nor less."*
> —Lewis Carroll
>
> *Through the Looking Glass, and What Alice Found There, (1871)*

Earlier, we promised that we would explain the difference between exercises and problems. Some writers will claim there is no difference between these two terms — or three terms, if you include puzzles, as we have in the section header. That is their prerogative.

We are choosing to identify three disparate concepts, and attach existing words which more or less fit those clusters of ideas, regardless of other writers' or teachers' use of those words. At this point, it's worth calling your attention back to the point earlier in the chapter, where we mentioned that while you are experimenting in mathematics, you are free to use ideas however they are most useful to you, as long as they are useful.

In defense of this approach, consider natural languages. Without much effort, you should be able to find a word in English (or any language you're comfortable speaking) that is related to a concept, yet surprisingly distant from it. For instance, a folder (in English) is a rigid paperboard container for holding loose paper. The name comes from the property of folding in half, and is a sufficiently well-established term that native speakers rarely think about the etymology of the word while making reference to it.

Meanwhile, in French, the equivalent word is *chemise*, which is the same word used for a shirt. French speakers decided that the principal property that should identify a rigid container for paper is that it *covers* its contents, like a shirt. In a classroom setting, no one confuses the two; the etymology of the word is forgotten as students open their *chemises* to pull out their homework for the day.

There are countless other examples like this. *Stand*, in English, means either to be upright (*stand up*), or to set up a final defense against a threat (*make a last stand*), a stationary piece of furniture that can hold objects (*microphone stand*), or a small temporary building where a vendor works (*ice cream stand*). As English speakers, we manage not to confuse these terms in their proper contexts. Nevertheless, there is a clear connection between the root word and all the other connotations.

There is precedent, then, for reusing an existing word in a new context, especially when there is some similarity between the two.

We will (and you should) feel at liberty to use this same process to put names to concepts whenever necessary.

Unlike Humpty Dumpty, however, we will always use the one word to mean the same thing!

What, then, are we choosing to call "exercises," "puzzles," and "problems?"

Exercises

Mathematical exercises are similar to physical exercises. When you wish to strengthen your abdominal muscles, you can perform sit-ups. Every exercise must be performed until a certain point to cause a training effect, and must be repeated to build results. Doing one sit-up per day will not make you strong, and doing a thousand at a time carries a high risk of injury. The optimal workout is somewhere in between: schedule 100 sit-ups in groups of 20 to begin with, and increase the number and difficulty day by day until you achieve your goals.

Mathematical exercises are self-contained examples demanding the application of a single concept (or a group of related concepts) that can be performed repeatedly over time to strengthen a single mental pathway. To learn the rules of exponentiation, you can solve a dozen or a hundred exercises involving those rules until the answers come easily to you. At some point, further exertion will bring smaller and smaller rewards.

Note that, just as with physical exercises, when performing mathematical exercises, we need not concern ourselves with the big picture. A professional athlete doesn't need to think about the big game on Saturday to perform at his best in the gym; similarly, we don't need to be thinking about the role of mathematics in the universe to benefit from doing a dozen mathematics exercises.

If students spent all their time doing exercises, and never got to see "the big game," it would be no wonder when they would complain that mathematics is tedious and pointless. Fortunately, there is more to mathematics — much more.

Puzzles

Puzzles are self-contained, like exercises, and like exercises promise a well-defined solution. However, the similarities stop here. Puzzles tend to demand much more time and focus than exercises, and there is often more than one solution, or no solution at all — which, in a way, is also a solution.

There are two defining characteristics of puzzles, according to the definition that we are using. First, puzzles demand nonlinear thinking: the application of some principle not obviously connected to the problem at hand, or the application of an obvious principle in a profound way. Second, puzzles typically (and intentionally) incorporate context which can confuse or mislead.

To summarize, a mathematical puzzle is a type of problem that is self-contained, and has at least one intended answer, but whose solution demands nonlinear thinking or the clever interpretation of facts.

Compare an ideal exercise to a puzzle. In the former case, the exercise will be presented in a chapter or section where an astute reader can infer what approach he is to take. Flipping backward in the text should reveal some example or approach that can be adapted to the current task. In the latter case, even when presented in a specific chapter where examples might abound, a puzzle can taunt us by substituting the familiar with the obtuse.

Successfully solving a puzzle can demand interpreting information, bringing in cultural or other outside knowledge, or recognizing the pattern of a familiar approach hidden in an unfamiliar form. Puzzles train what we have termed meta-skills, most notably the traits of persistence and experimentation. The best puzzles invite us to attempt them over and over again, before finally yielding to the application of a simple principle that, in a different form, would barely be worthy as a single exercise in the middle of a homework assignment.

Puzzles act as a bridge between practice and application. As we will see, recognizing, filtering, and interpreting information are critical skills in successful problem solving, and solving puzzles gives us the chance to build up to these skills.

Also, note that our definition excludes some common types of puzzles. One example would be Sudoku. While completing a Sudoku game might fit the definition of a puzzle in certain cases, a reader experienced with the game will neither be surprised by the content nor the presentation of a given board. Indeed, with practice, solving a Sudoku game becomes akin to a pleasant exercise, demanding no great energy or focus from the reader.

Problems

In our taxonomy, and in this book, we will spend the most time on *problems*. Mathematical problems are the broadest possible type of mathematics, and where all the skills we have learned come together. Whereas puzzles are often crafted to be deliberately confusing, mathematical problems need not be intrinsically confusing; a problem should be no more complicated than the world itself.

The goal of problem solving is, not to put it too mildly, solving problems. Problems are frequently related to a real-world concern: constructing a budget, maximizing returns on investments, or designing a building. Where a problem is different than the previous two categories is how success is defined. At the end of this chapter, we will discuss some caveats around the topic of what makes a solution "good," but for now, we will simply point out that a single problem can easily have multiple contradictory yet correct solutions.

There is a lot of hand-waving happening here, so let's examine a specific example. You're tasked with putting together a meal. Given no other constraints or specifications (time, money, materials, etc.), what would you do?

One solution would be to order food from a restaurant. It would satisfy the lone constraint (offering a meal), but might be more expensive than other alternatives.

Another solution would be to assume that your audience will prefer a variety of tasty foods, and copy a meal plan from a famous chef, executing it exactly.

A third solution would be to look closely at the definition of "meal," and realize that serving boiled potatoes, however boring, constitutes a meal.

All three responses solve the problem, yet each one brings different values and assumptions into play to produce a distinct result. Further, note that in the second instance, we used the word "copy." This is no mistake: pragmatic problem solving often prioritizes finding a result over understanding every step of the process. There are major downsides to this approach, but be aware that there are times that it is nevertheless the correct approach.

While the subject of this book is *Understanding Mathematics Through Problem Solving*, we will present a variety of exercises and puzzles as a gentle introduction to our topic.

Solving problems

Now that we have a working concept of what makes something a problem, how can we define a "solution?"

Some solutions are simple. A mathematics problem can have any solution, such as the number 4.

What can you do with this information? Not much; we haven't provided so much as a problem to go with this number, much less an explanation of what the solution means or why the solution is correct. Other times, though, a number is all we need.

Some solutions are just numbers

In practical problem solving, a number can be enough. Think about the example of assembling a meal from the previous section. If your goal is to make a dish according to an exact recipe, accomplishing that task successfully doesn't require you to understand the calculations or principles of cooking that went into producing the recipe. Sometimes simply executing prescribed steps is sufficient to achieve a good result.

Similarly, the chef who created the recipe may not have been thinking of the precise rules of cooking that went into producing that dish. Her job demands that she make delicious, edible food, no matter the manner which she arrives at that product. That process may involve modifying another chef's work, or combining two smaller components into a new creation; whatever her process, the sole value necessary to "prove" her work is whether or not it tastes good.

Some students get stuck trying to apply the "right" steps when problem solving. They may not yet have gained the courage to produce solutions by any means necessary, demonstrating to themselves (and only themselves, for a time!) that they're on the right track when needed, and only cleaning up at the end once they are satisfied that the "dish tastes good."

Solving problems in either of these two roles (as a cook following a recipe, or a chef creating food from scratch), and in many other areas of life, does not require knowing or even remembering every principle underlying the solution. But when you think of our chef, ask yourself: do you believe that she can make elaborate dishes without knowing the basics of cooking? Do you think she can serve dozens or hundreds of dishes per night safely without having ingrained the concepts of food-borne illness?

Similarly to a dish served without a recipe, a mathematical answer provided without any context might satisfy your immediate needs, while leaving you hungry later. You will be able to solve a narrow class of mathematical problems by memorizing particular recipes, but sooner or later you may desire to solve a type of problem you've never seen before — to make a mathematical "dish" from scratch — at which time your deep understanding of mathematics will pay dividends.

The characteristics of good solutions

Good mathematical solutions have several properties. First, every solution should be correct no matter where or when you look at it. The arguments presented in Euclid's *Elements*, first compiled in the 3rd Century B.C.E., form the basis for modern geometry. Although they have been translated and refined over time, the essence of Euclid's arguments have passed from one reader to another for thousands of years. The longevity of his geometry is a testament to the quality of the ideas contained therein. Entire scientific theories have come and gone in the centuries since Euclid.

Second, a good mathematical solution should be understandable to a reader on its own. Earlier, we suggested that you do not let notation or terminology stand in the way of your *internal* understanding of mathematics. While the language of mathematics[8] has a direct relationship to the mathematics itself, your understanding of mathematics can be wholly separate from how you write it down!

As a demonstration of this idea, Chapters 2 and 3 emphasize achieving mathematical understanding through written words. As we progress through the subsequent chapters, we will increase the use of more traditional mathematical notation until you are comfortable using that notation to support and enrich your reasoning. Reading mathematics is an essential part of learning new mathematical concepts, after all, but what is most important is how you understand something!

This leads into the third property of good solutions. A good solution should shed insight on the problem. There are caveats, of course: the prerequisite knowledge necessary to understand some problems is so difficult that understanding the solution must be even more difficult! Shedding insight, in these cases, is a relative term.

Sometimes problems are easy to understand, but the best solutions we know of are incredibly complicated. One of these is the four-color theorem. It states that any map divided into countries that satisfy certain properties (for instance, every country must be contiguous) can be colored with no more than four colors such that no two neighboring countries (sharing a common border) are colored the same.

It's a simple problem to understand, and one that can be explored by young children. (Simply print out some maps of the United States, Europe, or Africa, hand them four crayons, explain the rules, and come back a half hour later.) It is also a theorem rather than a conjecture, and thus we know it to be true. The most well-known proof of the four-color theorem, from 1976, required that the authors[9] catalog all possible maps which could possibly be the smallest possible counterexample to the proof.

After exhaustive analysis, both hand-written and by computer, the authors concluded there were 1,936 possible counterexamples (that

---

[8] That is, numbers, symbols, and formulas — what we call variously "school mathematics" or "textbook mathematics."

[9] Kenneth Appel and Wolfgang Haken.

is, maps that *might* require five colors), that they had correctly enumerated all of them, and that none of these did in fact require more than four colors. But many people did not find their proof satisfying!

Surely, if needed, another mathematician could use their result with confidence, but the hundreds of pages in printed material produced in arriving at their conclusion did not shed any insight on the reason the four-color theorem was true.

Good solutions show us something novel about a problem. Seeing that novelty for the first time is like discovering a new word for a phenomenon you always knew existed, but never knew how to express.[10] Each new insight offers a higher-order abstraction than what came before, a way to sweep up many small ideas into something larger that can be used as a tool, shaped to fit new situations, and most importantly: remembered.

This, then, suggests our goal and yours throughout this book: to find the kernel of each problem, to isolate it and present it in a way that can be remembered for a long time. By learning mathematics in this manner, you will gain a new understanding of not only mathematics, but also how it relates to real-world problems.

And who knows? Maybe with your new-found view of the world, you will be able to identify a problem that no one else had seen before, one that you can solve. Above all, the elegance, or cleverness, of a solution is perhaps one of the most rewarding aspects of problem solving. It exhibits the power and beauty of this subject!

Mathematical problems (and solutions!)

In all this discussion of mathematical problems and all this high-minded talk about what mathematics can achieve, you might start to believe that it can do anything.

---

[10] There are countless examples of these — *saudade*, a Portuguese word meaning a deep nostalgia for something loved; or *l'esprit de l'escalier*, a French term for the feeling one gets when they can't think of a clever joke until after the moment has passed — but in English perhaps the most familiar example is the German *schadenfreude*, or "taking pleasure in others' misfortune."

Mathematical problems are unique among all problems in that they are constrained to only what information we allow through the door. One theme that runs throughout this book is that, when problem solving, it is our duty to interpret facts and separate relevant and irrelevant information in order to find a useful solution.

Exercises and puzzles are more forgiving in this regard. Neither of these classes of problems will demand that you bring in truly new information in order to arrive at a solution (although again, they demand some familiarity with existing concepts).

An applied problem — that is, one whose solution we wish to apply to some real-world situation — might be mathematically correct, while the answer it provides is misleading. There are many possible reasons for this state of affairs. For instance, we might forget to account for some important factor, or our initial values might be incorrect. Either way, the end result is the same: the mathematics is correct, but does not model the world correctly!

Introductory physics students become intimately familiar with this situation. First-year physics textbooks leave out inconvenient facts, like the existence of friction or air resistance, to make their core lessons more clear. So while those students might study the trajectories of ideal cannons firing ideal cannonballs, woe to any military that tries to use those solutions to win a war!

If you think this is a silly example, you should be aware that in order to compute a ballistic trajectory[11] accurately, some of the factors that must be taken into account are: the flight characteristics of the projectile, air density, the angle and cant of the cannon, wind speed (both perpendicular and parallel to the cannon), the amount of propellant, the speed and direction of the target, and even (depending on how accurate you wish to be) the rotation of the earth!

It is possible to derive formulas that account for some or all of these factors, each of which can be solved using correct mathematics, that will differ in their results.

---

[11] That is, a trajectory that is determined solely by launch angle and speed, and not corrected via engines or fins after launch.

"Of course that's true," you might be saying. Consider this, though: the results of mathematical equations can be used to drive the goals of political or economic policy, but where do the values these equations point to come from? Mathematics, *per se*, carries no moral or ethical values; thus any moral conclusions derived from mathematical formulas must come from somewhere else. If a cannon ball falls far from its mark, we say "oh, they must have forgotten to take wind speed into account." Similarly, if some country applies the results of a model to its economy, and it leads to unfavorable outcomes, it's very possible that the mathematics are correct, but the model itself, or the inputs chosen, are wrong.

Problem solving is only as powerful as our ability to translate the world, in all its infinite complexity, into the language of mathematics. While mathematical problem solving can often get us close enough to the target to be useful, the universe will always be wilder than our attempts to measure it.

Understanding Mathematics Through Problem Solving!

We have spent this chapter engaged in what we hope is a useful dialogue about what it means to problem solve. It's a lot to take in, though!

As a reminder, here are the highlights:

- Understanding mathematics on a deep level demands a different approach than mere memorization
- Understanding comes from generating (correct) mental models
- There is no "right way" to understand — use whatever techniques or tricks make sense to you!
- To explain yourself to others and understand what others have written, you will eventually have to learn the language of mathematics
- No matter what approach you take, practice is essential to understanding
- By understanding mathematical problem-solving, we can discover applications in surprising places!

Like all advice, we fully expect that most of you won't see what we mean until you look back on your journey. But, from time to time, we hope that you think about what you read in this chapter and say "oh, now I see…"

Now, take some time to stretch your legs and sharpen your pencils and get ready to be entertained, because we're going to dive in to Chapter 2 with some entertaining problems!

Chapter 2

Entertaining Mathematical Problems

"Pick A Card, Any Card..."

It's a classic opening in any magic act: the magician, looking to engage his audience, pulls a volunteer out of the crowd and asks him or her to select a card, memorize it, and replace it in the deck. The volunteer follows along with our magician as he shuffles, deals, flips, and counts until, with a flourish, the magician delights the audience by producing the correct card.

There are countless possible variations on this trick, no two quite the same. Some require feats of memory, or digital dexterity; in keeping with the theme of this book, we will now present a trick that requires no more than a minor feat of math.

(At this point we entreat you to find a deck of playing cards or, should that prove difficult, any 21 pieces of paper, such as index cards. If you're using index cards, simply number them 1 through 21. As tempting as it may be to try to read straight through the text without performing each problem, we ask that you resist that temptation.)

Start by counting 21 cards off the top of the deck; ignore the rest of the cards. At the beginning you can act as both magician and volunteer; once you've gained some confidence in this trick you can invite

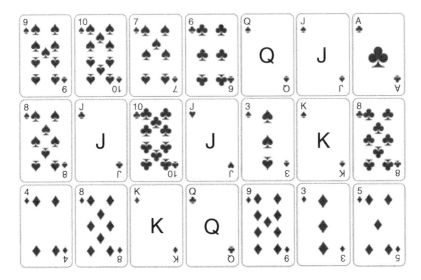

Figure 2.1. The standard layout for the 21 card trick.

others to participate. Thus, select one of the 21 cards and memorize its face.

After replacing the chosen card, shuffle the deck. Deal three rows of seven cards each, as shown in Figure 2.1.

Prompt your volunteer to point to the row that contains their card. Stack the cards in a consistent manner, from one side to the other; at this point you should have three stacks of seven cards each. Now, place the stack corresponding to the row your volunteer indicated between the other two stacks (see Figure 2.2).

At this point, you will deal the cards back out (without shuffling), making sure to deal one three-card column at a time. Repeat the previous step: have your volunteer indicate the row that contains his or her card, stack the cards into three piles from right to left or left to right, and place the indicated stack between the other two. Repeat this process one more time.

After this third cycle, you will once again have all 21 cards in your hand. At this point the magic is ready. Deal off the top of the deck, silently counting as you go. When you arrive at the eleventh card,

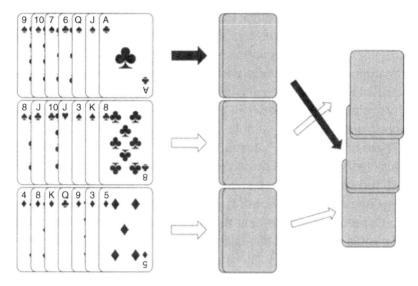

Figure 2.2.   Stacking the cards and placing the indicated row in the middle.

say "is this your card?" and, triumphant, turn it over to reveal the magic card!

Mathematical Entertainments

Two questions you may be asking at this point are, "What does this trick have to do with math?" and "How can mathematics be entertaining?" We will address these questions in reverse order.

Some people enjoy all kinds of problem solving. Indeed, it is one of our hopes that after completing this book, you join their ranks! Were that already the case, we could simply call each chapter "Entertaining Mathematical Problems" and leave it at that.

However, we will somewhat dodge the question of what makes an entertaining mathematical problem by parsing the phrase differently. Instead of wondering, "How can mathematics be entertaining?" we invite you to think of this as a chapter on *how mathematics can be used to entertain others*. That is, the problems in this chapter focus on tricks and puzzles that can be used to enliven a party or break the ice.

This is no mere theoretical exercise. Solving problems with others can open up new perspectives and deepen your understanding by forcing you to put your hypotheses into words.[1] Further, when exploring mathematical problems with others, explaining yourself with physical objects — such as playing cards — gives you a greater opportunity to close the loop between the real world and the abstract world of pure math.

These mathematical entertainments are meant to be shared, and by sharing them you will be using other parts of your brain — verbal, social, and spatial — to support you on your journey into learning.

The question remains: how is this card trick mathematical?

Let's pretend for a moment that we opened this book with a problem on rates of change. (We will, in fact, address these sorts of problems in a later chapter.) The archetypical rate of change problem begins something like this: "A train carrying coal leaves Tulsa at 60 miles per hour..." Your eyes may already be glazing over.

A goal of this book is to address common mathematical bad habits head on. Otherwise competent students often confuse the purpose of mathematics exercises with the purpose of mathematics itself. That is to say, these students are comfortable reading a problem, translating it into a known form (in the example above, something involving *distance equals rate times time*), working for a number of minutes with a pencil, and then writing down an answer ("4"). A particularly stringent teacher might demand at this point that the students write down units — 4 *miles per hour*, 4 *apples* — but the demands of the classroom often make it difficult to tell if the students are simply parsing out a likely unit from the problem statement, or actually understand (and can demonstrate) why that unit of measure makes sense, and relate it back to the original problem.

Useful mathematics — the kind of mathematics that can make predictions about the future — demands one further step: there must

---

[1] In computer programming, there is a term for solving problems by voicing them aloud to anything, even an inanimate object: "rubber duck debugging." The general idea is that by forcing yourself to state your givens and assumptions out loud, you force yourself to confront facts that your brain may have previously glossed over.

be real-world consequences, or a way to otherwise relate an answer back to the real world. This kind of practical mathematics forms a loop: one must be able to extract information from a situation, convert the situation into a mathematical form, find some answer, and convert that answer back into information (through actions or the interpretation of results).

We chose to open this chapter with a card trick precisely because the mathematics are shallow, so that we can always keep our eyes on the problem while performing manipulations. We will use logic — an essential part of any mathematics — and the bare minimum of numbers to make our point.

There is one final point we wish to emphasize. In the previous chapter we noted that it is possible to use mathematics without understanding it. Precisely because we will be demonstrating this card trick without extensive recourse to mathematics, you will probably be able to replicate it without understanding the ideas we are about to demonstrate. This can be fun and rewarding! But do try to keep in mind the purpose of our approach here, as a gentle introduction to problem solving, and try to get as much out of it as possible.

Keep Your Eye on the Card

Do you have a guess how this trick works? How might you begin to figure out the trick on your own?

One way to start is to keep an eye on the cards that could potentially be the magic card. For the sake of illustration, we will highlight the cards that could potentially be the magic card, step by step.

First, deal out all 21 cards in any order. For the sake of argument, say that the volunteer indicates his card is in the top row, as shown in gray in Figure 2.3.

Now, if you follow the instructions in the first section, once you deal the cards out again, they will be arranged as shown in Figure 2.4.

Do you notice anything peculiar? All the cards that were in the top row are now in the center of the array, meaning that you *know* the magic card must be one of the center seven cards.

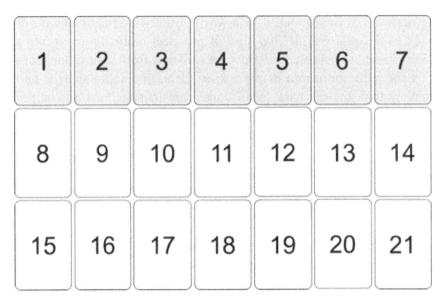

Figure 2.3.   Numbered cards, with the indicated row highlighted.

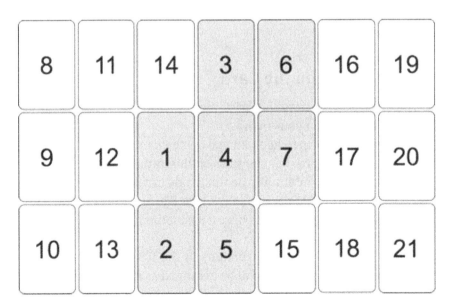

Figure 2.4.   After one iteration. Notice the former top row is now in a block in the middle.

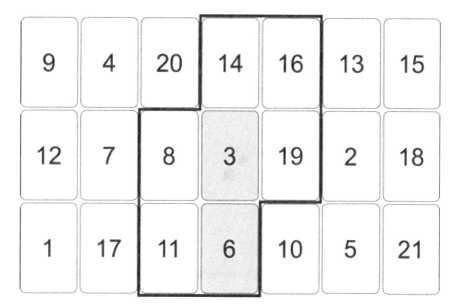

Figure 2.5. After two iterations. The cards that were previously in the top row are circled; the only two possible magic cards are highlighted.

Again, let's suppose that the volunteer indicates his card is in the new top row. Scooping up the cards into rows and dealing them out again by columns, you should arrive at the arrangement shown in Figure 2.5.

There are only two possibilities here: the magic card is the one numbered 3, or the card numbered 6. Why? Because these are the only two cards that were in both rows selected by your volunteer! These two cards are both in the center column, and you can predict that the volunteer can't say that his card is now in the top row!

Whatever his choice, when you pick up all the cards, the magic card will be the fourth card in the middle set. Let's assume the volunteer indicates his card is the bottom row. You stack the rows from left-to-right, and put the stack formed from the bottom row between the two others. There were seven cards in the top row, and thus, there are seven cards above the middle set. The middle set consists of seven cards, and the magic card has, at this point, been moved to the fourth position; there are a further ten cards below it. By addition, you can

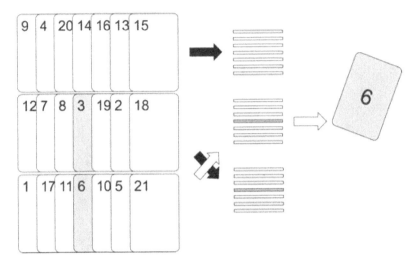

Figure 2.6. How the magic card finally moves to the middle of the deck. Numbers have been moved to the corner for clarity.

see that there are ten cards above the magic card and ten cards below it (Figure 2.6). When you begin counting, that card *must be* the eleventh card, which just so happens to be the middle card in the deck!

One Trick, Countless Possibilities

Breaking this trick down step by step will help you build a great deal of insight into the problem. This is a skill that shows up again and again in mathematics. Like the old joke says, "How do you eat an elephant? One bite at a time."

What other things can you do with this knowledge? How can you relate this newfound understanding back to the original trick?

Here we will stop to pose a number of questions. Spend some time seeing if you can answer them before moving on.

- Which of our original instructions don't affect the final outcome?
- As we saw, the magic card appeared in the middle of the deck. Can you make it appear in other places instead?

- Does this trick work with different numbers of cards? Which numbers, and why?
- Finally, are there any flourishes you can add to increase the appearance of magic?

We ask these questions because we already have answers prepared for them. However, when pursuing a mathematical goal, it is not always clear which questions lead to new solutions, and which questions lead nowhere. Asking good questions is as important a skill in problem solving as "getting the right answer;" some of the most frustrating problems and insightful results in mathematics have come about because someone stopped to ask themselves "what happens when we change this problem?"

Learning to ask these questions, and answer them, takes practice. Learning to ask *good* questions requires experience and judgment; these are traits that build over a period of years. It's possible that you're reading this book without ever having come face-to-face with a problem that wasn't answered in the "back of the book," so if you're one of these readers, this is an opportunity to learn something truly new.

Magic explorations

Having had time to experiment with the previous questions, you should have been able to reach some conclusions. What did you discover?

In response to the first question, note that the first time you gather up the three rows, it doesn't matter how neat you are: whatever you do, the seven cards from the indicated row will end up in the center of the array in the next round. On the other hand, each subsequent time you collect the cards, your goal is to move the magic card closer and closer to the center of the deck as a whole; thus, it is important to collect these cards in an orderly fashion.

Further, it is clearly essential to collect by row, and deal by column. At each iteration, your goal is to extract a bit more information from the volunteer; if you were to, say, collect by row and deal by row, you could be assured that the magic card is in the middle row. In this situation, the only way to gather "new" information would be by asking the volunteer which column their card is in — and anyone who has seen Cartesian coordinates will immediately see through the trick!

An interesting exception to this is if you were to reverse the directions, collecting one column at a time before dealing three rows of seven. We will illustrate this example in Figures 2.7 and 2.8 to make the forthcoming explanation easier to follow.

Notice that, while the indicated row isn't concentrated in the center, there is definitely a pattern to the cards! The original trick could be performed step-by-step without understanding how it works. However, this new variation would require more than mechanical competence; it demands memory and skill.

The purpose of this demonstration was to show that not all questions that can be asked are immediately useful. While we won't pursue this particular line of questioning any further for now, we will

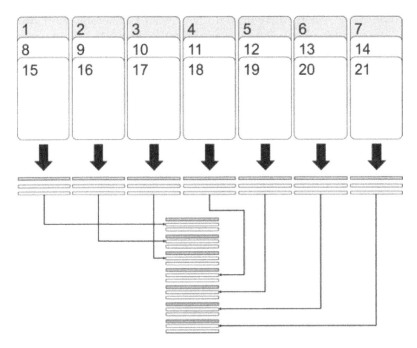

Figure 2.7. Collecting cards by column.

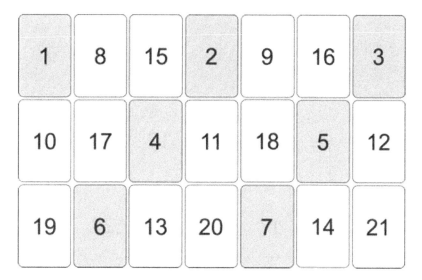

Figure 2.8.   Card layout after picking up by column and dealing by row.

return to it once we've added some particular tools to our mathematical toolbox.

There remains one major step in the trick we haven't questioned: the order of the stacks. If instead of putting the row indicated by the volunteer in the middle, you put it on top as shown in Figures 2.9 and 2.10, what would happen?

Notice that with this minor change in process the magic card moves further and further to the top left corner, until it's the first card. The second question we posed was whether it is possible to cause the selected card to move to other positions, and clearly it is! This provides an answer to our second question — you can move the card to the middle, the top, or (as you could demonstrate on your own) the bottom (see Figures 2.11 and 2.12)!

Moving a card anywhere

In fact, through some clever manipulations, it is possible to move the selected card almost anywhere in the deck. A question remains, though: how can you formalize this intuition?

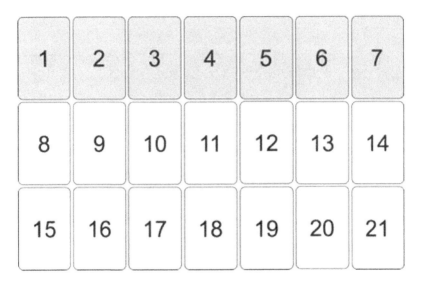

Figure 2.9. The volunteer indicates their card is in the top row.

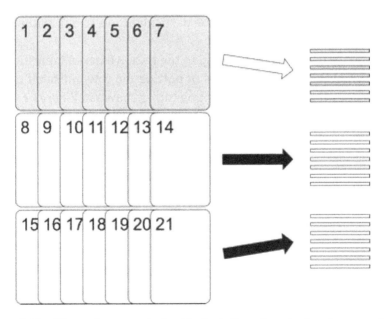

Figure 2.10. The cards are stacked, with the indicated stack placed on top of the deck.

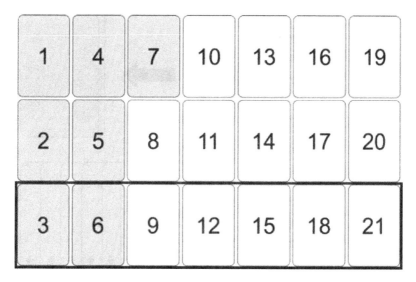

Figure 2.11.   After one iteration, the top row is dealt to the left side. The volunteer selects the bottom row this time.

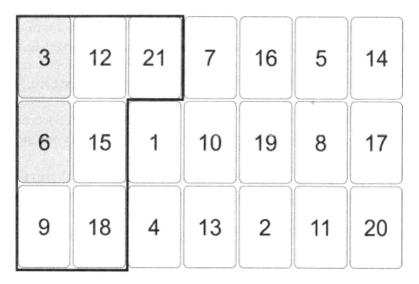

Figure 2.12.   After two iterations. The previous bottom row has moved to the top left, and the magic card must be in the top left corner.

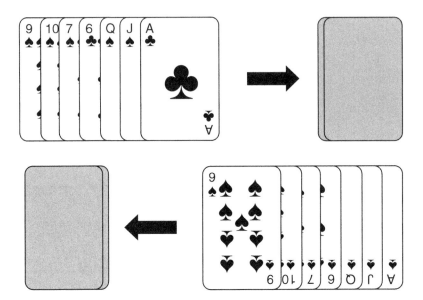

Figure 2.13. (Top) Stacking left-to-right. (Bottom) Stacking the same cards right-to-left.

First, we'll define some terms. There are two natural ways to collect all the cards in a row: from left-to-right and from right-to-left (Figure 2.13). Right-to-left involves picking up the rightmost card in the indicated row and putting it on top of the next rightmost card, and so on; left-to-right is, of course, the opposite.

To simplify this portion of the chapter, we will always collect cards left-to-right, so that the leftmost card will be at the top of the stack when dealing.

Next, we'll define *candidate cards*. Candidate cards are those cards that we have been highlighting throughout this chapter; that is, the set of cards that *must* logically contain the volunteer's selected card. Each time the volunteer picks a row, you gain a little more information, and that information reduces the number of candidate cards until there is only one left: the selected card. We have been highlighting candidate cards with the color gray in our diagrams.

A *block* or *candidate block* like the one shown in Figure 2.14 is the group of seven candidate cards that remain after the first pass.

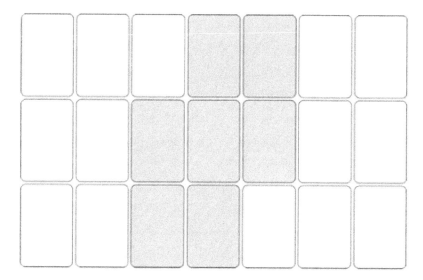

Figure 2.14.   A candidate block.

This suggests a natural extension, and, indeed, we will use the term *candidate column* for the collection of cards that must contain the magic card after two iterations.

Finally, we will use the terms *top-first, middle-first,* and *bottom-first* to refer to how the indicated row is stacked.

In the previous section we demonstrated that by stacking top-first, middle-first, or bottom-first, the selected card moves to the top, middle, or bottom of the entire deck, respectively, by the end of the trick.

Next, note that when the first stacking operation is performed top-first, the candidate block consists of the seven top- and leftmost cards. If you collect rows left-to-right, then on the next iteration the candidate column will consist of at most the first three cards of that respective block. (This column will not necessarily be contiguous, as you can see in Figure 2.15, but there will be at most three cards, one in each row, in the containing block.)

At this point, there remains one iteration to perform, but you already know (if you've been following the cards to this point) which column contains the magic card. Thus, when collecting the cards, you

Figure 2.15. A candidate column. Candidate columns can be contiguous or non-contiguous.

can make the magic card appear in the first seven, second seven, or last seven cards at the end of the trick; since you know which column the card appears in, you can simply add the number corresponding to the magic card's column to either 7 (if collecting middle-first) or 14 (if collecting top-first) to "foresee" which card will be magic!

All this suggests a new trick: ask your volunteer to pick a card *and a number*, and then stack top-, middle-, or bottom-first to manipulate the position of the magic card to that number (Figure 2.16). For instance, if your volunteer's number is seven, you can stack the cards bottom-first (pushing the candidate block to the far right), bottom-first again (ensuring that the candidate column appears on the far right), and then top-first. Voila! The selected card will appear in the seventh position.

If you're off by one at the second-to-last step (due to a non-contiguous candidate column), use a bit of stagecraft to trick your volunteer. Say you've gotten their card to the twelfth position in the deck, but their number is thirteen. Then, turn over twelve cards (including the magic card) and then say, with confidence, "I bet you the next card I turn over will be your card." Your volunteer might smile knowingly,

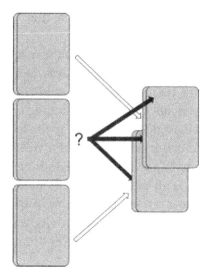

Figure 2.16. Three different ways to stack the rows: top-first, middle-first, or bottom-first.

but you're already a step ahead of them: flip over the twelfth card again, so it's face down. The thirteenth card you turned over was indeed their card!

Before we move on, note that defining terms allowed us a great deal of flexibility in analyzing this problem. Defining terms, even experimentally, gives you a great deal of leverage over a problem. Thinking in terms and phrases allows you to manipulate ideas as a whole rather than as a collection of parts, and gives you the opportunity to recognize new patterns and structures inside an otherwise complex situation!

Sizes and shapes

We've laid a lot of groundwork so far, so answering the next question — "does this trick work with different numbers of cards?" — should be much easier. This magic trick works with different numbers of cards; there is nothing special about the number 21!

Take note, though: when using the original rules (always stacking middle-first), it is impossible to perform the trick reliably using an *even* number of cards. Why is that? Because there is no "middle card" when dealing with an even number!

On the other hand, you can easily move the magic card to the top or bottom of the deck using any number of cards, as long as that number forms a rectangle with two or more rows and columns. For instance, using thirty-six cards dealt into a 6×6 rectangle, after one iteration (stacking top-first) the magic card must be in the leftmost column. This thirty-six-card variant of our trick works after two iterations.

Only two iterations? Indeed, there is nothing magical about any part of our trick — it is all mathematical. It should be clear that the variant of this trick using three cards requires only one iteration; with a little effort, you should recognize that two iterations are enough for six or nine cards; and finally, using three iterations is adequate all the way up to twenty-seven cards.

Why is that? What is special about those numbers?

Note the following:

$$3^1 = 3$$
$$3^2 = 9$$
$$3^3 = 27$$

This is no coincidence! The maximum amount of knowledge you can extract from three pieces of information (i.e., the volunteer pointing at one row out of three, three times) is one of 27 possibilities! If this statement is unclear to you, don't worry; we will attend to it in due time, when we explore a visual cousin of this problem in Chapter 5.

This principle holds for other shapes, as well: if you elect to use four rows of cards instead of three, you should predict that you can find the magic card in three iterations for any number of cards up to $4^3 = 64$ cards, which is more than enough for an entire deck!

What is magic?

By dissecting this magic trick mathematically, we have stripped away a lot of what makes it magic. It's worth spending a few paragraphs asking what, exactly, about the original trick makes it engaging.

We have conclusively demonstrated that there is nothing special about the number of cards used, or any one of the steps we take to arrive at the "prestige" — the reveal of the magic card at the end of the trick. However, there must be some justification for why the creator of this trick made these decisions.

One strategy when entertaining an audience is to keep things long enough to become engaged, but not so long as to get bored. By using 21 cards (instead of, say, 51 or 9), the trick builds an aura of mystery until the very end, delighting the audience.

Moving the magic card to the middle (rather than one of the ends) infuses just enough visual "randomness" to the trick that the audience is, again, engaged. Magicians say that you should never perform the same trick twice; however, the process here is mechanically simple, yet visually confusing enough that you could perform it two or even three times without the audience noticing that the magic card is gradually moving to the middle of the deck. It's doubtful that you could make the same claim about one of the variations where the card is moved to the top or bottom of the deck.

Finally, the potential embellishments (for instance, turning over the twelfth card twice) can help engage your audience, taking advantage of their overconfidence only to subvert it a moment later.

Why, in a book about mathematics are we spending time talking about magic? To paraphrase E.B. White, "explaining a magic trick is like dissecting a frog. You understand it better, but the frog dies in the process."

Mathematics allows us to "dissect the frog" — to strip away the magic from a trick, and to see it for what it is. Effective problem solving allows us to strip away magic and see to the heart of a problem, only looking at the information that matters. Unlike the frog, though, we can use our problem-solving abilities to put that same information back together in different ways!

A Different Kind of Card Trick

"For our next trick, we will again need a volunteer…"

A					B					C					D					E			
1	3	5	7		2	3	6	7		4	5	6	7		8	9	10	11		16	17	18	19
9	11	13	15		10	11	14	15		12	13	14	15		12	13	14	15		20	21	22	23
17	19	21	23		18	19	22	23		20	21	22	23		24	25	26	27		24	25	26	27
25	27	29	31		26	27	30	31		28	29	30	31		28	29	30	31		28	29	30	31

Figure 2.17.   Five magic cards.

At fun fairs or carnival prize booths you will often see small magic tricks of different kinds. One that you might notice is a small stack of cards with numbers printed on them; we reproduce an example in Figure 2.17.

While this is a card trick of a sort, these are not playing cards. In time, we will show you how to construct a set of your own. Until then you can refer to the above diagram.

The way this trick works is as follows: tell your volunteer to choose a number between 1 and 31. Then, without telling you what the number is, indicate which of the cards contains that number. A moment later (perhaps after some theatrics, such as closing your eyes and holding your hand to your head dramatically), you tell her which number she chose.

The trick is as follows: add up the numbers in the top left corner of each of the indicated cards, and that sum is the "magic number." Try it and verify it works!

To explain this trick, we will need to take an aside into number systems.

Binary, ternary, and decimal, oh my!

There are countless ways to represent numbers. You are almost certainly most familiar with the decimal, or base-10 positional, numeral system, wherein the digits 0 through 9 are used in a sequence to represent numbers.

For instance, the number twenty-four is written in decimal as 24, which we can interpret as "two tens, and four ones." No one familiar with the decimal system stops to think about numbers in that sort of detail, but it is illustrative for our purposes.

If decimal is only one of many number systems, what others can we think of? Well, the most trivial of all number systems is a simple tally. You might use a tally when counting people walking through a door at, for instance, a sporting event.

A tally representation of 24 would look like the following:

There are alternative methods for tallying; you can group by fives in the Western style, as shown in Figure 2.18.

Or gradually draw the five-stroke Chinese character 正 over and over, shown in Figure 2.19, as is typical in East Asian cultures.

Can you think of other ways to represent numbers? Another system, which is not positional, is that of the Roman Numerals. In Roman Numerals, 24 is written as XXIV.

Roman Numerals utilize positions, but not in the same way as decimal numbers. A full Roman Numeral number consists of four groups, each one optional, ordered by magnitude: thousands, hundreds, tens, and ones. The decimal number 2097 is written as MMXCVII, and written in groups is [MM][][XC][VII]. The word "positional," as it's used to refer to decimal numbers, means that each position in the final number *must* be occupied by *only* one symbol — something we can see doesn't happen with Roman Numerals!

Figure 2.18.   One type of tally marks; groups of five are indicated by a slash through four lines.

Figure 2.19.   East Asian-style tallying. (Top) The progression from 1 to 5. (Bottom) The number 24, marked in this system.

All this is to say that there are many ways to represent the same number beyond base-10 positional notation!

Using the metacognition skills you've been developing, one of the first questions you should be asking is "if I'm most familiar with base 10, what other bases can I use?"

Now using the number 324, observe that

$$324 = 3 \times 100 + 2 \times 10 + 4 \times 1$$

But also

$$324 = 3 \times 10^2 + 2 \times 10^1 + 4 \times 10^0$$

This suggests a natural extension of the idea of a base, using different numbers instead of ten. If you wanted to write a number in a different base, say, base 5, you could use any of the digits between 0 and 4 (that is, the number that is one less than the base), and you would find that all the standard rules of arithmetic that you know from the decimal system still apply.

But how can you convert a number from one base to another?

We will demonstrate a brute-force method to convert numbers between bases. It's a cumbersome process, but we want you to recognize that these numbers are the same regardless of representation, and not just the output of some formula.

Write down all the powers of your new base (in this case, 5) until you find the first number that is greater than or equal to your original number.

Thus,

$$5^0 = 1$$
$$5^1 = 5$$
$$5^2 = 25$$
$$5^3 = 125$$
$$5^4 = 625$$

The number 324 is certainly less than 625, so you can stop here. Note that whatever the base 5 representation is of 324, it must be a four-digit number (since $125 < 324 < 625$).

Subtract the largest possible number from this list from 324, adding 1 to the appropriate column, until you arrive at 0. Thus,

$$324 - 125 = 199 \Rightarrow 1???, \text{base } 5$$
$$199 - 125 = 74 \Rightarrow 2???$$
$$74 - 25 = 49 \Rightarrow 21??$$
$$49 - 25 = 24 \Rightarrow 22??$$
$$24 - 5 = 19 \Rightarrow 221?$$
$$19 - 5 = 14 \Rightarrow 222?$$
$$14 - 5 = 9 \Rightarrow 223?$$
$$9 - 5 = 4 \Rightarrow 224?$$
$$324_{10} = 2244_5$$

(We use subscript numbers at the end to indicate the base of the respective numbers.)

As we said, this procedure is quite cumbersome! It might help to understand the motivation for this approach by looking at a visual representation of the same process. You can follow along with this process step-by-step by looking at Figure 2.20.

We hope to drive home the point that the numbers haven't changed, only their representation has. Still, there must be an easier way to perform this conversion — and, of course, there is.

By repeatedly dividing by the target base and taking note of the remainder R, you can speed up the previous process as follows:

$$\frac{324}{5} = 64 \, R \, 4$$
$$\frac{64}{5} = 12 \, R \, 4$$
$$\frac{12}{5} = 2 \, R \, 2$$
$$\frac{2}{5} = 0 \, R \, 2$$

Reading off the remainders in ascending order gives you the same result as before[2] — $324_{10} = 2244_5$.

---

[2] Why ascending? If you divide by the same number repeatedly, what remains has a greater "weight" to it than what has already been removed. If you're struggling to convince yourself of this fact, consider the number 125_{10}. Clearly, we can divide this number by five three times, leaving no remainder each time; from the previous discussion we know that $125_{10} = 1000_5$.

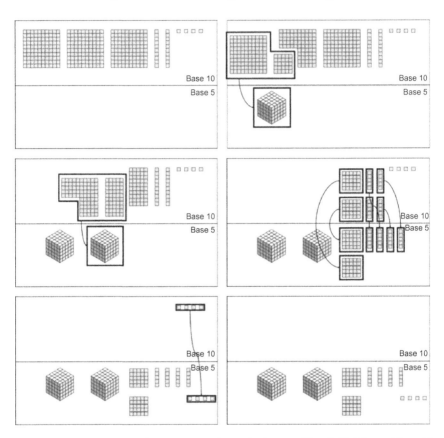

Figure 2.20. Converting bases using physical objects. (Top left) Three 10×10 squares (hundreds), two 1×10 sticks (tens), and four ones. (Top right) Making a $5 \times 5 \times 5$ cube. (Middle left) Repeating this process. (Middle right) Using what's left to make 5×5 squares and 1×5 sticks. (Bottom) Moving the ones over together results in two cubes, two squares, four sticks, and four ones, representing 2244_5.

To convert back, you can multiply each digit of the converted number by its corresponding power of 5 (in decimal notation):

$$2 \times 5^3 + 2 \times 5^2 + 4 \times 5^1 + 4 \times 5^0$$
$$= 2 \times 125 + 2 \times 25 + 4 \times 5 + 4 \times 1$$
$$= 250 + 50 + 20 + 4$$
$$= 324$$

Great! You have verified that our process works backwards and forwards, and can be reasonably confident that you're working with a definition that works for any integer (i.e., whole number) base.[3]

The most common number system used today is called *binary*, derived from the Latin prefix *bi-*, meaning two. We can say with the utmost confidence that binary is the most widely used number system today because the world is well on its way to converting almost all human output (whether it be words, photos, videos, or math), into a form that is understandable by computers. Computers represent everything — from computations to data — with electronic circuits that can be in one of two states, on or off, corresponding to the binary digits 1 and 0.

Without going into too much detail about the implementation of binary computers,[4] we will note that part of what makes binary so useful is that many processes can be described as either-or. A light switch is on or off; a poet at a fork in the road can go right or left; a statement is either true or false;[5] "you're either with us or against us." Binary numbers can represent these dichotomies just as easily as they can act as numbers, and binary computers are much easier to design and implement than computers using, say, base 10.

For practice, we will show you how to convert the number 324_{10} into binary and back; this is meant to give you a chance to gain initial familiarity with the number system. If you don't get it right away, don't worry — you will have many more opportunities to practice with it throughout this book!

---

[3] We are choosing to gloss over the demonstration of this fact, but feel free to play with this definition and convert numbers between bases until you're comfortable with it.

[4] There were a number of ternary computers in the early days of computing. Most notable of these was the computer Setun, built in Moscow in 1958.

[5] This isn't actually true, for reasons that are well outside the scope of this book. Suffice it to say, it can be proven that there are statements that are unprovable.

$$\frac{324}{2} = 162 \, R \, 0$$

$$\frac{162}{2} = 81 \, R \, 0$$

$$\frac{81}{2} = 40 \, R \, 1$$

$$\frac{40}{2} = 20 \, R \, 0$$

$$\frac{20}{2} = 10 \, R \, 0$$

$$\frac{10}{2} = 5 \, R \, 0$$

$$\frac{5}{2} = 2 \, R \, 1$$

$$\frac{2}{2} = 1 \, R \, 0$$

$$\frac{1}{2} = 0 \, R \, 1$$

Putting this all together, you should have $1 \, 0100 \, 0100_2 = 324_{10}$.

Reversing the procedure, you can see that

$$1 \times 2^8 + 0 \times 2^7 + 1 \times 2^6 + 0 \times 2^5 + 0 \times 2^4 + 0 \times 2^3 + 1 \times 2^2 + 0 \times 2^1 + 0 \times 2^0$$
$$= 1 \times 2^8 + 1 \times 2^6 + 1 \times 2^2$$
$$= 256 + 64 + 4$$
$$= 324$$

Demonstrating once again that this conversion process works.

The mathematical magic cards

That was quite the aside, but you finally have the tools needed to see how the previous trick was performed!

For reference, we've copied the cards from earlier in the chapter to Figure 2.21.

Do you have a guess where all this is headed? If so, congratulations! If not, pay close attention to the following demonstration.

First, we will write the numbers 0 through 31 in binary in Table 2.1, with enough leading zeroes that all the numbers are the same length.

Do you observe any patterns here? As tedious as it might sound, it can be worthwhile to write out these numbers yourself; if you don't see any patterns, writing these numbers by hand will surely show you something.

Recall that any number can be written in binary as a sum of the powers of 2 corresponding to its digits. Thus,

$$13 = 1 \times 8 + 1 \times 4 + 1 \times 1$$

Note also that the number 13 appeared on the cards starting with 1, 4, and 8. Surely this isn't a coincidence!

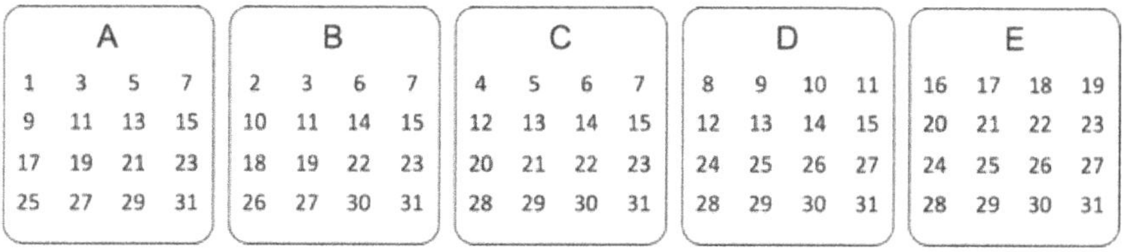

A					B					C					D					E			
1	3	5	7		2	3	6	7		4	5	6	7		8	9	10	11		16	17	18	19
9	11	13	15		10	11	14	15		12	13	14	15		12	13	14	15		20	21	22	23
17	19	21	23		18	19	22	23		20	21	22	23		24	25	26	27		24	25	26	27
25	27	29	31		26	27	30	31		28	29	30	31		28	29	30	31		28	29	30	31

Figure 2.21. The five magic cards from before.

Table 2.1. The numbers 0 through 31 and their binary representations.

0	00000	8	01000	16	10000	24	11000
1	00001	9	01001	17	10001	25	11001
2	00010	10	01010	18	10010	26	11010
3	00011	11	01011	19	10011	27	11011
4	00100	12	01100	20	10100	28	11100
5	00101	13	01101	21	10101	29	11101
6	00110	14	01110	22	10110	30	11110
7	00111	15	01111	23	10111	31	11111

This trick works precisely because a number is the sum of its powers of 2; there is one, and only one, way to represent each number using binary digits, and thus the "magic" here corresponds perfectly to the process you learned in the previous section!

If you're still confused, consider this: any number can be written in a unique binary form. Choosing again the number 13, and writing it in binary, yields

$$13_{10} = 01101_2$$

Card A contains only the numbers with a 1 digit in the rightmost position, so 13 should appear on that card. Card C contains only those numbers with a 1 in the third position, so 13 should appear on card C. Finally, card D lists all numbers that have 1 in the second position from the left, so 13 is on card D.

We told you that we'd show you how to make cards like this on your own, and the previous paragraph suggests how you might start to do just that. We will illustrate two methods of making "magic cards" of your own; this method is by far the more tedious of the two. Write down every number between 1 and 31 in binary form, and then perform the procedure above: look at each digit of each number in turn; if a 1 appears in position 1 for that number, write the number on card A, repeating for all five digits in all 31 numbers.

A simpler method involves first noticing that each binary digit alternates periodically. Thus, the rightmost digit alternates between 0 and 1, so on card A write down the (odd) numbers 1, 3, 5, 7... The next digit alternates between 0 and 1 every four numbers, thus you would skip 0 and 1, write down 2 and 3, skip 4 and 5, ... The results of this process are highlighted in Table 2.2.

In Table 2.3, we provide an alternate version of the Table 2.2, before extending the principle.

You can quickly verify that all of the binary representations on a given card have one digit that remains fixed. For instance, every number on card C, when written in binary, has a 1 in the third position. By virtue of this fact you can reliably state that if a number appears on card C, then in order to convert that number from binary to decimal,

Table 2.2.   Patterns of 1s in successive binary numbers.

1	0	0	0	0	1
2	0	0	0	1	0
3	0	0	0	1	1
4	0	0	1	0	0
5	0	0	1	0	1
6	0	0	1	1	0
7	0	0	1	1	1
8	0	1	0	0	0
9	0	1	0	0	1
10	0	1	0	1	0
11	0	1	0	1	1
12	0	1	1	0	0
13	0	1	1	0	1
14	0	1	1	1	0
15	0	1	1	1	1
16	1	0	0	0	0

you must at some point add 4 — the top digit on that card — to the total.

More observant readers may have noticed other patterns. For instance, each card contains sixteen numbers — this is no coincidence — but if you think a bit about the nature of binary numbers, there is a very good reason for this.

Consider the restrictions characterizing each card: one digit must remain fixed, while the other four can take on any value. Exactly sixteen distinct numbers can be represented using four binary digits: 0000, 0001, 0010, 0011, and so on; thus there are sixteen different numbers of the form XXXX1 (or XXX1X, or XX1XX...).

Ignoring for a moment the fixed digit on each card (cover it with a small strip of paper, or perhaps a pencil), note that at each step, the other four digits are identical, no matter which card you look at!

Table 2.3. The numbers that appeared on the original cards, alongside their binary representations.

Card A		Card B		Card C		Card D		Card E	
1	00001	2	00010	4	00100	8	01000	16	10000
3	00011	3	00011	5	00101	9	01001	17	10001
5	00101	6	00110	6	00110	10	01010	18	10010
7	00111	7	00111	7	00111	11	01011	19	10011
9	01001	10	01010	12	01100	12	01100	20	10100
11	01011	11	01011	13	01101	13	01101	21	10101
13	01101	14	01110	14	01110	14	01110	22	10110
15	01111	15	01111	15	01111	15	01111	23	10111
17	10001	18	10010	20	10100	24	11000	24	11000
19	10011	19	10011	21	10101	25	11001	25	11001
21	10101	22	10110	22	10110	26	11010	26	11010
23	10111	23	10111	23	10111	27	11011	27	11011
25	11001	26	11010	28	11100	28	11100	28	11100
27	11011	27	11011	29	11101	29	11101	29	11101
29	11101	30	11110	30	11110	30	11110	30	11110
31	11111	31	11111	31	11111	31	11111	31	11111

Table 2.4. The fifth row of the preceding table. Note the pattern of the digits.

9	01001	10	01010	12	01100	12	01100	20	10100

For instance, look at the fifth row, reproduced in Table 2.4 for your convenience.

If you ignore the highlighted digits, the remaining digits all read 0100. Fascinating!

This mind reading trick has given us the chance to learn about a powerful tool: the binary number system. However, it's worth spending a little time thinking about how this trick would work under other number systems.

The world's worst magic trick

Let's say that you want to perform this magic trick, but you are having trouble remembering how to construct the appropriate cards. What else could you try to do?

If you recall the trick itself — there's something about adding up the smallest number on each card — you could attempt to construct cards that correspond to base 10 numbers.

You might begin by writing the number 1 on a card, followed by other numbers that end in that same digit:

1	11	21	31	41
51	61	71	81	91

If your volunteer picks 1, then this card would be enough, but there are nine other numbers to account for on this card alone. To be able to guess any of these numbers would require at least one more card, and, remembering the "mindreading rule" that says in order to guess a number you add up the top left number on each selected card, you reason that some other cards will be necessary.

10	11	12	13	14
15	16	17	18	19

20	21	22	23	24
25	26	27	28	29

Continuing this process, you end up with eighteen different cards. You can see what these look like in Figure 2.22.

This isn't good at all! There are nearly four times as many cards as were necessary for the previous trick, and the secret should be obvious to all, because the numbers are staring right at them!

Nevertheless, you can take some lessons from this, the world's worst magic trick, to draw conclusions about the general case.

First of all, note that it's natural to write down all numbers up to, but not including, some whole number power of the base. In the

A	B	C	D	E
1 11 21 31 41	2 12 22 32 42	3 13 23 33 43	4 14 24 34 44	5 15 25 35 45
51 61 71 81 91	52 62 72 82 92	53 63 73 83 93	54 64 74 84 94	55 65 75 85 95

F	G	H	I	J
6 16 26 36 46	7 17 27 37 47	8 18 28 38 48	9 19 29 39 49	10 11 12 13 14
56 66 76 86 96	57 67 77 87 97	58 68 78 88 98	59 69 79 89 99	15 16 17 18 19

K	L	M	N	O
20 21 22 23 24	30 31 32 33 34	40 41 42 43 44	50 51 52 53 54	60 61 62 63 64
25 26 27 28 29	35 36 37 38 39	45 46 47 48 49	55 56 57 58 59	65 66 67 68 69

P	Q	R
70 71 72 73 74	80 81 82 83 84	90 91 92 93 94
75 76 77 78 79	85 86 87 88 89	95 96 97 98 99

Figure 2.22.   A base 10 version of the previous magic trick.

original trick, that number was $2^5 - 1 = 31$. But that number is neither special, nor magical: we could have just as easily stopped at 15 (i.e., $2^4 - 1$) or 63 (i.e., $2^6 - 1$). As before, there's a balance to be struck between engaging your audience and keeping things complicated enough to hide the "trick." The most natural number to use for the base 10 variation is then 99 ($10^2 - 1$) — but as you just saw, it doesn't make for a very *interesting* trick.

Is there an easy way to determine the number of cards necessary to perform this kind of trick in any base? Base 2, binary, presents a good starting point. No other numerical base should require fewer cards than binary, and using the principles you saw earlier it is quite straightforward to compute the number of cards necessary to for any variation of this trick in binary. In order to perform this trick with numbers up to 31, you need five cards; similarly, you need four cards for numbers up to 15 and six for numbers up to 63.

Let's do a brief experiment. How many cards are necessary to perform this trick, in base 10, for numbers up to 9? Nine, clearly. How many cards are necessary to perform this trick in base 5, for numbers up to 4? It should be clear that the answer is four. Slightly complicating matters, can you determine how many cards are necessary, in base 5, for numbers up to 24?

Table 2.5. The numbers 0 through 24 and their base 5 representations.

Base 10	Base 5	Base 10	Base 5	Base 10	Base 5	Base 10	Base 5	Base 10	Base 5
0	00	5	10	10	20	15	30	20	40
1	01	6	11	11	21	16	31	21	41
2	02	7	12	12	22	17	32	22	42
3	03	8	13	13	23	18	33	23	43
4	04	9	14	14	24	19	34	24	44

Table 2.6. Base 10 and (in parentheses) equivalent base 5 numbers, organized according to their locations on each card.

Card A	Card B	Card C	Card D	Card E	Card F	Card G	Card H
1 (01_5)	2 (02_5)	3 (03_5)	4 (04_5)	5 (10_5)	10 (20_5)	15 (30_5)	20 (40_5)
6 (11_5)	7 (12_5)	8 (13_5)	9 (14_5)	6 (11_5)	11 (21_5)	16 (31_5)	21 (41_5)
11 (21_5)	12 (22_5)	13 (23_5)	14 (24_5)	7 (12_5)	12 (22_5)	17 (32_5)	22 (42_5)
16 (31_5)	17 (32_5)	18 (33_5)	19 (34_5)	8 (13_5)	13 (23_5)	18 (33_5)	23 (43_5)
21 (41_5)	22 (42_5)	23 (43_5)	24 (44_5)	9 (14_5)	14 (24_5)	19 (34_5)	24 (44_5)

Again, we will create two tables. Table 2.5, lists each number in both decimal and base 5; Table 2.6, shows which numbers appear on each card.

You can clearly observe the same kinds of patterns you saw on the binary cards earlier. Each card has one digit fixed, while the other digit is free to count from 0 to 4; what differentiates this case from the binary one is that there are four possible non-zero values for each card's digit (1, 2, 3, or 4), rather than one (1). This suggests a method to count the number of cards necessary to perform this trick.

Do you see it? In order to perform this trick with numbers strictly less than some number $n = b^x$, where b is our chosen base and x is an integer exponent, we need $x(b - 1)$ total cards. Since we promised we wouldn't use excessive mathematical notation in this early chapter, we will make this explicit: in order to craft magic cards for 80 possible numbers in base 3, first calculate $3^4 = 81$, and then recognize that 80 is one less than 81. Then, note that there are two non-zero digits in base 3: 1 and 2. Thus, this form of the magic trick would require $4 \times 2 = 8$ total cards.

Try it and you will see!

Overthinking, or the right amount of thinking?

If you found the previous example exhausting, you are right to feel this way! We covered a great many topics in a short period of time, but as always we had a goal in mind.

After all that explanation and elaboration, you should feel more capable of recognizing and making use of patterns. You also now know what binary is! The former is a skill, or even a meta-skill, that will serve you well throughout this book (and life); the latter is a tool that will crop up over and over, to the point that you will start to see it coming before we even mention it.

Finally, you learning to use tools — and learning to recognize new problems as simply variations of skills already in your existing toolkit — is a fantastic way to gain competence in mathematical problem-solving!

Perfect Shuffles and Moving Cards

As a break from the intensity of the previous problems, we will spend a little time practicing memory and logic. The next problem should be solvable without any great insights or tricks; its primary purpose is a warmup for the next major demonstration.

Suppose you are given a stack of eight cards. You reveal the top card and place it in a separate pile, face up, then move the next card to the bottom of the deck, repeating this process until all eight cards are face up. When you've revealed all eight cards, they are in the order Ace, King, Ace, King, Ace, King, Ace, King. What was the original order of the cards?

The simplest way to solve this problem is to reverse the steps. Thus, picking up the last King and putting an Ace on top of it, remove the bottom card from the two-card deck (the last King) and place it on the top. Then, place the second-to-last King on top of the deck, move

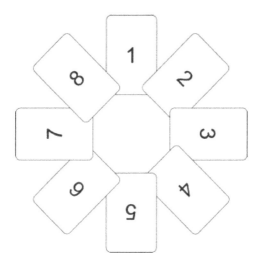

Figure 2.23. Dealing a deck of cards in order into a circle.

the bottom card to the top, and repeat until all the face-up cards are back in the deck.

Turning them over, you should see that the original order was (using appropriate abbreviations) AAKAAKKK.

That was pretty easy, right? However, the procedure we provided is error-prone. Is there a more reliable way to verify this solution?

Here's a hint: think of the transformation we performed earlier, at the beginning of the 21 card trick. Is there any way you can do something similar here?

Indeed there is! Take eight index cards and write the numbers 1 through 8 on them. Arrange them in a circle, as shown in Figure 2.23.

Pretend that this circle is a deck of cards, and whatever card you point at is the top card. In this representation, the card immediately counterclockwise from your finger represents the bottom of the deck. Moving your finger clockwise, then, is equivalent to taking the top card off the deck and moving it to the bottom.

Now, set aside the card numbered 1 in a separate pile, skip the next card, set aside 3, skip 4, and repeat this process until you have set aside all eight cards, as shown in Figure 2.24. What order are they in?

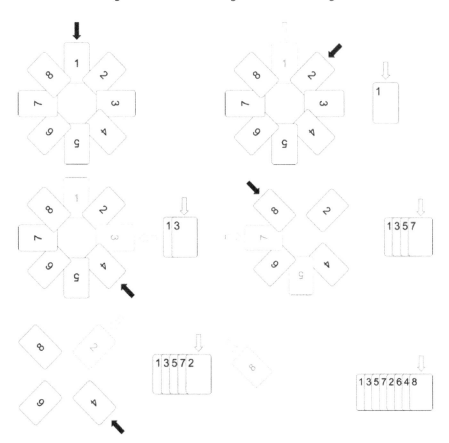

Figure 2.24.   Remove one, skip one. (Top left) Card number 1 is at the "top" of the deck. (Top right) Setting aside card 1, card 2 is now at the top. (Middle left) Skipping 2 and removing 3, card 4 is now on top. (Middle right) Continuing in this fashion with cards 5 and 7. (Bottom left) Skipping card 8, card 2 is removed next. (Bottom right) The result of this process.

You can see that the cards are now in the order 1, 3, 5, 7, 2, 6, 4, 8. Substituting in the values of the playing cards described in the beginning of the problem, you can see that cards 1, 5, 2 and 4 were Aces; the rest must be Kings. Mapping the cards back to their starting order, following the arrows backwards in Figure 2.25, will tell you what order they appeared in the original deck: AAKAAKKK. And indeed this is exactly what you saw before, confirming that the original ordering you found was correct!

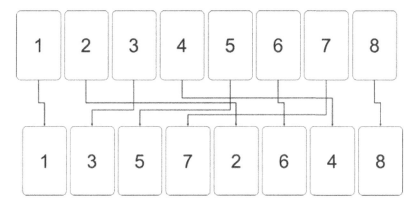

Figure 2.25. The order of the cards in the original deck, mapped to their final order.

This problem was much shorter than the first two in this chapter. You can see that this brought together several skills and tools — for instance, working backwards from the solution, transforming a problem into a different, more useful representation, and looking at the cards themselves in a different way — which allowed you to find the same answer two different ways. This, believe it or not, is a skill in itself: if you have the time, solving the same problem twice (and finding the same answer twice) is an excellent way to check for errors.

For the next trick, we will combine two of the tools you've seen so far. As you are reading, try to guess what those two tools are.

Faro shuffles

What would you call a "perfect shuffle?"

Two answers spring to mind. When playing card games, shuffling should create randomness, enough to make the next game unpredictable. There's nothing worse than playing a card game and recognizing a string of (poorly shuffled) cards that survived, unchanged, from the previous game. From this perspective, a perfect shuffle is a means of incorporating randomness into the game, making it more fun.

The other kind of perfect shuffle implies something quite different. A typical shuffle is performed by cutting a deck more or less in half and then letting down one or more cards from each side.

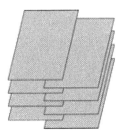

Figure 2.26.   A perfect faro shuffle.

By contrast, the other variety of "perfect" shuffle involves cutting the deck exactly in half, and interleaving the cards from the left and right sides *perfectly*, like a zipper.

There are two points worth noting about this kind of shuffle: it's beyond the skill of most amateurs, but an absolutely *fundamental* component of stage magic; and the results of this kind of shuffle are *completely predictable*. (The first point arises naturally out of the second.)

This type of "perfect shuffle" is so fundamental that it even has a name: the faro shuffle.[6] While we don't expect that most readers will be able to perform the faro shuffle, understanding it can strip away the magic of some tricks, and reveal some surprising insights about mathematics that go beyond mere playing cards.

Begin by cutting a standard deck exactly in half.[7] Take each half in hand, and interweave the cards from one half into the other, ensuring the cards alternate between the left and right halves. We depict a simplified version of this, with only eight cards, in Figure 2.26.

You can simulate this by manually layering the cards one at a time, left then right, until the deck is remade.

---

[6] You may also encounter the terms *riffle shuffle* or *weave shuffle*. They are different names for the same process.

[7] Any book on card tricks will offer more advice on how to achieve this than we have room for. However, one way to start practicing this would be to take a standard deck and count to the 26th card, taking note of its value. Then, try to cut the deck exactly in half; you will know you've succeeded when the bottom card is the one you memorized.

It can be cumbersome to do this for all 52 cards, especially if you don't have the knack for it. So, for the next portion, we will substitute a smaller deck consisting of only eight cards in order to try to identify useful principles.

Using eight cards (we suggest using cards from Ace through 8, or, again, index cards with the numbers 1 through 8 on them), split the deck into two piles of four cards each, interleaving them until you are again left with one deck. How does this affect the order?

This is a trick question. Do you see why?

It turns out that there are two distinct ways to perform a faro shuffle. In one, the card that began on the top of the deck remains on top at the end of the shuffle. In the other, the card that began on the top of the deck is shuffled into the rest of the cards. In either case, the cards are shuffled together like a zipper, one side after the other, as illustrated in Figure 2.27.

The first of these shuffles is called an *out shuffle*, because the top and bottom cards remain on the outside of the deck, as you can see on the left side of Figure 2.27. The latter is naturally called an *in shuffle*.

At this point you could attempt to formalize how different shuffles alter the ordering of the cards, but at best you would be feeling

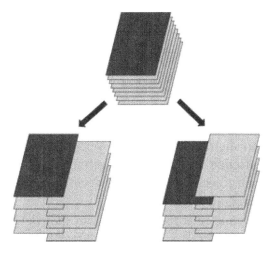

Figure 2.27. Two possible ways to perform a faro shuffle. (Left) In an out shuffle, the top card remains on top. (Right) In an in shuffle, the top card enters the deck.

Table 2.7. The order of the cards in an eight-card deck after successive out shuffles.

Position	1	2	3	4	5	6	7	8
Start	1	2	3	4	5	6	7	8
Shuffle 1	1	5	2	6	3	7	4	8
Shuffle 2	1	3	5	7	2	4	6	8
Shuffle 3	1	2	3	4	5	6	7	8

Table 2.8. The order of the cards in an eight-card deck after successive in shuffles.

Position	1	2	3	4	5	6	7	8
Start	1	2	3	4	5	6	7	8
Shuffle 1	5	1	6	2	7	3	8	4
Shuffle 2	7	5	3	1	8	6	4	2
Shuffle 3	8	7	6	5	4	3	2	1
Shuffle 4	4	8	3	7	2	6	1	5
Shuffle 5	2	4	6	8	1	3	5	7
Shuffle 6	1	2	3	4	5	6	7	8

around in the dark. You could shuffle a small deck multiple times, each time writing down the order of the cards that results. We have done the work for you in Table 2.7; look at it and see if anything stands out to you.

After only three out shuffles, the deck is back in the original order!

Table 2.8 shows the same kind of data, but this time the result of multiple in shuffles.

After three in shuffles, the deck is reversed, and after six shuffles it's back in the original order. How interesting!

What patterns do you see? Do you have any hypotheses about the relation between shuffles and ordering?

Any patterns you see at this point may or may not hold for decks of different sizes. The goal for now is to identify the behavior of a single card, rather than the whole deck. You may be able to identify a general principle that applies to any card, but it may not yet be clear how to start down that path. For now, then, use Table 2.9 to follow the position of a single card as it changes over multiple shuffles.

Note that, after each of the first three shuffles, the index of the card in which you are interested doubles in value — it moves first from 1 to 2, then from 2 to 4, then from 4 to 8. The reason for this should be self-evident: for a card at position n, n cards from the opposite half of the deck are shuffled into positions above it.

By contrast, when card 1 is in the bottom half of the deck it moves up by the same amount, as you can see in shuffles 4 through 6 of Table 2.9: it moves to the second-to-last position, then the fourth-to-last, then the eighth-to-last position (that is, the first position).

When performing out shuffles, you can see a similar phenomenon occurring (Table 2.10), with one minor exception: the cards on either end don't move.

Table 2.9.   The position of a single card relative to the others in an eight-card deck after successive in shuffles.

Position	1	2	3	4	5	6	7	8
Start	1							
Shuffle 1		1						
Shuffle 2				1				
Shuffle 3								1
Shuffle 4							1	
Shuffle 5						1		
Shuffle 6	1							
Position (from bottom)	8	7	6	5	4	3	2	1

Table 2.10.   During successive out shuffles, the cards on either end do not move, while the cards in the middle do.

Position	1	2	3	4	5	6	7	8
Start	1	2						8
Shuffle 1	1		2					8
Shuffle 2	1				2			8
Shuffle 3	1	2						8

From this observation, you can construct a hypothesis: if a deck of a given size returns to its original order after a certain number of **in** shuffles, a deck with two more cards returns to its original order after the same number of **out** shuffles.

You have enough information to build on, but still don't have the whole picture. In order to flesh out your understanding, you will need to test the hypotheses you have derived so far against other examples, and you can start that by considering a deck of nine cards.

Shuffling oddly

Let's spend some time examining what happens when you perform the same actions as before, this time with an odd number of cards.

The first thing you should notice is that there are two different ways to cut a deck containing an odd number of cards: one way leaves a larger number of cards in the top half of the deck, and the other does the opposite. How will you choose to name these?

In the case of an eight-card in shuffle, the top card goes *into* the deck. Starting instead with nine cards, if the bottom half of the deck has one more card than the top half, the only way to interleave the cards necessitates that the top card go into the deck.

As we originally defined it, an in shuffle moves both the top and bottom card into the deck. Since you can't possibly meet both of these criteria with an odd deck, you will have to eliminate one or the other to make it apply to all possible deck sizes. Ultimately, this decision is arbitrary — you could decide it via a coin flip — but for reasons of consistency, we will suggest the following definition: an in shuffle is a perfect interleaving shuffle that results in the *top* card of the deck moving to the second position.

Note that this doesn't change any of the behavior you identified earlier. This is a good thing! Changing the definition has maintained its descriptive power, while extending it to more cases. Since all whole numbers are even or odd, this should be the last change you need to make.

Faro shuffles have been addressed in other texts, and much more concisely than our treatment here. As we will continue to emphasize, the subject of this book is *Understanding Mathematics Through Problem Solving*, and so if you will indulge us for a moment, we ask you to activate your metacognition to consider the purpose of the past few pages.

When you imagine the day-to-day work of mathematics, what do you think of? Is it neat, exact, and the result of genius arriving at the perfect answer via established rules? Or is it messy, tentative, and the result of trying things, not ever knowing whether they will arrive at a result?

The purpose of the past few pages is to give you a sense for what real mathematics is like, and to "give you permission," as it were, to be messy. Just as the artist doesn't paint her masterpiece without making preliminary sketches, and the musician doesn't write his hit song without writing and rewriting music until everything works, the mathematics you read in textbooks is not representative of the bloody mess that is discovering new math. We won't belabor this point, but it is worth remembering from time to time and whenever you feel frustrated.

Returning to the topic of shuffling with a working definition of an in shuffle, you can now examine how cards behave in the context of odd-numbered decks. Go through Table 2.11 step-by-step, and again look for patterns.

Curious! Comparing Table 2.11 to Table 2.8, you can see that repeated in shuffles with nine cards produces the exact same result as the case with eight cards, except for the addition of the (fixed) ninth card on the end!

As you may have already surmised, something similar happens when performing repeated out shuffles. Out shuffling an

Table 2.11. Repeated in shuffles of a nine-card deck result in a similar pattern as the eight-card case.

Position	1	2	3	4	5	6	7	8	9
Start	1	2	3	4	5	6	7	8	9
Shuffle 1	5	1	6	2	7	3	8	4	9
Shuffle 2	7	5	3	1	8	6	4	2	9
Shuffle 3	8	7	6	5	4	3	2	1	9
Shuffle 4	4	8	3	7	2	6	1	5	9
Shuffle 5	2	4	6	8	1	3	5	7	9
Shuffle 6	1	2	3	4	5	6	7	8	9

Table 2.12. Repeated out shuffles of a nine-card deck.

Position	1	2	3	4	5	6	7	8	9
Start	1	2	3	4	5	6	7	8	9
Shuffle 1	1	6	2	7	3	8	4	9	5
Shuffle 2	1	8	6	4	2	9	7	5	3
Shuffle 3	1	9	8	7	6	5	4	3	2
Shuffle 4	1	5	9	4	8	3	7	2	6
Shuffle 5	1	3	5	7	9	2	4	6	8
Shuffle 6	1	2	3	4	5	6	7	8	9

odd-numbered deck is similar to in shuffling an even-numbered deck, with an extra card on top.

At this point, you have seen how decks containing odd and even numbers of cards behave under the two different shuffles. There are other mathematical curiosities hiding in this problem, which we will address in the next chapter. For now, it's worth maintaining our focus on the original goal of this exercise, which is to find some way of predicting the order of cards in the deck, and from there figuring out a way to control the results.

Magic moving cards

"Pick a card, any card…"

Your volunteer draws a card out of the deck, looks at it, and then hands it back to you. You place it on the top of the deck, and, to create the illusion of magic, shuffle the deck repeatedly. Unbeknownst to him, the card is now located in a precise position within the deck. How can you make this happen?

You don't quite have the mathematical machinery to answer this question, but there are a lot of pieces lying around. Let's see if there's a way to make it all fit together.

Up to now, you have been using small decks to identify movement patterns. You can extend this machinery, but more examples will help you to identify the concrete patterns.

Recall from our discussion of Table 2.9 *why* each card moves the way it does. When a card is in the nth position in the deck, the same number of cards from the opposite half of the deck enter into positions above it, shifting its position down by n positions. We illustrate the case, following the fourth card in the deck, in Figure 2.28.

You can track any card from the top half of the deck in the same manner. During an out shuffle, one fewer card enters the deck above any given card. You can calculate the resulting change in position of any card in the top half of the deck using these two rules; Table 2.13 shows how cards in each position move after different types of shuffle. This table assumes that, whatever the total size of the deck, the current card is in the top half of the deck.

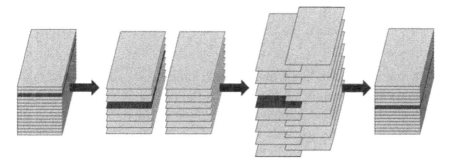

Figure 2.28.   The fourth card in the deck moves to the eighth position after cutting the deck in half and performing an in shuffle.

Table 2.13.   How the position of a card changes under two types of shuffle.

Position	Out	In
1	1	2
2	3	4
3	5	6
4	7	8
5	9	10
6	11	12
7	13	14
8	15	16

There is certainly a pattern on display here, although it does seem a bit awkward. Excusing our use of algebra, we can express the two above patterns as follows:

$$\text{out}(x) = 2x - 1$$
$$\text{in}(x) = 2x$$

For reasons that are difficult to motivate directly, we are going to suggest a change. Instead of calling the top card "position 1," we will call it position 0. You can adjust Table 2.13 to reflect this change in numbering by subtracting 1 from each value. Table 2.14 reflects these changes.

Adjusting the previous formulas for this change of base, you should find:

$$\text{out}(x) = 2x$$
$$\text{in}(x) = 2x + 1$$

This might not look like a meaningful change, but consider this: take a binary number, shift all the digits to the left by one space, and add either a 0 or 1. What happens to the value of the number in each of these cases? As you can see in Table 2.15, adding a 0 onto the right side of a binary number is equivalent to multiplying by 2, whereas

Table 2.14. How the position of a card changes under two types of shuffle, starting with the top card as position 0.

Position	Out	In
0	0	1
1	2	3
2	4	5
3	6	7
4	8	9
5	10	11
6	12	13
7	14	15

Table 2.15. Adding either a 0 or a 1 to the right of a
binary number produces a predictable result.

Number	Binary	Add 0	Value
		Add 1	Value
1	1	10	2
		11	3
2	10	100	4
		101	5
3	11	110	6
		111	7
4	100	1000	8
		1001	9

adding a 1 to the right side is equivalent to multiplying by 2 and adding 1.

Do you see it now?

We're ready to wrap up this trick with a bang.

When your volunteer places their card on top of the deck, you can move it from that position to any other by determining the target position's value, in binary form, and then performing the series of in and out shuffles corresponding to that number. So, let's say you want to move the volunteer's selected card to the exact middle of the deck, card number 26:

$$26_{10} = 11010_2$$
$$0 \cdot 10 + 1 = 1$$
$$1 \cdot 10 + 1 = 11$$
$$11 \cdot 10 = 110$$
$$110 \cdot 10 + 1 = 1101$$
$$1101 \cdot 10 = 11010$$

One in shuffle will move the card from the top of the deck to position 1. Another in shuffle will move the card to position 3 (11, in binary form). Now, out shuffle, and the magic card moves to position 6... hold on a second! The binary representation of the target number is practically a mnemonic for how to get there!

That is, the shuffles necessary to move a card to position 26 are In, In, Out, In, Out… IIOIO… that looks a lot like 11010!

By breaking a complicated problem down into smaller and smaller parts, and then building the patterns we found back up into a formal description, we have discovered a way to move a single card anywhere in the deck. This method — converting the target position into binary, then reading it from right to left as a series of in and out shuffles — works for decks of any size, and makes all sorts of magic tricks possible.

Taking Stock

Believe it or not, we have only solved four problems so far. What powerful problems they have been!

It's understandable if your head is reeling right now. The mathematics you are accustomed to is probably much more well-defined, whereas we have been helping you create mathematical tools from scratch, while building problem solving skills step-by-step.

By now we hope you understand, by following our examples, that mathematical problem solving is a process of experimentation, play, and testing and creating tools to discover how (and if) they're useful.

So, from one point of view, we have only solved four problems. From another point of view, we have used four examples to demonstrate the process of breaking large problems into dozens of small pieces, each of which can be tackled on its own.

To extend an earlier analogy, most mathematics problems you will ever see are like recipes. A recipe might tell you to "dice two medium potatoes," which is an adequate instruction, if you already know your way around a kitchen and have all the tools you need. However, that brief line is loaded with implicit knowledge! If you're trying to follow that recipe for the first time, with no experience in the kitchen, you might not know how to choose the right potatoes, how to identify a medium one, what kind of knife to use, how to hold it, what size a "diced potato" is, and so on.

We have shown you how to analyze new problems, and how to quest for your own solutions. We have also shown you what the process of building up and testing your own tools — models, definitions, transformations — looks like. And we have shown you that making mistakes, trying approaches that might lead nowhere, and struggling with new concepts are not only allowed in math, they are completely normal.

From here on, though, we will focus on building tools and demonstrating connections between old and new problems. The responsibility for taking the time to understand a solution, using any technique that works, is now yours.

A Logical Card Trick

We present to you one final card trick. In truth, it was developed as a psychological test of deductive reasoning; in light of this, we are using it as a first demonstration of how re-contextualizing information can make it easier to reason about it. It is counter-intuitive at first, but easy to demonstrate, which makes it an excellent candidate for entertaining you or your friends.

The Wason Selection Task was first demonstrated in 1966. There are several variants, but the easiest to demonstrate quickly (assuming you still have blank index cards available) is the following.

Someone presents you a rule: "Each card has a letter on one side and a number on the other. All cards with an even number on one side must have a vowel on the opposite side." They then show you the cards in Figure 2.29.

What is the smallest number of cards you must turn over to determine whether the rule holds?

Figure 2.29.   Example cards for the Wason Selection Task.

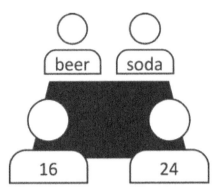

Figure 2.30.   Four people sitting at a table enjoying their drinks.

The way the question is framed suggests there is some trick; the number of flips is probably smaller than you initially believe, or requires that you flip different cards than you might first guess.

But how can you verify your suspicions?

Let's reframe the problem this way: in a certain country, the legal age to drink beer is 18. There are four people sitting at a table (Figure 2.30). You know the ages of two of them, but can't see what they're drinking. You don't know the ages of the other two, but you can see that one is drinking beer and the other is drinking soda. What do you have to check to ensure the law is not being broken?

The situation here is easier to reason about: it doesn't matter what the 24-year old is drinking, nor does it matter how old the soda drinker is. (The law says that people older than 18 can drink beer, not that they have to!) However, you should ensure the 16-year old isn't drinking beer, and that the beer drinker is at least 18.

To connect this to the motivating example, "drinking beer" here is equivalent to the condition "has an even number on one side," and "is at least 18" is equivalent to "has a vowel on the opposite side." In order to check for violations of the original rule, you would need to make sure the one "beer drinker" (even number, 4) has a vowel on the opposite side, and that the one visible "underage customer" (consonant, T) does not have an even number on the opposite side. Figure 2.31, below, shows the same information as Figure 2.30, but presented like the cards at the beginning of this section.

Figure 2.31.   An equivalent representation of the four people above, this time in the language of cards.

Reframing the question ("if someone is drinking beer, then they must be 18 or older"; "if a card has an even number on one side, it must have a vowel on the opposite side") doesn't necessarily make the problem easier to solve. Something about framing the problem in the context of social rules leverages a part of the brain that is much more intuitive!

If you don't want to rely on your intuition and translation, consider the problem case-by-case.

- The card is displaying an odd number. The rule doesn't tell you anything about what kind of letter is on the back.
- The card is displaying an even number. Then, the reverse side must have a vowel.
- The card is displaying a vowel. The number on the opposite side could be either odd (which our rule doesn't care about) or even (which is allowed by our rule), so you don't need to check it.
- The card is displaying a consonant. Then, to check for a violation, you must verify that there isn't an even number on the back.

This approach to logic is ultimately more robust,[8] but takes time and practice to master, and even then, it's easy to make mistakes.

Entertaining Brain Teasers

In the beginning of this chapter, we invited you to share tricks with an audience. You may not have access to playing cards or other props at

---

[8] Students of formal logic should recognize this pattern as contraposition. The rule, as usually stated, is: "if P then Q implies if not-Q then not-P," or $(P{\rightarrow}Q) \rightarrow (\neg Q \rightarrow \neg P)$.

every gathering, so here we present a series of problems that are challenging and entertaining while demanding no particular preparation. The problems themselves are easy to remember (or look up on the internet, as long as you remember certain key words), so even if you don't recall the solution, you can share in the fun of debating solutions as you and your friends close in on the answer!

Now, let's consider the following problems.

The vexing poisoner

A nobleman, who is known for his wine collection, is planning to host a party for a large number of distinguished guests next week. In an attempt to humiliate him, one of his enemies has poisoned one of his 1000 barrels of wine. The poison is slow-acting but effective, causing no effects at first, before killing its victim on the third day no matter how small the dose.

The nobleman has hired you to find the poisoned wine. He has provided you with ten mice, and given you permission to take samples of wine from as many barrels as you need. Can you use the mice to identify the poisoned barrel before his party?

Direct testing will almost certainly be too slow. If you were to test ten barrels at a time, in the worst case scenario, it would take almost a year to find the tainted wine. You might be able to test multiple wines at a time — giving one mouse a taste of wine from the first barrel on the first day, the eleventh barrel on the second day, the twenty-first barrel on the third day, and so on — to get through them more quickly, but you wouldn't be assured to find the poison before the party.

The next strategy you might suggest is to test multiple wines at a time. If you mix a single drop from each of 100 barrels and give the wine to the mice, you can identify a set of 100 barrels that contains the poison. After that, it is a matter of repeatedly narrowing down candidate barrels: you can use the nine remaining mice to test those 100 barrels, then test the 11 (or 12) candidates that remain with the eight mice that have survived the first two tests. However, this process is still too slow to find the poisoned wine before the party.

You should eventually realize that there must be some way to mix the wines that "encodes" the value of the poisoned barrel onto the mice. How, though?

Begin by writing the numbers 1 through 1000 on the barrels, then place the mice in individual boxes labeled 1 through 10. Next, *convert the numbers on the barrels to their binary equivalent*, and, identifying all the barrels that have a 1 in a given position, mix those wines together and give the mix to the corresponding mouse. For instance, all odd-numbered barrels must end in the binary digit 1, so, take a drop of wine from each of those 500 barrels and give the mix to the mouse in box 1. Two out of every four barrels will have the digit 1 in the second position from the right, so mouse 2 will receive wine from barrels 2 and 3, but not 4 and 5; 6 and 7, but not 8 or 9; 10 and 11, but not 12 or 13, and so forth. (You should recognize this process as similar to the method for constructing magic cards identified in Table 2.2.)

After three days, some number of mice will have perished (RIP). You can now read off the numbers on their cages, and use them to construct the binary number corresponding to the poisoned barrel!

To demonstrate, let's say that mice 2, 4, 7, and 9 have died. Mouse 2 would have consumed wine from all barrels with a 1 in the second position from the right; mouse 4 would have consumed wine from only barrels with a 1 in the fourth position, and so on. Whatever barrel contains the poison must have 1s in the positions corresponding to these mice, and the digit 0 in all other positions. The only wine that all these mice, and only *these*, would have consumed comes from $01\ 0100\ 1010_2 = 330_{10}$, so you can tell your patron that the poison is in barrel 330!

You might be interested to learn that a variation of this method applies in real-world medical testing.

First, you might ask how many barrels we can test with ten mice. At the end of the 21-card trick on page 49, we showed how, each time your volunteer points to one row out of three, you can distinguish between three times as many cards as before. The same principle applies here. Each additional mouse allows you to test twice as many barrels as before, as each mouse can be in one of two states (alive or dead).

You can motivate this intuition by starting from a small case and building up. If the nobleman has two barrels of wine and you know that exactly one is poisoned, you can give wine from one of the barrels to a single mouse; if the mouse is alive after three days, the poison was in the other barrel.

With three or four barrels of wine to test, there is no simple way to find the poison with only one mouse. Dividing the barrels into two groups, two different ways, will allow you to test up to four barrels with two mice. The most straightforward way to divide four things into two groups two ways is by numbering them in binary, as you have seen before and as we illustrate in Table 2.16.

Proceeding in this manner, you can see that for a given number of mice m, we can test 2^m barrels of wine for poison. Thus, with ten mice you can test up to $2^{10} = 1024$ barrels of wine.

Finally, you might ask if there is some way to test more barrels or to use fewer mice, and there is! In the original problem, we stated that the poison kills with 100% lethality after three days, and the deadline for testing the wine is a week in the future. If you run a test starting on day one, by day four some or none of the mice will have died. Since you know that only one barrel has been poisoned, you can overlap tests: start one test on day one, finishing on day four; another test, with a different set of barrels, would start on day two and finish on day five; and so on.

You now have four possible starting dates and ten mice, which will allow you to test up to $4 \cdot 2^{10} = 4096$ wine barrels!

Table 2.16. A simple case with only three barrels demonstrates the structure of the tests.

Group 1	Group 2
Barrel 1 (01_2)	Barrel 1 (10_2)
Barrel 3 (11_2)	Barrel 3 (11_2)

Making an omelet the hard way

As the saying goes, "you can't make an omelet without breaking some eggs."

You are an employee at a new restaurant at the ground floor of a fifty story building, and your boss wants to set up a spectacular opening attraction to stir-up business: he wants you to drop eggs, without breaking them, from the highest possible floor.

Your boss is also cheap, so he's only given you two eggs with which to find the target floor. What is the smallest number of drops you'd need to perform to guarantee you've found the correct floor?

The first process that might come to mind is go up one floor at a time, dropping an egg then fetching it, repeating until it breaks. The worst-case scenario, though, would require fifty drops — certainly more effort than you wish to expend! How can you reduce this number?

You might start by dropping the first egg from the fiftieth floor, and if it breaks, dropping the second egg one floor at a time, starting at the first floor. This doesn't reduce the worst-case scenario, but it does hint at an approach that might work better.

Next, you consider what would happen were you to drop the first egg from the twenty-fifth floor... Aha! Now you're getting somewhere!

Here, we will suggest a testing procedure. If you know that an egg will break if dropped from floor x, but that it won't when dropped from floor y, your only recourse is to try each floor from just above y to one less than x, one floor at a time. You can drop one egg — the first egg — from floors with a common distance between them to cover distance more quickly. So, you could drop the first egg from the twenty-fifth floor, and if it survives, drop it from the fiftieth floor; if it ever breaks at either of these floors, continue testing with the second egg one floor at a time, starting from the next highest floor that didn't break the first egg.

Under this procedure, the worst case scenario requires twenty-six drops (drop the first egg from the twenty-fifth floor, then the fiftieth, where it breaks; drop the second egg from each of the floors from twenty-six to forty-nine), but the most likely scenarios all save time compared to the prior approaches.

We will refer to the jumps in height taken with the first egg as the testing *step size*. So if you drop the first egg from floors ten, twenty, thirty, and so forth, the step size is equal to ten.

Continuing along this line of thinking, you can vary the step size you take with the first egg to identify which one produces the best results. For instance, using a step size of ten, and assuming an egg will break when dropped from the twenty-seventh floor, you would end up dropping an egg from the following floors: ten, twenty, thirty (whereupon the first egg breaks), then floors twenty-one, twenty-two, twenty-three, ... up to twenty-seven, where the second egg would break, for a total of ten drops.

You can reason through the worst-case scenario with any step size. (We examine several of these in Table 2.17.) If the first egg is dropped at regular intervals, the maximum number of drops you would ever need to perform would be equal to the most drops you would ever need to break the first egg, plus the greatest possible number of drops of the second egg.

Hold on a moment! The number of drops needed in the worst-case scenario went up at the end!

There is an easy way to understand this behavior. A step size of one and a step size of fifty result in the same worst-case performance: both situations have a case that would require fifty drops.

Table 2.17. Worst-case testing sequences for different step sizes. Drops conducted with the first egg are in bold.

Step Size	Maximum Drops	Longest Series
50	50	**50**, 1, 2, 3, 4, 5, 6, 7, 8, 9, 10, ..., 49
25	26	**25, 50**, 26, 27, 28, 29, 30, 31, ..., 49
20	21	**20, 40**, 21, 22, 23, 24, 25, 26, ..., 39
15	17	**15, 30, 45**, 31, 32, 33, 34, 35, ..., 44
10	14	**10, 20, 30, 40, 50**, 41, 42, 43, ..., 49
5	14	**5, 10, 15, 20, 25, 30, 35, 40, 45, 50**, 46, 47, 48, 49
3	18	**3, 6, 9, 12**, ..., **42, 45, 48**, 46, 47

The latter, you can see in Table 2.17; while the former is the same strategy first identified in this section, wherein you would drop one egg from each floor starting at the first floor. If your eggs could survive a forty-nine story drop, testing one floor at a time would necessitate fifty drops.

After examining other potential step sizes, it should become clear that the necessary number of drops falls, but you know it must rise again at the end, suggesting the graph in Figure 2.32. Notice that the minimum and maximum worst-case performance are the same, and that there is a minimum somewhere between them. The point corresponding to a step size of ten must be somewhere near the minimum of the graph, but it remains to be seen where that minimum point is in relation to this value.

The minimum point could be either to the right or left of ten. By testing other values in this neighborhood, you should be able to find the true minimum, as in Table 2.18.

Take note of the method here for narrowing down the minimum values (that is, best worst-case performance) of the graph here: first, test values somewhat to the right and left of the currently known minimum; then decrease the step size; and finally test values outside

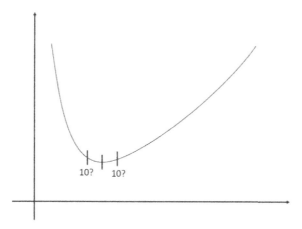

Figure 2.32. The sketch of a graph of step size (x-axis) versus worst-case (y-axis). If the worst case for a step size of 1 and 50 are the same and maximal, the graph of worst cases versus step size should curve down between them.

Table 2.18. Testing additional values near the minimum known value for worst-case performance.

Step Size	Maximum Drops	Longest Series
15	17	**15**, **30**, **45**, 31, 32, 33, 34, 35, ..., 44
10	14	**10**, **20**, **30**, **40**, **50**, 41, 42, 43, ..., 49
12	15	**12**, **24**, **36**, **48**, 37, 38, 39, 40, ..., 47
8	13	**8**, **16**, **24**, **32**, **40**, **48**, 41, 42, ..., 47
9	13	**9**, **18**, **27**, **36**, **45**, 37, 38, 39, ..., 44
7	13	**7**, **14**, **21**, **28**, **35**, **42**, **49**, 43, ..., 48
6	13	**6**, **12**, **18**, **24**, **30**, **36**, **42**, **48**, 43, ..., 47
11	14	**11**, **22**, **33**, **44**, 34, 35, 36, 37, ..., 43

the current window to find other nearby values that have the same minimum.

Somewhat surprisingly, the graph doesn't have a single minimum value. Instead, there are four different step sizes that all share the same minimum. Any of these solutions would suffice if your task were simply to find a "pretty good method" of testing the eggs' durability. Can you improve on these results, though?

Once again, the best way to demonstrate the trick here is by working backwards.

If you could only drop an egg once, you could only test a one-story building. (see Figure 2.33, left building.)

If you could perform two drops, you could test a building up to three stories tall. Think about it like this: drop the first egg from the second story; if the first egg breaks when dropped from the second floor, drop the second egg from the first story (whether or not it breaks, you're done); if the first egg doesn't break when dropped from the second floor, drop it from the third story. (Figure 2.33, center building.)

The three-drop case becomes somewhat harder to reason about, so think about it as an expansion of the two-drop case. If the first egg doesn't break on the first drop (whatever floor you eventually drop it from), then repeat the process from the previous paragraph on the last three floors. If it does break, you can test at most two floors in two drops with the second egg. (Figure 2.33, right building.)

Each time you allow one more drop, the graph that results contains the next smaller graph. Thus, if you are allowed four drops, if the first egg survives its fall, you can test at most six higher floors in accordance with the procedure on the right in Figure 2.33. If the first egg doesn't survive its fall, then you can test at most three floors with the second egg. Repeating this process again and again, you should produce a diagram similar to Figure 2.34.

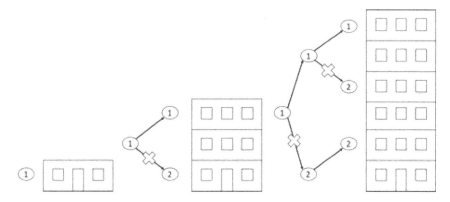

Figure 2.33. Maximum possible building sizes for one, two, and three drops. If egg 1 breaks, use egg 2 on a lower floor.

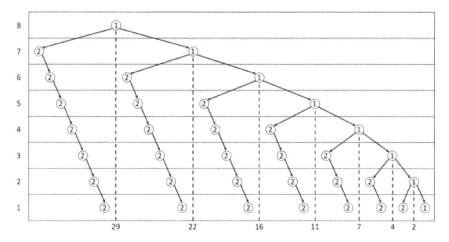

Figure 2.34. The previous diagram, turned clockwise and extended. Horizontal rows indicate the drop number; vertical dashed lines indicate the number of floors down from the top of the building.

By reading information off Figure 2.34, you can begin to observe patterns. If you are permitted one drop, you can test a building that is one floor high, starting from the first floor from the top. If you can make two total drops, you can test a building three stories high. If the first egg survives its first drop, two stories from the top, drop it again from one floor higher.

We have catalogued the results of this process in Table 2.19.

From this table you can see that the answer to the original puzzle is ten drops. That is, you can find the maximum height an egg can be dropped without breaking from a fifty-story building by performing no more than ten drops.

This is not the only conclusion to be drawn from Table 2.19, though — there is certainly some pattern emerging! One final transformation should make the pattern absolutely clear. If you ignore the floor numbers and instead look at the egg-drop problem from the perspective of "how much information can be drawn from different numbers of drops using only two eggs," you will produce a graph similar to Figure 2.35.

Here, the question you're trying to answer is "how large of a building can I test using two eggs?" You can determine the minimum

Table 2.19. Increasing the number of allowed drops increases the size of the tallest building that can be tested by a predictable value.

Max drops	Max Floors	Starting Floor (from top)	Step Size on Success
1	1	1	N/A
2	3	2	1
3	6	4	2
4	10	7	3
5	15	11	4
6	21	16	5
7	28	22	6
8	36	29	7
9	45	37	8
10	55	46	9

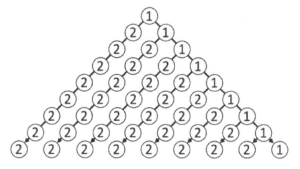

Figure 2.35.   A different way of looking at Figure 2.34. Here we can see the "quantity of information" that can be extracted using two eggs.

number of drops necessary to test a building of a given size by looking for the smallest triangle that has at least as many circles in it as the building has floors. This is the same kind of solution we found in the "Vexing Poisoner" problem, but because of a difference in perspective, the connection was not initially obvious!

You can formalize this result, and if you are familiar with triangle numbers you should already see where this is headed. With two eggs and d drops, you can test buildings up to a height h of

$$h = \frac{d(d+1)}{2}$$

Other questions you can ask include how large a building can you test with three, four, or more eggs? We will leave these as exercises to you, the reader, with one exception. If there is no limit to the number of eggs you can break, how many drops would it take to test the height of a fifty-story building?

Start at the middle floor of the building (an example building can be seen in Figure 2.36). Drop an egg. If it breaks, start over with a new egg, dropping it from the halfway between the middle floor and the ground. If, on the other hand, the first egg survives its fall, drop that same egg from the floor that is halfway between the starting point and the top of the building. After each drop, the range of floors you need to test is cut in half: when an egg breaks, you certainly don't

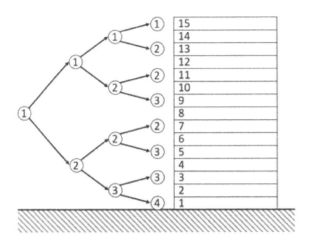

Figure 2.36. A protocol for testing egg strength using an unlimited number of eggs.

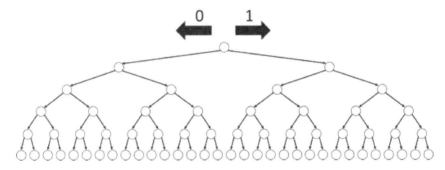

Figure 2.37. The tree that is produced when you can break any number of eggs. At each step, you discover one new piece of information: did the egg break at this floor (0) or did it survive the fall (1)?

need to test any of the higher floors; when an egg survives, you certainly don't need to test any of the lower floors.

Note that adding one more drop doubles the number of floors you can test. At each step, you determine a new piece of information, as we show in Figure 2.37. If an egg dropped from some floor doesn't survive its fall, write down a 0; if it does survive, write down a 1.

This should seem familiar! If, every time an egg breaks, you record a 0, and every time an egg survives a fall, you record a 1, this problem reduces *exactly* to a type of problem you've seen before! That is to say,

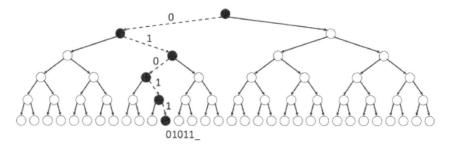

Figure 2.38. An example of a chain of drops from different floors of a sixty-four story building. Each drop generates a new piece of information — did the last egg break (0) or survive (1) — and stringing those digits together corresponds to the maximum height an egg can be dropped from, in binary.

the digits you record with each drop tell you the maximum height an egg can fall from without breaking.

If an egg can survive a drop from the twenty-second floor, and the building you're testing is sixty-four stories tall (Figure 2.38), the first egg will break when dropped from the thirty-second floor (0), the second egg will survive its fall from the sixteenth floor (1) but break when dropped from the twenty-fourth floor (0), and so on. Reading off the list of digits, you find the binary number 010110, which, converted to decimal, yields 22!

Falling pressure

We will take a break from mathematics with a humorous variation of the previous problem.

Simple problems, and many puzzles, are worded in a way that points the reader toward information that is important. The poisoned wine problem, for instance, is usually stated with 1000 barrels (or bottles) of wine. Were we to use the number 1024, the answer might be too obvious; were we to use some other equally valid number, like 537, the answer might be too obscure to find without frustration.

Real-world problems and puzzles are often misleading. Extraneous information may be included that causes you to get lost in a blind corner, futilely striving to find a way out. The problems we've demonstrated so far are meant to help you get familiar with the process of extracting a limited amount of information from a narrowly defined scenario; starting in Chapter 3 we will spend more time learning how to recognize and handle these traps.

Too much school-style mathematics can cause students to develop bad habits. We have repeatedly alluded to looking things up "in the back of the book," but there is a habit that is even more insidious than this. One way to get good at the kind of mathematics seen in school and on tests is to become good at identifying "what the question is asking." In 1979, French researchers documented one problem that illustrates this type of thinking. Students between the ages of nine and ten were presented the following question:

On a boat, there are 26 *sheep and* 10 *goats. What is the age of the captain?*

Some students correctly recognized that the problem does not provide enough information to be solved. Others students saw that a question was being asked, and simply scanned backward, looking for numbers to manipulate. Fully three-quarters of students responded that the captain must be $26 + 10 = 36$ years old!

Recognizing that assignments are written in the context of a chapter or subject, and thus must have an answer that is relevant to the current material, can be a good heuristic for narrowing down which approach the teacher or test-writer "expects." In some problems, and in real life, questions are rarely so simple, so while this can be an excellent strategy for getting good grades and succeeding on tests, too much of it can result in the inability to handle truly novel problems.

The opposite scenario is one you very well may see amongst your friends. You decide solving a puzzle will be a fun diversion; one of your friends — the class clown, the joker — deliberately misinterprets the rules you've established, or jumps on some relevant omission, to disrupt your fun.

In the moment, you'll have to find a way to handle your friend on your own. In the meantime, we offer this apocryphal example for your consideration.

In a physics class somewhere, the story goes, a student reads the following question on an exam: "Using a barometer, how can you find the height of a very tall building?"

Per the previous discussion, you can understand that the teacher's intended solution involves measuring air pressure at ground level (using the barometer), again at the roof of the building, and then performing some calculation involving the difference in pressures (presumably from the current chapter of the book!) to determine the change in vertical distance.

The wise-guy student, however, suggested the following answers:

- Using a stop watch, drop the barometer off the roof, timing it as it falls. Using the formula for a falling object and plugging in the time measured between release and impact, you can then calculate the height of the building.
- Tie the barometer to a long rope, and let it out down the entire height of the building. When you feel it touch the ground, mark the rope and pull the barometer back up. Then, simply measure the length of rope between your mark and the barometer to determine the building's height.
- Using the same rope, drop the barometer down until it's almost touching the ground. Set it swinging and, using the stopwatch from before, measure the time taken by each swing of this *ad hoc* pendulum. Plugging this value into the pendulum equation, you can determine the length of the pendulum and thus the height of the building.
- Find the office of the building superintendent. Knock on his door and, when he answers, say, "hello, sir; could I interest you in this gorgeous barometer? In exchange, I was wondering if you could tell me the height of this building?"

(In the story, the student admitted he knew what the intended answer of the question was, he simply didn't feel like always answering what he was "supposed to" answer.)

As you progress through this chapter, and the rest of the book, be mindful not to spend too much time trying to figure out how you are "supposed to" think, nor ignoring signposts that point you toward the right answer!

Problems you can solve in the dark

The next four problems are about information that you cannot see, but have to use logic to manipulate. How can you think about problems when you can't see the answers?

Lights in the attic

You just purchased a house, and there are three switches on the top floor corresponding to three different light bulbs in the attic. You want to know which switch corresponds to each bulb, but you don't want to have to climb up and down the attic ladder to do so. How can you determine which switch corresponds to each bulb in a single trip?

At this point, your joker friend might ask if he can see into the attic from where the switches are located. The setup for this puzzle is usually more convoluted, specifically to prevent this sort of thinking ("you have been captured by an evil villain, who has locked you in a room containing three switches..."). This kind of contrived setup can turn readers off to problem solving, so we're choosing to avoid it.

Assuming that there is no possible way to see the lights or any of their effects from where the switches are located, how might you solve this problem?

Based on the lessons of this chapter, you should notice that if you rely on the state of the bulbs themselves (on or off), there is no way to distinguish between more than two bulbs based on light alone. You would need at least one more piece of information to distinguish between three bulbs.

Luckily, lightbulbs do not only produce light! Turning on a light switch heats up the bulb (this even applies to modern lightbulbs, based on compact fluorescent tubes and LEDs); recognizing this fact points toward a solution.

Turn one switch on, leaving it on for several minutes. Then, turn it off, and turn a different switch on. Climb the ladder into the attic and check which of the two dark bulbs is hot — that bulb corresponds to the first switch you flipped — and which bulb is lit — corresponding to the second switch. The cold, dark bulb of course, corresponds to the third, untouched, switch.

The sleeping spouse

You are getting ready for work early one morning when you realize you forgot to set aside socks for your outfit the previous night. Trying not to wake your spouse, you slip back into the bedroom and, without turning on the light, find your way to the dresser. Knowing that you own socks of three different colors, how many individual socks must you pull from the drawer to ensure that you have at least one pair that matches?

This question is relatively straightforward, but the insight needed to solve it will eventually lead us to lead us to some surprising insights in Chapter 3.

Two socks are not *guaranteed* to be sufficient, since you own three different types of socks. Similarly, the worst-case scenario if you only take three socks from the drawer is that you have one of each type, and thus have to wear two different socks to work, making yourself a potential target for embarrassment.

If you take four socks out of the drawer, you must have at least one matching pair. You can enumerate all possible collections of four socks to drive this point home.

Suppose that you have white (W), gray (G) and black (B) socks. Writing out every possible arrangement, as in Table 2.20, you can verify, row-by-row, that every row contains at least one matching pair. While making this kind of list is an effective check on your logic, it's much more difficult to list all possible sets of four socks, and confirm that all of them are listed, than it is to trust and verify your own logic!

A coin-flipping game

This puzzle can be acted out literally as described, or solved via discussion and experimentation. While nothing prevents you from trying

Table 2.20. All possible collections of four socks of three different colors: white (W), gray (G) or black (B).

W	W	W	W
W	W	W	G
W	W	G	G
W	G	G	G
G	G	G	G
G	G	G	B
G	G	B	B
G	B	B	B
B	B	B	B
B	B	B	W
B	B	W	W
B	W	W	W
W	W	G	B
W	G	G	B
W	G	B	B

it out exactly as described, you may provoke a more enjoyable social experience by talking it through with friends first.

Your friend empties her pockets, pulling out sixteen coins. She proposes a challenge to you: she will put all of the coins flat on a table, then ensure exactly six of them have the head-side facing up (leaving ten head-side down), before turning off the light and leaving the room.

You are to enter the room, and manipulate the coins so that they form two distinct groups with the same number of heads in each group. You won't be able to see the coins, and it is not possible to feel the difference between heads and tails by touch alone. How can you outsmart your friend?

What a tricky situation! Your first attempts to solve this problem may leave you feeling shortchanged. However, we assure you that this puzzle is solvable by logic alone!

A careful reading of the prompt suggests two things. First, your friend did not tell you that the groups had to be of equal size. Second, she did not say that you couldn't flip coins yourself. Put a mental bookmark here for now, because we'll perform some investigations using smaller sets of coins before revealing the answer.

If your friend used two coins and said both were heads, there would only be one possible division of the coins into groups, and it is obvious that the two groups would each contain one head.

If both coins were tails, the same thing would be true: dividing the coins into two "groups" of one coin each would result in zero heads per group.

If, however, only one coin were heads, you would have no way of knowing which coin was which. However, as long as you flip one of them, as in Table 2.21, you have two groups with the same number of heads — zero, or one!

If she told you there were three coins, one of which was heads, the same approach would apply, with one caveat: were you to flip a coin in the group of two coins, there is a chance that you wouldn't end up with the same number of heads in both groups. However, as you can see in Table 2.22, it seems that if you flip the coin in the group of one coin, you will always end up with the same number of heads in both groups.

What property makes this work? Or, is it simply a coincidence?

It's impractical to make a table for a large number of coins (simply enumerating all possible arrangements of the coins would be difficult enough!), so you'll have to find another way to explore this theory.

Suppose for a moment that you have a large number of coins, exactly half of which are showing heads. By a stroke of luck, when you

Table 2.21.   If you split two coins, one heads and one tails, into two groups, then no matter which coin you flip, there will be the same number of heads in each group.

Coins	Group 1	Group 2
HT	H → T	T
HT	H	T → H
TH	T → H	H
TH	T	H → T

Table 2.22: The previous method fails in some cases where you flip some or all the coins in the larger group. When you flip only the coin in the one-coin group, both groups always contain the same number of heads.

Coins	Group 1	Group 2	Valid?
HTT	HT → HH	T	Invalid
HTT	HT → TT	T	Valid
HTT	HT → TH	T	Invalid
...			
HTT	HT	T → H	Valid
THT	TH	T → H	Valid
TTH	TT	H → T	Valid

split them into two groups, you happened to put all the heads on the right side of the table and all the tails on the left. Now, were you to flip all the coins in one group, both groups would be all heads or all tails. (This is essentially identical to the problem with only two coins.)

Now comes the tricky part: choose one coin at random from both sides, flip both, and switch their positions. There are still equal numbers of heads on both sides, because if all tails were showing, you have just added one head to each side; if all heads were showing, you simply subtracted one head from each side.

Between the two groups of coins, you can imagine an invisible wall: whenever a coin passes completely through this wall, it flips. If the group on the left was the one you flipped, then passing the invisible wall returns the coin to its original orientation, whereas moving a coin from right-to-left would reverse the coin relative to its starting position.

Note that no matter which coins you move nor how many you move, as long as you trade them one-for-one and flip them when you cross the centerline of the table, there will always be the same number of heads on both sides.

To put some numbers to this situation, suppose you start with 200 coins, 100 of which are showing heads. If you move 100 coins from the right side of the table to the left side, 37 of which are heads-up, then the right side of the table would have $(100 - 37)$ heads, and the left side would have 37 **tails**. Since there are 100 coins on the left side of the table, there must be (100 − 37) heads on that side — the same as on the right!

Now, you can put all these pieces together. In Table 2.22, when working with three coins and one head, flipping the sole coin by itself was sufficient to guarantee the same number of heads in both groups. Looking at the previous paragraph, you can just barely see the heart of an idea talking about all groups of coins. Let's see if there's a way to refine it.

If this explanation is confusing, feel free to follow along with the example provided in Figure 2.39.

You have *c* coins, *h* of which are heads. Move *h* coins at random to a second group. There are now *c* − *h* coins remaining in the first group. Some of the *h* coins in the second group are heads, but you do not know (and will never know) how many there are. Call this number *r*, for "removed."

The first group contains *r* fewer heads than it did at the start, so there are *h* − *r* heads there total.

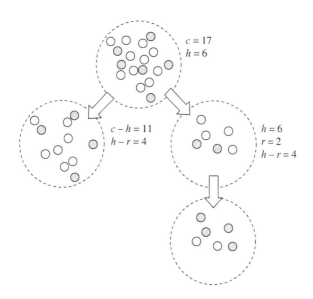

Figure 2.39. Splitting *c* = 17 coins containing *h* = 6 heads into two groups. Some number *r* of heads are moved at random to the second group; when those coins are flipped, both groups contain the same number of heads.

In contrast, the second group contains r heads and $h - r$ tails. By flipping *every* coin in the second group, you reverse these totals: after flipping, there are r tails and $h - r$ heads.

This provides a solution not only to the original puzzle, but also a rough proof allowing you to solve all possible puzzles of this type, no matter what number of coins your friend uses. You can freely plug in the numbers from the original statement, if that helps you reason about the last three paragraphs; you will discover that they hold true, and you can demonstrate this irrefutably by locating some coins and seeing the process work for yourself.

Thinking about yourself

This next puzzle doesn't involve darkness, but it does involve hidden information that may leave you in the dark, until you find the solution!

The king of the land, on hearing of your success in dealing with the case of the Vexing Poisoner, invites you to a test to see if you are wise enough to become his advisor. You have competition in the form of two other candidates.

He places the three of you in a room, facing one another, and then places a skullcap on each of your heads. You can see that the other candidates' caps are black, but try as you might, you cannot see your own. The king then makes the following pronouncement:

"Each one of you is wearing a cap that is either black or white. At least one of your caps is black. The first person who finds out the color of their own cap, and can explain how they know it, will be my next advisor."

After several moments, neither of the other two candidates seems to have an answer. Can you figure out the solution?

There are two ways to entertain your friends with this tricky logic puzzle. The first is to simply read them the puzzle as we have written it, discussing it as you go. However, you can also turn this into a party game, assuming you already know the answer: you can use five hats of two different colors, or, if you don't have hats, you can stick a red or black playing card to each volunteer's head!

To figure out this puzzle, you will have to learn to see yourself through another person's eyes. We used logic to reason about numbers in prior puzzles; this puzzle demands that you use logic to reason about logic.

Even with that hint, the solution may not jump out at you. So, you can begin as you have in the past, enumerating possible configurations of caps, as in Table 2.23.

Is that the whole table? The king stated that there is at least one black cap; thus, the table does not contain a row with three white caps. Was it right to omit that row, though?

Consider the counterfactual case where the king had placed white caps on each of your heads (Figure 2.40). You could quickly deduce

Table 2.23. All possible arrangements of hats, given that at least one is black.

Candidate 1	Candidate 2	You
White	White	Black
White	Black	White
Black	White	White
White	Black	Black
Black	White	Black
Black	Black	White
Black	Black	Black

Figure 2.40. If only one cap were black, one of the candidates would know their cap's color immediately.

that each candidate would see two white caps and, lacking the original restriction, none of you would be able to tell what color your own cap was.

You know that there is at least one black cap, though. If you again test a counterfactual situation, this time asking what would happen if there were two white caps, you should quickly realize that one of the candidates would be able to look at the other two and, seeing only white caps, determine that their own cap must be black.

You know that both your fellow candidates are wise, thus you can surmise that they would be able to draw the same conclusions as you.

Now, assume that you are wearing a white cap. The other two candidates would see one white cap, one black cap, and not be able to see their own cap. Critically though, each of you would know that if any one of you sees two white caps, that person would immediately know their cap's color. Since you all know this, and all of you know that none of you has answered yet, none of you must be wearing a white cap.

Your hat must then be black!

Reasoning about the process of reasoning can be incredibly difficult (see Table 2.24). Until now, most, if not all, the information you have used to determine solutions has been drawn directly from the problem statement (albeit often several steps removed). The critical piece of information needed to solve this puzzle came from thinking

Table 2.24. Successive phases of reasoning about reasoning.

I am...	I see...	If I am wearing...	Then I know...
Candidate 1	White, Black	White	Another candidate would have answered immediately.
Candidate 1	White, Black	Black	No one else answered immediately, thus no one saw two white caps, thus I must be wearing black.
You	Black, Black	White	One of the other candidates would have figured out the previous line already.
You	Black, Black	Black	No one would have answered yet.

what information you had, what information others must have, and then inferring a conclusion from the assumption that our opponents would be rational — not an easy task!

We will not spend a lot of energy on this type of problem. It is however worth knowing that approaches like this are possible, because thinking about others' thinking transcends the math, cropping up wherever humans compete: from games, to economics, to politics.

Thinking laterally

It's no accident that many of the problems so far have explored similar themes. By now, you have seen that each of the tools you build or discover can be used in unexpected ways. To avoid giving the impression that these are the *only* tools you can use in mathematics, for the remainder of the chapter we will explore some short problems that are no less entertaining than those that came before, but will exercise different parts of your brain.

Filling buckets

You are at a cabin in the woods, far from civilization, mixing concrete to build a barbecue pit. The instructions on the bag say that for optimum strength, you should mix the entire contents of the bag with exactly four gallons of water. Luckily, you bought a five-gallon bucket when purchasing the concrete; unluckily it doesn't seem to have any intermediate markings on the inside. Scrounging around the cabin, you find another bucket that holds three gallons.

Using nothing but these two buckets and a water faucet, how can you measure precisely four gallons of water for your project?

This problem[9] does lend itself to creative half-solutions. Practically speaking, many people in this situation would guess at how much water they needed ("yeah, that looks like about four gallons") and leave it at that; some of you reading this may feel the same way.

---

[9] Actually, class of problems, as we will see.

Here are three possible rebuttals. First, we could contrive a situation where guessing is impossible or would lead to disastrous results ("if you use even a drop too much or too little water, the concrete will explode!"), but this lends an unrealistic air to many mathematics problems. The second would be to resort to something we identified earlier: suspension of disbelief. Under this paradigm, we assume you're interested in the puzzle itself, and thus treat the problem statement as a means to present information, independent of its particular details.

We will take a third approach here. While humans can usually estimate quantities in containers that are regular (that is, having straight sides) with a high degree of accuracy, we are often overconfident about our abilities to estimate volumes in containers with irregular shapes. Even ignoring contrived questions such as "how much water is there in all the pipes in your house," or "how much sand can fit in the passenger compartment of a car," shapes with easily described geometry — cylinders, spheres, pyramids, and so forth — can demonstrate unexpected relationships between the height of liquid inside and the volume.

The most telling real-world example of this phenomenon can be seen anywhere drinks are served. Beverage containers — paper and plastic cups, pint glasses, and so on — generally have a tapering design to aid in stacking (Figure 2.41). As a side effect of this design, the upper part of these glasses make up a disproportionate percentage of the total volume. Every few years, some restaurant or bar gets in trouble with the authorities for short-pouring drinks, depriving customers of 10–25% of their purchase with every drink served. These scams take time to detect, because most customers see a glass that is a half-inch short, or with a half-inch of extra foam, and don't realize how much they're getting ripped off!

When you aren't sure if you can rely on your estimating skills, applying logic can help.

Within a few minutes, you should be able to find a solution to this problem. At any moment, there are a limited number of moves you can make: fill one of the two buckets, empty one of them, move water from the larger bucket to the smaller, or move water from the smaller bucket to the larger. With perseverance, you can combine these steps in some order to arrive at a solution.

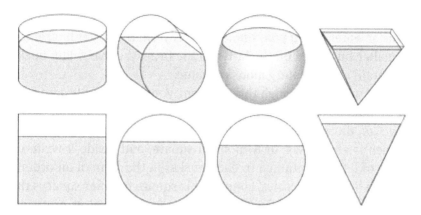

Figure 2.41. (Top) Three-dimensional representations: a vertical cylinder, horizontal cylinder, sphere, and square pyramid containing varying amounts of liquid. (Bottom) Cross-sections show where the water line would be in each shape when 75% full by volume.

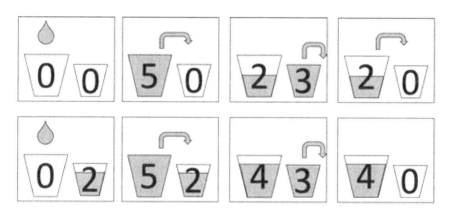

Figure 2.42. A successful strategy to end up with exactly four gallons.

Figure 2.42 shows a solution you could arrive at *via* this kind of trial-and-error. Fill the larger bucket from the faucet. Next, using the larger (five gallon) bucket, fill the smaller bucket to the brim; the larger bucket will now contain two gallons of water. Empty the smaller bucket, then pour the entire contents of the larger bucket into it. Again fill the five gallon bucket from the faucet. At this point, the five gallon bucket is full, and the three gallon bucket contains two gallons of water. If you pour enough water from the larger bucket into the smaller bucket to fill the latter to the brim, the five gallon bucket will contain exactly four gallons of water.

When working through this problem, you may stumble across the idea of making a diagram of transitions between possible states instead of drawing pictures of buckets. The general idea of this approach is as follows: draw a small square, and inside it write the amount of liquid contained in each bucket in the form of an ordered pair. With lines and arrows, connect this square to other squares that can be reached by filling one of the buckets from the faucet, or moving water from one bucket to another.

For instance, the buckets begin in the state (0, 0). If you fill the larger bucket, you will move to the state (5, 0). By pouring water from the larger bucket to the smaller one, you move to state (2, 3). If you're not careful, your diagram can quickly become cluttered and unreadable, like the complete state diagram in Figure 2.43.

What a mess! Each transition encapsulates a single action — filling one bucket, emptying it, or pouring from one bucket to another — but the results get quite hard to follow!

After a little time tinkering with this approach, you might realize that there is some underlying order to this method. What is it, though?

First of all, the choice of notation is natural, but it so happens that it's the same as the notation we use for Cartesian graphs. How can you use this fact?

Draw points on a plane corresponding to each of the possible quantities of water the two buckets can hold, and then connect valid transitions with arrows. You should notice that, when translating the original solution to this graphical format, there are three valid movements: horizontal movement, indicating filling or emptying the larger bucket; vertical movements, representing the smaller bucket; and movements that go up and left or down and right, corresponding to pouring water from one bucket to another.

Trace the route taken by the arrows in Figure 2.44, and compare the transitions there to the pictures of buckets in Figure 2.42.

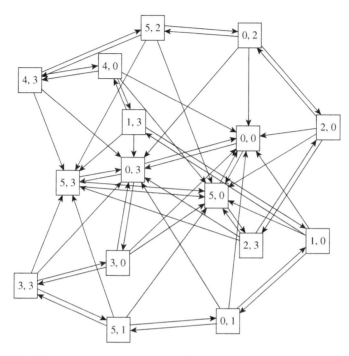

Figure 2.43.   The complete state transition diagram for this problem. Arrows represent possible routes between states; for instance, emptying a bucket in (4, 3) can carry you to (0, 3) or (4, 0).

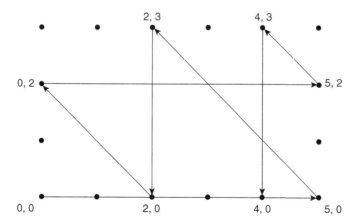

Figure 2.44.   Viewing volumes as points on the plane, and actions (filling, emptying) as arrows between points.

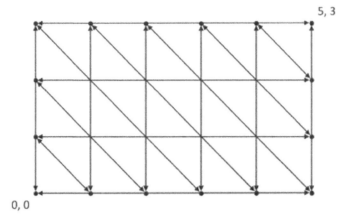

Figure 2.45. All possible states (dots) and transitions (arrows) using buckets of a given size.

Note, too, that there is no way to stop partway across on a line. From any valid point, you are only able to draw a line that stops at one of the outside edges. This makes sense, given the restriction that you can only fill or empty buckets.

If you repeat this process over and over, eventually you will discover the map in Figure 2.45, illustrating all possible transitions between states.

Three conclusions that can be drawn from this diagram follow:

- In the case of a five-gallon and three-gallon bucket, we can reach any whole-number state where at least one bucket is full or empty.
- This process can be extended to any problem involving two buckets, regardless of their sizes.
- We have stumbled on a method for marking the interior measures of both buckets! Each bucket can be filled to a precise value, at which point you can make a mark at the waterline in the interior of the bucket corresponding with that value.

The first conclusion includes several qualifiers. Are all of these necessary?

As an experiment, try this same method using two buckets of size two and four gallons. Is it possible to fill either bucket to the one-gallon mark?[10]

Matchstick puzzles

The final class of problems for this chapter is one that invites engagement from everyone, young and old alike. The most common name for them is "matchstick problems," but since it's unlikely in the current age for your fellow party-goers to be carrying boxes of matches, toothpicks often suffice.

A typical puzzle looks like this.

Puzzle 1. Given matchsticks in the arrangement shown in Figure 2.46, can you move five sticks to leave exactly four squares?

There are a few general rules. When a problem asks you to move a certain number of matchsticks, you must move that many sticks exactly. "Dangling" matchsticks, like the one shown in Figure 2.47, generally aren't allowed.

When a puzzle specifies that you must *move* matchsticks, removing them from the puzzle is not allowed. And, unless permitted by a specific puzzle, you are not allowed to break matches. These puzzles originated with people sitting around salons and bars, passing time — if they broke all of their matches, they'd never have a chance to use them!

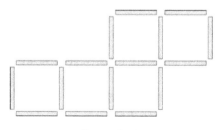

Figure 2.46.

---

[10] If you attempt this problem on your own, you will discover that there is no way to fill *either* bucket to either the one- or the three-gallon mark. Strange!

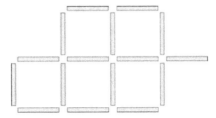

Figure 2.47. A dangling match. Most puzzles do not allow you to leave matches dangling.

We will work our way through a number of these problems, testing your ability to "think outside the box" in increasingly surprising situations!

You can work through these problems on your own or with others; if you enjoy them, you can find hundreds more of these puzzles online, or in a multitude of books available on puzzles or recreational mathematics.

To get truly good at matchstick puzzles, it can be helpful to create a mental vocabulary for describing different moves. For instance, we will refer to the removal of a single matchstick from the side of a square as *breaking* the square. Further, since it is almost never permissible to leave dangling sticks, we refer to a series of steps that removes a square, along with all the dangling edges, as *removing* the square. (Your terminology doesn't have to be complicated, just useful!)

Puzzle 1 solution

The solution to the first problem involves breaking the two outermost squares, while re-using one of their sides, and breaking an additional square — the third one on the bottom row — to get the third stick for the top left square (see Figure 2.48).

Because of the nature of these problems (it's too easy to accidentally peek ahead and ruin the problem), for the remainder of this section we have moved the answers further back in the chapter.

The next puzzle is a variation of the previous one.

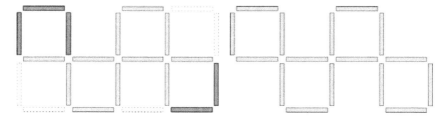

Figure 2.48. The solution to Puzzle 1.

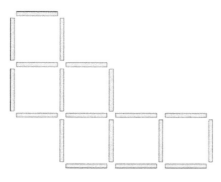

Figure 2.49.

Puzzle 2. Starting from the same arrangement as in Figure 2.46, how can you move two matchsticks to leave four squares?

Puzzle 3. Given the arrangement in Figure 2.49, we ask you to move two sticks to turn six squares into seven.

Puzzle 4. Move three sticks from the arrangement in Figure 2.50 to make five squares.

Puzzle 5. Move two sticks in Figure 2.51 to make eleven squares.

Puzzle 6. In Figure 2.52, move four sticks to make three squares.

Puzzle 7. There are four squares shown in Figure 2.53. Move two sticks to make two squares.

Puzzle 8. There are fourteen squares in Figure 2.54. **Remove** (don't move) four matchsticks to leave five squares.

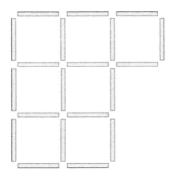

Figure 2.50.

Figure 2.51.

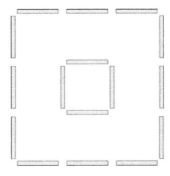

Figure 2.52.

Puzzle 9. There are fourteen squares shown in Figure 2.55; remove three sticks to leave seven.

Puzzle 10. Remove six sticks from Figure 2.56 to leave three squares.

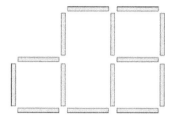

Figure 2.53.

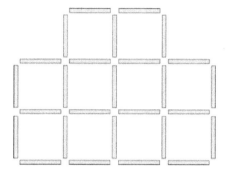

Figure 2.54.

Figure 2.55.

Figure 2.56.

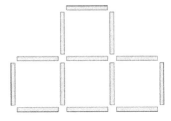

Figure 2.57.

Figure 2.58.

Figure 2.59.

Puzzle 11. In Figure 2.57, move four sticks to change four squares into six.

Puzzle 12. Move one matchstick in Figure 2.58 to make a square. (Careful, though; this is a bit tricky!)

Puzzle 13. Add three matchsticks to those shown in Figure 2.59 to make four equilateral triangles.

Solutions to matchstick puzzles

Puzzle 2 solution

The result in Figure 2.60 is a mirror-image of the first solution, seen in Figure 2.48.

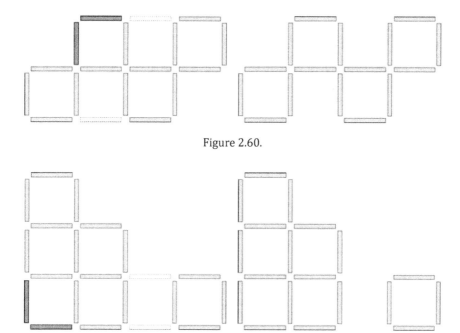

Figure 2.60.

Figure 2.61.

Puzzle 3 solution

Here, there are two different types of squares you can remove by moving two sticks; anything on a corner, and the square that belongs to the long bottom section. There is not an obvious way to increase the number of squares from six to seven unless you look at squares in a different way.

Moving the sticks in the manner shown in Figure 2.61, you maintain the number of small squares while creating a single *large* square (2×2) in the left-hand block.

Puzzle 4 solution

Now that you know that squares of different sizes are allowed, start by counting the number of squares. Do you see all nine of them?

Using the same method as before, there are several ways to identify candidate sticks. You can remove three sticks from the lone square

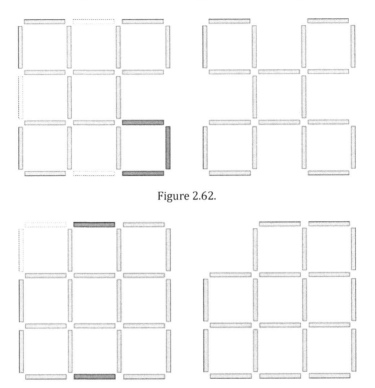

Figure 2.62.

Figure 2.63.

in the top right corner, removing that square; you can remove two sticks from a corner square and one from a side square; or you can break three small side squares. The last approach, applied in Figure 2.62, produces the correct answer.

Puzzle 5 solution

This puzzle is similar to the previous one (see Figure 2.63).

There are no surprises here, so you should be able to figure this puzzle out by applying techniques you have already practiced.

Puzzle 6 solution

This is a new twist! Using the counting method we have been using, four sticks suggests that you should either move the entire center

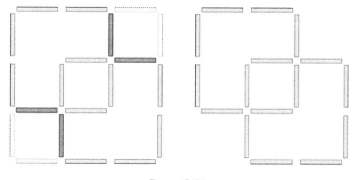

Figure 2.64.

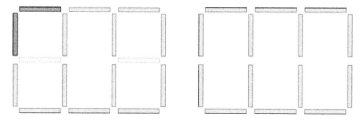

Figure 2.65.

square, or the two corner squares. And indeed, by *inverting* two corners, as seen in Figure 2.64, you end up with three squares, but in a surprising configuration!

Puzzle 7 solution

There are fourteen matchsticks in this problem. If the resulting squares were small, you would need at most eight sticks; if the squares were 2×2, you would need at most sixteen sticks to draw them. Fourteen is certainly closer to sixteen than it is to eight, which suggests that the solution involves two overlapping large squares, as pictured in Figure 2.65.

Puzzle 8 solution

The goal here is to "ruin" as many squares as possible with the fewest removed matchsticks. Coming from the previous problem, you have

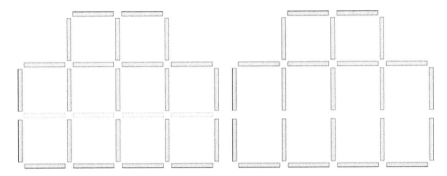

Figure 2.66.

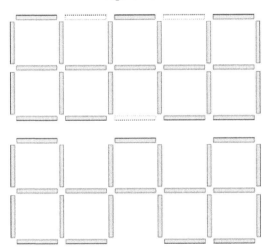

Figure 2.67.

already seen a strategy that reduces the number of small squares while resulting in large squares. Some experimentation will produce the result depicted in Figure 2.66.

Puzzle 9 solution

This is another case where the goal is to ruin as many squares as possible with the fewest moves. Removing a single stick from the outside of the figure can ruin as many as two 2×2 squares and one small square. Do this three times, and you'll end up with seven squares, as in Figure 2.67.

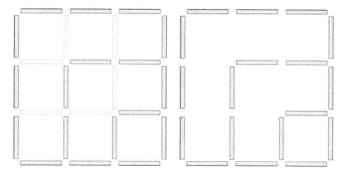

Figure 2.68.

Puzzle 10 solution

By now, you should be a little suspicious when we don't provide a count of existing squares. You're already accustomed to counting 2×2 squares, so the inclusion of the 3×3 outer square, without drawing attention to it, is both an attempt to mislead and a hint.

The solution shown in Figure 2.68 is not the only solution (and it's possible you found other solutions to earlier problems); there are four equivalent solutions that are simple rotations of the one provided, and it's possible you found a solution where only the small square and the medium square share a corner.

Puzzle 11 solution

This one is particularly tricky. Analyzing this puzzle under the rules we've established so far doesn't lead anywhere: there are no two corner squares to remove, and while there are three candidates for removing three matchsticks, it's not clear that there's a "good" way to remove a fourth.

Failing to find a solution in this way is a hint to try other approaches. Each problem we've shown so far is tricky, but in a particular way; this trickiness is new, and you shouldn't be disappointed if you didn't see the approach shown in Figure 2.69. Here, we overlap two matchsticks to create four squares half the normal size, one 1×1 square, and one large 2×2 square. Nothing in the rules said this approach isn't allowed, but it does require "thinking outside the box!"

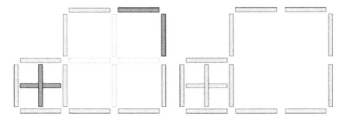

Figure 2.69.

Figure 2.70.

Puzzle 12 solution

By now, the rules you know are completely breaking down. They may have guided you to think "wait, there is no way to build a normal square in one movement!" But that's no guarantee that you would have seen this coming.

In this problem, move the top matchstick up slightly, as shown in Figure 2.70. Do you see the square? It's in the middle of the sticks, as wide as a match itself.

This is the only puzzle we demonstrate that does *not* work with toothpicks, if you cannot find wooden matches when posing these puzzles to your friends, you can omit this one. This is a particularly tricky puzzle; however, this kind of thinking will help keep your mind sharp and open to new ideas.

If you want to make a particularly cruel puzzle of your own, you might combine this puzzle with some of the ideas you've seen in other puzzles. Don't blame us, though, when your friends no longer want to talk to you afterward!

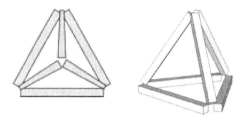

Figure 2.71.

Puzzle 13 solution

We include this puzzle because it is quite possibly the simplest puzzle that uses the third dimension.

None of the other puzzles in this chapter use triangles, so this puzzle is doubly unfair: you have neither had time to consider rules relating to triangles, nor have you had the chance to think about solving puzzles in 3-D! Nevertheless, as you can see in Figure 2.71, it is quite solvable.

To demonstrate this solution in real life, it may be necessary to rest a finger on top of the point where the three angled sticks meet.

Matchstick puzzles are an excellent way to upend your intuition, forcing you to consider problems from different perspectives. Hopefully, this series of escalating challenges provided you with ample opportunity to think further and further outside the box, since this is the central theme of Chapter 3.

Entertaining Problems

The ostensible theme of this chapter was "Entertaining Mathematical Problems." After reading this far, we hope you agree that *Understanding Mathematics Through Problem Solving* can be quite entertaining and challenging! More importantly, we hope you understand that mathematics doesn't always require extensive formalism, and that problem solving is something that anyone can begin to learn and enjoy.

You have seen examples of how to solve problems by a direct approach, how to think in unconventional ways, how to ask questions, how to identify meaningful information and work with it… and above all, how to think, and think about thinking. These skills are foundational. Every problem that we demonstrate (and that you solve) from this point forward will demand you apply these skills without our explicit guidance.

The de-emphasis of these topics is a concession to the limitations of print. We have a lot more to cover, and only a limited number of pages in which to cover it! So while we will be addressing more topics, and more advanced tools, never forget what you have learned in these pages!

Now, prepare to have your intuition challenged, as we begin to explore Counterintuitive Problems in Mathematics.

Chapter 3

Counterintuitive Solutions to Simple Mathematical Problems

Consider the following riddle:

> *As I was headed to St. Ives, I met a man with seven wives,*
> *The seven wives had seven sacks,*
> *The seven sacks had seven cats,*
> *The seven cats had seven kits,*
> *Kits, cats, sacks, wives,*
> *How many were headed to St. Ives?*

What do you think the answer is?

Many listeners start by adding 7 to 7 times 7, and then adding…

Read the first and last lines again. How many were headed to St. Ives? Just one — the poet!

What Is Your Intuition?

Counterintuitive problems are like little mathematical riddles.

Humor often works by subverting expectations. You follow each step, listening carefully, nodding along, and suddenly the joke veers in an unexpected direction, in a way that, somehow, still makes sense!

In the same way, the "counterintuitive" part of a counterintuitive problem or solution tends to sneak up on you. If you see the "trick" on your own, you feel stupid for not seeing it earlier; if you don't see it, the problem itself might seem impossible. In either case, the most productive approach is generally not visible from the start!

Some mathematics problems succumb to a direct approach. Chapter 2 was filled with these; while we did demonstrate many connections (and some that were a bit counterintuitive), most of the perspectives and solutions we described could be discovered by anyone through thorough experimentation and a little bit of insight.

Other mathematics problems are routine extensions of the previous work, practically copying and pasting some existing solution. One phrase you may have seen in mathematics books is "this is left as an exercise to the reader." While often vexing, it is, like many other phrases in mathematics, a stock phrase that is heavy with connotations. The first time you read it, you might think "why is this an exercise for the reader? You're the one writing the book; you should explain it!" After seeing it often enough, occasionally humoring the author by solving the suggested problem, you may come to realize what is actually being said: these "exercises" are a natural extension of work you have already seen, or have the same structure as something that has already been demonstrated. There are, in a word, no new insights to be discovered or applied; nothing in these problems' solutions will challenge your intuition, only train it.

Far more rare are the mathematical problems that demand a particular twist of perspective. These solutions form the core of mathematics texts and academic research, and especially important results and methods are given important names in capital letters ("The Quadratic Formula"), or named after their creator ("The Pythagorean Theorem"). As you learn these and your mathematical toolbox expands, problems that you once found challenging can become glaringly obvious.

Between these extremes there is an entire world of mathematics. Most problems you ever solve will require a little bit of cleverness, and a little bit of experience.

While we cannot "trap lightning in a bottle" and teach you how to be brilliant, we can do something similar: demonstrate different types of counterintuitive problems, using these to develop your intuition about when a problem might challenge your intuition.

Counterintuitive problems need not be long. Frequently, the "counterintuitive bit" of a problem is quite short, sandwiched between layers of comparatively simple applications of known math. When stripped of all context, these twists may not appear counterintuitive at all. The best examples, then, are like riddles, similar to the one we used to open this chapter: short, to the point, with just enough detail to hide a twist that subverts your expectations (yet seems obvious in retrospect).

If solving problems is about identifying the meaningful information and applying the established rules to that information, counterintuitive problems (or steps) are about testing and ignoring invalid assumptions, or bringing in new information whose relationship to the problem may not have been obvious. As you can surmise, the reason for practicing these problems is to build your mathematical flexibility, and to train you to recognize when a direct assault on a problem will be unproductive.

The most concise examples of counterintuitive problems trend toward the definition of "puzzle" we described in Chapter 1. They are short problems, with a clear (often contrived) setup. With a little experience, you will begin to recognize problems that have been designed to mislead you, and by reflex learn to throw away your first, "instinctive" response.

Throwing away a first intuition: A case study

A well-known example of challenging intuition comes to us from the Second World War.

The unprecedented scope of global industrial warfare drove many countries to extremes of research and development, trying to carve out the slightest advantage over their adversaries. Rather than enlisting all able-bodied men as soldiers, most of the powers in conflict sought to identify talent and apply it to the problems of war — designing weapons and machines, breaking codes, precomputing the trajectories of bombs and artillery rounds to simplify the work of soldiers in the field, and so forth. Physicists, engineers, and mathematicians worked around the clock, and around the world, to end the war more quickly.

The names of some of these groups are well known. The Manhattan Project plucked the most brilliant scientists, engineers, and mathematicians from both American universities and the ranks of European scholars, many of whom were displaced by war, to develop the first atomic bomb. On the other side of the Atlantic Ocean, the Bletchley Park codebreakers were men and women tasked with breaking the German military codes, incidentally laying the groundwork for the age of computing. Other groups' duties were not as well defined, and their accomplishments are less well known, yet their work was no less important to the war effort and post-war research than their more famous cousins.

One of these groups, the unassumingly named Statistical Research Group, or SRG, gathered together some of the brightest minds of the day in the field of statistics to solve problems related to warfighting. The SRG undertook to answer many mathematical problems related to the efficacy of weapons and, as an extension, the survivability of ships, aircraft, and men. The first question given to the SRG was whether it was better to arm aircraft with four large machine guns or eight smaller guns; to answer this problem, they learned about dogfighting techniques, and the accumulated learning from this investigation led to further questions about how to find and engage with (or avoid) enemies. The knowledge of how planes maneuvered led to questions about targeting with dive bombers; the knowledge of the performance of dive bombers led to questions about how to most effectively aim anti-aircraft weaponry. For over 3 years, the SRG continued in this manner, applying and developing techniques that would turn out to have wide-ranging applicability beyond the war.

The alumni of the SRG include some of the most notable names in economics and statistics of the post-war period. In this story, we concern ourselves with one of them in particular, Abraham Wald.

One day, a set of data came across Wald's desk. The military had provided the SRG with detailed data on the locations of bullet holes in bombers returning from missions, and tasked them with improving the planes' survivability. There was, the story goes, a belief within the military that the planes should have protection added in the locations where more bullets had hit.

Wald turned that assumption on its head. He assumed, instead, that bullets should hit a bomber anywhere with equal probability, and that the data provided by the military indicated which parts of a plane could be hit *without* impeding the ability to return to base.

If no surviving plane, he reasoned, had bullet holes in a particular location (such as the engine), then all the planes that had been hit there crashed or were otherwise destroyed.

Wald's counterintuitive deduction proved correct, and provides a remarkable example of what is now called *survivorship bias*. While counterintuitive when first described, this principle has become a standard tool in most fields of scientific research. Consider, the following:

- When researching cancer survival rates, you might discover that the people who survive five years post-diagnosis all have some trait in common. But, before trumpeting your discovery, you should check: did those who did *not* survive also have this trait?
- When investing for retirement, you could compare the performance of several mutual funds to determine where you should put your money. Most of them will have impressive rates of return over the previous 10 years. But you should ask yourself: how many mutual funds had to shut down in the same timeframe?
- Older buildings are often more attractive than newer buildings. Is this because *all* buildings were more attractive hundreds of years ago, or because buildings that are not attractive are more likely to be torn down?

Learning to see the counterintuitive

Wald's discovery, and those of other wartime researchers, were the result of people asking the question: "what if this assumption is wrong?" Real-world problems don't always cry out for this type of approach. In fact, there is almost certainly something in your day-to-day life that could be improved by asking counterintuitive questions and measuring the results of change.

The results brought to us by history are no doubt important, but they are not the whole story. Habit or tradition can easily conceal unquestioned assumptions about what is right and wrong, or what is correct and incorrect. There is no formula or equation to reveal these unstated assumptions, but by the end of this chapter, you will have had many chances to practice the mental strategies that should serve you well in life.

Let's Make a Deal

By far the most well-known counterintuitive result in popular mathematics comes to us by way of a game show.

Let's Make a Deal first aired in the early 1960s. The show's iconic host, Monty Hall, gives us the name of this first problem: *The Monty Hall Problem*.

A simplified version of the game goes as follows. There are three curtains. Behind one of them is a desirable prize — a vacation, electronics, or, in this example, a new car — while the other two are hiding dummy prizes ("goats"). The host prompts the contestant to choose one of the curtains and, without revealing what's behind it, opens one of the other two curtains to reveal one of the goats, since Monty Hall knows where the goats are located. Monty Hall then gives the contestant a choice: will he stick with his original choice, or will he instead choose to switch to the other curtain?

At this point, alarm bells should be ringing in your head. Everything — from the title of the chapter, to the introduction, to the setup of this problem — says that your intuition will mislead you here. Yet, given all these facts, the show has run on and off for almost 60 years, and contestants (and puzzle-solvers) keep guessing wrong!

Without reading this book further or having seen this problem before, you could be forgiven for thinking there are even odds that the car would be behind either curtain. That is, you may believe there's a 50–50 chance that the car will be behind either of these two curtains.

We will demonstrate, in three different ways, why that intuition is incorrect.

A wrong way to count

It will be instructive to demonstrate one process by which people are misled.

We begin by spelling out all the implied constraints.

1. The initial location of the car is random; there is a one-in-three chance that it is behind any curtain.
2. Monty will always open one of the curtains that has not been selected by the contestant.
3. Since he knows where the car is, he will always show a goat, and never the car.
4. The host will always offer the contestant the opportunity to switch curtains.

With all these constraints before you, you have enough information to try to calculate the odds that switching curtains will win a car. How do you begin, though?

One method we've demonstrated is to enumerate all possibilities. There are three possible places the car could be located (Table 3.1), and to the best of our knowledge, all are equally likely.

There are also three possibilities for which curtain the contestant selects. However, there is symmetry in the contestant's choice: since you don't know where the car is, your analysis should be the same no matter which curtain is selected. We will assume, for the sake of this demonstration, that the contestant selects Curtain 2.

Table 3.1.   Three possible locations for the "good" prize.

Curtain 1	Curtain 2	Curtain 3
Car	Goat	Goat
Goat	Car	Goat
Goat	Goat	Car

Finally, there are three possible curtains Monty Hall could open. Combining this idea with the previous three possibilities, you can generate a table of potential scenarios, as we have in Table 3.2.

Not all of these nine scenarios are possible. Monty Hall cannot reveal the prize behind the curtain selected by our contestant, nor can he reveal the car. Under these restrictions, the number of possible scenarios is reduced to the four in Table 3.3.

This is the point where your intuition betrays you. Counting all possible scenarios, you can see that there are only four possible Monty

Table 3.2.   Monty could open one of the three curtains, in bold.

Curtain 1	Curtain 2	Curtain 3
Car	Goat	Goat
Car	**Goat**	Goat
Car	Goat	**Goat**
Goat	Car	Goat
Goat	**Car**	Goat
Goat	Car	**Goat**
Goat	Goat	Car
Goat	**Goat**	Car
Goat	Goat	**Car**

Table 3.3.   If the contestant selects Curtain 2, and Monty will never reveal the car, there are only four ways he can open a curtain (highlighted).

Curtain 1	Curtain 2	Curtain 3
~~Car~~	~~Goat~~	~~Goat~~
~~Car~~	~~Goat~~	~~Goat~~
Car	Goat	**Goat**
Goat	Car	Goat
~~Goat~~	~~Car~~	~~Goat~~
Goat	Car	**Goat**
Goat	Goat	Car
~~Goat~~	~~Goat~~	~~Car~~
~~Goat~~	~~Goat~~	~~Car~~

Hall options: two where the contestant wins if he sticks with his choice, and two where he loses. From this perspective, you might believe that there is a 50–50 chance the contestant is justified sticking with their original choice. This, however, is incorrect.

Three curtains, counted

Most people stop here, saying "aha! I was right! It's a fifty-fifty chance!" To upend your intuition, you must look at the problem from a slightly different perspective: every time you consider an option, compute the likelihood that you are in that situation.

There is a one-in-three chance that the car is behind any curtain at the beginning of the game. When the contestant chooses Curtain 2, that chance does not change. Moreover, in two scenarios (the top and bottom scenarios in Figure 3.1), Monty Hall's choice is no choice at all: he *must* open one specific curtain. The top and bottom scenarios are equally likely to happen, and each of these happens with the probability of one-third.

Meanwhile, if the contestant happened to choose the car on their first guess, there are indeed two possible branches, each happening with a 50% probability. Given that there is a one-third chance that the

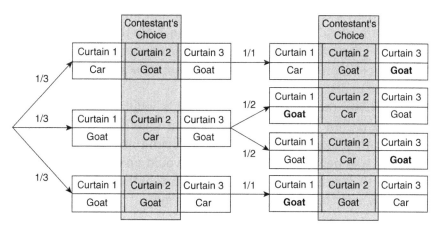

Figure 3.1. Conditional probabilities. The contestant is wrong 2/3 of the time, and in these cases, Monty has no choice in which curtain he opens.

contestant's initial choice was correct, there is a one-in-six chance — that is, $\frac{1}{3} \cdot \frac{1}{2} = \frac{1}{6}$ — that the contestant is in each one of these situations.

The contestant can indeed be in one of four different situations, but *not all of these are equally likely*. The two situations where switching guarantees failure happen with a combined probability of one-third, and the two situations where switching guarantees success happen with a combined probability of two-thirds.

You can spend some time processing these numbers to verify that they make sense. However, the solution was staring at us before we even started into the previous two paragraphs: no matter what Monty Hall does, there is a two-thirds chance that the contestant's initial selection was incorrect. None of which is to say that understanding this explanation is easy — this is a problem that has misled countless people, even brilliant mathematicians! — only that, at this moment, you have all the information you need to guide you to correct understanding.

Countless curtains

If the previous demonstration did not change your mind, let's change the deal and see what happens.

You're watching a different version of the game show. Monty Hall presents the contestant with 100 curtains, and invites her to choose one of them. She does so, choosing Curtain 1. He then opens 98 curtains, revealing 98 goats. Do you think the contestant on this show should choose to stick with her original choice, or should she instead choose the undisclosed curtain?

In this case, the probability that the car is behind the other curtain is 99/100. If the contestant selected Curtain 1 and the car was behind Curtain 2, then Monty Hall would have to open curtains 3, 4, 5,...,100. If the car was behind Curtain 3, Monty would open 2, 4, 5,...,100,... and so forth, for each of 99 possible locations of the car.

Clearly, switching is the superior strategy!

Some readers may object to this approach, for one particularly good reason. The original game's rules specify one of the following two things, depending on how you look at it:

1. The host's job is to open *all but one* of the curtains concealing goats.
2. The host opens *only one* curtain that is concealing a goat.

Under the original game's rules, "all but one" and "only one" are equivalent. Either interpretation would result in the same situation: the contestant chooses one curtain, Monty Hall opens another (revealing a goat), and the contestant is given the opportunity to change her choice.

With the 100-curtain version of the game, option one (all but one) would involve Monty Hall opening 98 curtains, as in the previous example; however, option two (only one) would see him opening only one of them. Strangely, in this second case, the contestant would *still* have a slightly better chance of winning if she changed her selection!

The calculation to demonstrate this fact is tedious, but we include a highly abbreviated form in Table 3.4, below, for the sake of completeness.

To elaborate: The contestant's initial guess will be correct 1 out of 100 times. Monty Hall has the option to open one of the 99 possible curtains (column 3 of Table 3.4), and in each of these cases, the contestant will have the opportunity to switch to one of the 98 possible curtains (column 4), for $99 \cdot 98 = 9702$ different possibilities. Choosing any of these possibilities will, of course, be the incorrect choice, leading to a goat.

Table 3.4. More conditional probabilities. We count the number of ways switching results in finding the prize (correct) or a goat (incorrect).

First choice	Likelihood	Monty's options	Contestant's options	Correct	Incorrect
Car	1/100	99	98	0	9702
Goat	99/100	98	98	9702	941,094

If the contestant's initial guess is incorrect (99 times out of 100), Monty Hall has the option to open one of the 98 curtains. In each of these cases, when the contestant switches to one of the remaining 98 unrevealed prizes, they will be right $99 \cdot 98 \cdot 1 = 9702$ times, and wrong $99 \cdot 98 \cdot 97 = 941{,}094$ times.

Thus, if the contestant sticks with her choice, she will win a prize 1 out of 100 times, and if she chooses to switch, she will win 1 out of every 99 times.

$$\frac{9702}{9702+9702+941{,}094} = \frac{99\cdot98}{99\cdot98+99\cdot98\cdot1+99\cdot98\cdot97} = \frac{1}{1+1+97} = \frac{1}{99}$$

It might take some effort to follow each step of this demonstration, but you can confirm for yourself that, $\frac{1}{100} < \frac{1}{99}$, indicating that *even in this case*, it's in the contestant's best interests to switch to a different curtain![1]

Monty Hall's knowledge

One further justification for this counterintuitive result relies on a concept we touched on in Chapter 2: how much information is available.

In the beginning, the car's location is assumed to be random, as is the contestant's initial choice. You have no information about what lies behind any of the curtains, and thus, there is no reason to assume anything, nor any reason to make one choice over another. Monty Hall, however, knows the location of the car, and so whatever curtain he opens guarantees that no car will be there. His choice of curtain *adds* information to our decision-making process.

If, instead of revealing a goat, Monty Hall *removed one of the prizes from the stage at random*, he would not be adding any information to the problem. For all you know, the choice now might be between two goats! In this case, the intuitive answer — that switching is no better than staying — would be correct.

---

[1] These numbers are highly suggestive of a general-case solution that answers the question, "what is the contestant's chance of winning if there are *n* possibilities and Monty reveals *m* goats?" We leave this problem as an exercise to the reader.

Table 3.5.   A scorecard you can use to conduct your own experiment with the Monty Hall problem.

	Won	**Lost**
Switched		
Stayed		

It is difficult to translate this insight into numerical terms, but if you were still unconvinced at the end of the last section, we hope this final note was enough to help you understand *why* the previous solutions are true.

Finally, you can turn this problem into an entertaining problem. Take three playing cards — such as a Queen and two Jacks — and challenge a friend to guess where the Queen is. Each time she guesses, show her a Jack (by the same rules Monty Hall would use to reveal a goat), and ask if she wants to change her choice.

Tally the results of multiple repetitions in Table 3.5. After ten or twenty games, you should both be able to see the pattern emerging, proving our claim!

The Monty Hall problem is famous precisely because so many people have been stymied by it over the years. Even having read this section, some readers may remain unconvinced that our conclusions are correct.

There are plenty of counterintuitive problems in mathematics that are challenging without being as divisive at the Monty Hall problem! In time, we will examine a few others that are just as famous as this one, or nearly so; nevertheless, the well of counterintuitive problems is deep.

Getting Home in the Rain

The first step to handling counterintuitive problems is learning to recognize them. This is one area where puzzles excel. With just a little experience, you can begin to see a difference between puzzles that demand a direct approach and those that glance suggestively toward out-of-the-box thinking.

There are countless examples of the latter type of puzzle; we will demonstrate the general idea with one in particular.

A family of four is having dinner at a restaurant when it begins to rain. They eat slowly, hoping the rain will stop so they can walk home together, but by the time they finish dinner, the rain has become a downpour. Fortunately, they have an umbrella; unfortunately, it's only large enough to shelter two people from the rain at a time.

The two parents, A and B, can get home in one minute and two minutes, respectively. Their daughter, C, will take five minutes to make the trip. B's elderly father, D, is much slower; it will take him ten minutes to get back to the house safely.

Who should make each trip, and in what order, to get everyone home as quickly as possible without getting wet?

Your initial approach might be to assume that A can carry the umbrella back and forth between the restaurant and the house, making two and a half round trips, or five legs in total. By this reasoning, the legs of the trips will take (in no particular order) two minutes as A walks with B, then one minute for A to return, five minutes for A and C, another minute to return, and ten minutes as A walks with D, yielding a total of nineteen minutes.

A puzzle like this wouldn't be worth asking if the solution were so easy to find, so we ask: can you find a better solution?

You could exhaustively list all possible orderings in which the four people could make the trip, but this approach could be error-prone. So instead, ask yourself how to solve this puzzle through logical deductions.

No matter who carries the umbrella, it must travel an odd number of legs. If the umbrella took, for instance, four legs (two complete round trips) then it, and whoever had carried it, would end up back at the restaurant. Using more than five legs (three out and two returning) will always be suboptimal. There is also no way to move four people in fewer than five legs. For, if the umbrella traveled three legs, then there could be at most three people at the house, stranding the fourth person at the restaurant!

There is no way for grandpa D to get home in under ten minutes, but if you can combine his trip somehow with C, you can minimize the "expense" of the entire process. So, at some point, C and D should make the trip together.

How can the umbrella get back, though? Clearly, one of A or B must ferry it back to the restaurant, but in returning, whoever returned would be left behind and have to make the trip again.

If there is a combination of trips that takes less time than your first guess, it is probably one where C and D travel together. What if A and B make the trip first, then A ferries the umbrella back? These three legs — A travels with B, A returns to the restaurant, then C and D walk home — would take thirteen minutes. B, C, and D are all now at home, and B can return to the restaurant the fastest of these three, at which point A and B could once again travel home. Adding together these five segments results in an elapsed time of seventeen minutes, beating the first guess by a full two minutes!

Logically, this seems to be the best possible solution (and indeed it is, by the logic we spelled out earlier). If you don't want to believe it, you can exhaustively list all possible combinations of five legs[2] and compute the amount of time each takes. Using a computer program, we have calculated the results, which you can see in Table 3.6.

According to this table, there are two different ways to hit the seventeen-minute mark — A returns first, or B returns first — but it's clear that seventeen-minutes is the best possible time. Nineteen minutes isn't bad for a first attempt, though!

Some people might never detect that this problem has an optimal solution different from their first guess, but it does!

Since one of our goals is to help you learn to recognize when problems have an unexpected solution, we next offer a problem that doesn't immediately seem to have any solution at all!

---

[2] There are 108 different possibilities here. On the first leg, two family members travel, and there are six different combinations of two family members: AB, AC, AD, BC, BD, and CD. No matter who goes, there are two choices for the return journey: for, if A and B travel first, then either A or B can return the umbrella to the restaurant.

With three family members at the restaurant, there are three possible choices. For instance, out of A, B, and C, we can choose AB, AC, or BC.

One of the three family members can return, and then the two remaining people will make the final journey.

Multiplying these numbers together yields $6 \times 2 \times 3 \times 3 \times 1 = 108$ different combinations of five legs.

Table 3.6. All lengths of time possible using the optimal five trips, the number of different ways those trips can be constructed, and an example of each. Return journeys have been highlighted in bold.

Time (min)	Orderings	Example
17	2	AB, **A**, CD, **B**, AB
19	6	AB, **A**, AC, **A**, AD
20	12	AB, **A**, AC, **B**, BD
21	6	AB, **B**, BC, **B**, BD
23	6	AB, **A**, CD, **C**, AC
24	6	AB, **B**, CD, **C**, BC
26	8	AC, **A**, BD, **C**, AC
27	8	AC, **C**, BD, **B**, BC
30	6	AC, **C**, BC, **C**, CD
33	6	AB, **A**, CD, **D**, AD
34	6	AB, **B**, CD, **D**, BD
36	8	AC, **A**, BD, **D**, AD
37	10	AD, **D**, BC, **B**, BD
40	12	AC, **C**, BD, **D**, CD
50	6	AD, **D**, BD, **D**, CD

The Man with Three Daughters

A man and a woman are talking. "Tell me about your family," she says to him.

"Well, I have three daughters."

"That's nice. What are their ages?"

"You like puzzles, right? Maybe you can figure it out. The product of their ages is 36."

"I do enjoy puzzles," replies the woman. After a moment's thought, she says "hmm, I don't think that's enough information to answer my question."

"Okay, the sum of their ages is the same as my house number."

"I know where you live, but I still don't know how old they are," says the woman.

"One last clue, then: my eldest daughter has red hair."

"Okay, now I know how old they are!"

How did the woman figure it out?

This is another classic brain teaser, and it's a classic for very good reason. All the information you need to solve it is contained in the dialogue, but readers seeing it for the first time will often get tangled up by certain red herrings — misleading clues — and miss the mathematical heart of this puzzle!

After trying to solve it on your own, you may not see how to make any progress toward a solution. We suggest that you attack this puzzle deliberately, one step at a time.

First, the man states that the product of his daughters' ages is 36. How many ways can you break 36 into a product of whole numbers?

It is useful, but not necessary, to know that any whole number can be *uniquely* broken down into a product of prime numbers.[3] In this case, $36 = 2 \cdot 2 \cdot 3 \cdot 3$. We can also suppose that one (or more) of the daughters can be one year old.

There are many different combinations of ages that are possible given the product 36, as you can see in Table 3.7. We did our best to list them systematically; nevertheless, you may want to spend some time confirming that this table lists all possibilities.

Table 3.7. Possible ages of the man's daughters.

Daughter 1	Daughter 2	Daughter 3
1	1	$2 \cdot 2 \cdot 3 \cdot 3 = 36$
1	2	$2 \cdot 3 \cdot 3 = 18$
1	3	$2 \cdot 2 \cdot 3 = 12$
1	$2 \cdot 2 = 4$	$3 \cdot 3 = 9$
1	$2 \cdot 3 = 6$	$2 \cdot 3 = 6$
$2 \cdot 2 = 4$	3	3
2	$2 \cdot 3 = 6$	3
2	2	$3 \cdot 3 = 9$

---

[3] That is, numbers that can only be evenly divided by themselves and 1.

This may be as much information as can be obtained out of the father's first clue, so let's analyze the next one.

The man tells the woman a fact about the sum of his daughters' ages, but he doesn't tell her (or us) what that sum is. Regardless, you can calculate the possible sums given the previous table.

$$1 + 1 + 36 = 38$$
$$1 + 2 + 18 = 21$$
$$1 + 3 + 12 = 16$$
$$1 + 4 + 9 = 14$$
$$1 + 6 + 6 = 13$$
$$4 + 3 + 3 = 10$$
$$2 + 6 + 3 = 11$$
$$2 + 2 + 9 = 13$$

The woman knows where the man lives, but still doesn't know what his daughters' ages are. Why might this be? She knows the man's house number, but you don't need to; you only need to recognize that the woman can't tell what the solution is. This indicates that there is confusion amongst the age sums. What might that confusion actually be? There is one number that came up twice — 13 — leaving the woman baffled.

When the man says "my *eldest* daughter," the woman knows that he has a single eldest daughter and two younger daughters, rather than a pair of six-year-old twins and a one-year-old infant. Thus, the man's daughters must have the ages two, two and nine.[4]

Note, in this puzzle, that the *information* extracted out of the man's words is all logical or numeric. While it might have made some of the steps easier, you didn't need to know the man's house number to solve the problem. You certainly didn't need to know anything about anyone's hair color! The most important clue was that she was confused by knowing the house number, which indicated the double appearance of 13.

---

[4] While it is possible to have two children born in the same year that aren't twins (say, one born in January and the other in November), we choose to parse "eldest daughter" to mean "uniquely eldest." That the man assumes the final clue offers sufficient information to solve the riddle is enough for us to conclude that, as well.

However, learning that one of the daughters was older than the other two made the final piece of the puzzle slide into place.

Sometimes, a puzzle's setup makes it clear that the obvious solution is incorrect. Other times, it's not clear that a solution is possible. The next problem we will explore is, in a way, a lot more like the counter-intuitive problems you might encounter in real life.

Numbers with Meaning

We present, without comment, the following equation:

$$28x + 30y + 31z = 365$$

Find a single set of positive integer values of x, y, and z that make this equation true.

We will present three solutions to this problem. Two will be fast-paced, and we don't expect you to follow them in detail or understand every leap we will be making in the solution process. Skim through them for the counterintuitive reveal at the end.

Solution one: By any means necessary

To reassure ourselves that a solution is *at all* possible, we can begin by removing the requirement that all three numbers be positive.

We start, then, by breaking the elements of the equation into pieces. Thus,

$$28x + (28+2)y + (28+3)z = 365$$
$$28x + 28y + 2y + 28z + 3z = 365$$
$$28x + 28y + 28z + 2y + 3z = 365$$
$$28(x+y+z) + 2y + 3z = 365$$

One possible approach would be to let $x = 0$, and find values of y and z such that the entire leftmost term would disappear. If $y = -z$, then the value of the number inside the parentheses will be 0, and the equation will be reduced to the following:

$$28(0)+2y+3z=365$$
$$2y+3z=365$$
$$2y+3(-y)=365$$
$$2y-3y=365$$
$$-y=365$$
$$y=-365$$

One solution, then, would be

$$x=0$$
$$y=-365$$
$$z=365$$

We can use these values in the original equation to see if our solution is correct:

$$28\cdot0+30\cdot(-365)+31\cdot365$$
$$=0+365\cdot(-30)+365\cdot31$$
$$=365(-30+31)$$
$$=365(1)=365$$

Using similar reasoning, we can set x to have any value — such as 1 — and then find values for y and z that will nevertheless cancel out the entire left term.

$$(x+y+z)=0$$
$$1+y+z=0$$
$$z=-y-1$$

Again, plugging this value into the above equation, we have

$$2y+3(-y-1)=365$$
$$2y+(-3y-3)=365$$
$$2y-3y-3=365$$
$$-y=368$$
$$y=-368$$

$$z = -y - 1$$
$$z = -(-368) - 1$$
$$z = 368 - 1$$
$$z = 367$$

Finally, checking our work:

$$28 \cdot 1 + 30 \cdot -368 + 31 \cdot 367$$
$$= 28 + (-11{,}040) + 11{,}377$$
$$= 365$$

It seems this provides us with a mechanism for identifying any number of solutions, although many of them have large negative components.

Is there a way to find solutions that would meet the original restriction?

Solution two: Staying positive

Using the same grouping of terms that we used during the previous approach, we'll attempt to apply inequalities to put a cap on the maximum possible *positive* values of x, y, and z that result in valid solutions.

Ignoring for a moment all other terms in the equation, if all three variables are positive, then 28 times their sum must be less than 365.

$$0 < 28 \cdot (x + y + z) < 365$$
$$0 < (x + y + z) < 13.035\ldots$$

Since we're only searching for positive solutions, the minimum value each variable can take would be 1. Thus,

$$3 \le x + y + z \le 13$$

Recall (or convince yourself) that the product of an even number and any other number is always even (Table 3.8), and the sum of even numbers is always even (Table 3.9).

Looking at the earlier equation $28 \cdot (x + y + z) + 2y + 3z = 365$, we can assume that the leftmost product must produce an even number by virtue of the fact that 28 is even. Furthermore, since y is being

Table 3.8.   The parity of products of even and odd numbers.

Number 1	Number 2	Product
Even	Even	Even
Even	Odd	Even
Odd	Even	Even
Odd	Odd	Odd

Table 3.9.   The parity of sums or differences of even and odd numbers.

Number 1	Number 2	Sum/Difference
Even	Even	Even
Even	Odd	Odd
Odd	Even	Odd
Odd	Odd	Even

multiplied by 2, the center term must be even. Yet, the total, 365, is odd. The only way for that to be possible would be if $3z$ were odd; thus z must be odd.

Taking all these constraints into consideration, we can deduce all possible values of z: 1,3,5,7,9,11. This is because it must be positive (at least 1), it must be odd, and it must be 11 or smaller for the sum $x + y + z$ to be less than or equal to 13.

We can regroup the terms again in order to move all z terms to the right-hand side, producing the following:

$$28(x+y)+2y+31z=365$$
$$28(x+y)+2y=365-31z$$
$$14(x+y)+y=\frac{365-31z}{2}$$

In Table 3.10, we substitute each possible value of z one-by-one to calculate all possible values for the right-hand side of the equation.

We can work through the table to eliminate some possibilities quickly. Examining the last row in particular, if we choose minimal values for x and y ($x = 1$ and $y = 1$), the left-hand side of the equation equals 29, which is clearly not equal to 12. Since there are no smaller positive values to choose for x and y, z cannot be equal to 11.

Table 3.10. Using potential values of z to find all possible values for the right-hand side of the equation.

z	31z	Right side
1	31	167
3	93	136
5	155	105
7	217	74
9	279	43
11	341	12

Table 3.11. Finding upper bounds for the sum $x + y$ given various values of z.

z	$14(x + y) < ...$	$(x + y) \leq ...$
1	167	11
3	136	9
5	105	7
7	74	5
9	43	3

By similar logic, we can find upper bounds on the sum $x + y$ for other values of z. If z is equal to 9, then (reading from the previous table) the right-hand side of our equation equals 43. Looking at the left-hand side, $14(x + y)$ is the dominant value, so the sum of x and y must be strictly less than the previous value, 43, divided by 14. In other words, $(x + y) \leq \frac{43}{14} = 3.07...$. Since both values must be integers, their sum must be less than or equal to 3. Table 3.11 shows the upper bounds on the sum $(x + y)$ for different values of z.

To clean up some complexity, we can represent the right-hand side of the equation with the variable w. Then, re-arranging the equation:

$$14(x + y) + y = w$$
$$14x + 14y + y = w$$
$$14x + 15y = w$$
$$14x = w - 15y$$
$$x = \frac{w - 15y}{14}$$

The only way for the right-hand side of the equation to produce an integer is for the numerator of the fraction to be a multiple of 14, so certainly $w - 15y$ must be even. Therefore, if w is odd, y must be as well: the product of 15 and an odd number must be odd (see Table 3.8), and the difference of two odd numbers must be even (Table 3.9). For the same reason, if w is even, y must be as well.

To simplify matters, we can calculate all multiples of 14 up to the greatest possible value of w, 167, taken from the top row of Table 3.10. Table 3.12 shows these values, allowing us to look for candidate solutions for this equation.

Knowing that we're on the lookout for these values, we exhaustively compute all possible values of the right-hand side of the equation in Table 3.13, leveraging the fact that y must be even or odd depending on the parity of w.

Thus, we find two valid solutions. When $w = 74$, $z = 7$ (Table 3.11) and

$$x = 1$$
$$y = 4$$
$$z = 7$$

Meanwhile, when $w = 43$, $z = 9$ and

Table 3.12.   Multiples of 14. Precomputing these makes spotting them in the next step easier.

n	$14n$
1	14
2	28
3	42
4	56
5	70
6	84
7	98
8	112
9	126
10	140

Table 3.13.   For each possible value of *w* and *y*, we can calculate the entire right-hand side of the equation, looking for multiples of 14.

w	*y*	*w−15y*	Divisible by 14?
167	1	152	No
167	3	122	No
167	5	92	No
167	7	62	No
167	9	32	No
167	11	2	No
136	2	106	No
136	4	76	No
136	6	46	No
136	8	16	No
105	1	90	No
105	3	60	No
105	5	30	No
74	2	44	No
74	4	14	Yes
43	1	28	Yes

$$x = 2$$
$$y = 1$$
$$z = 9$$

Whew! We used a lot of tricks here — from algebra and number theory especially — to cut down on the amount of work we needed to do, that may have been tough for you to follow. You can check our work, and our results, to verify these solutions are correct — but you might want to read on to the following section before you spend all that energy!

Solution three: Stop and think

Stop and think: what do the numbers 365, 31, 30, and 28 mean to you?

$$28x + 30y + 31z = 365$$

Observe that there are 365 days in a year, and that there are three types of month: months that have 28 days, those that have 30, and those that have 31 days.

Count the number of each kind of month: February has 28 days; April, June, September, and November have 30 days; the other seven months of the year have 31 days apiece. Therefore,

$$x = 1$$
$$y = 4$$
$$z = 7$$

Since this problem asked for *one* solution, you are done!

Blinded by the ... algebra?

Did you see the trick right away? Why or why not?

If you did — congratulations, not many people do! You did, however, have a leg up: in a chapter on counterintuitive solutions, you are probably already thinking about how we might try to trick you. Consider what would happen if you saw mathematics like this on an algebra test, or in a news article. Would you have been on guard for a counterintuitive solution? Or, rather, would you have assumed the direct approach — something similar to the first two solutions — would be the appropriate one?

There are a couple of lessons to be learned here. The first is, of course, context matters. When and how you see a problem sets the tone for how you approach it. If you assume a problem that looks like an algebra problem will require an algebraic approach, you might miss something much simpler.

Second, the choice of notation matters. We have constructed alternate representations for many of the problems so far, all to make their solutions more comprehensible. But recall the discussion from Chapter 2 about the nature of magic: just as adding props and theatrics to what is essentially a mathematics problem can make it more entertaining (or confusing!), so too, using a confusing notation or representation can misdirect you, leading you to believe that a problem is harder than it is.

Finally, you will not always get road signs. Some problems are straightforward, and others require an indirect approach, and only rarely outside this book will you get clear guidance as to which is which. Some things that are mathematics problems do not look like mathematics problems, and some things which aren't, do.

At some future date, you may run into mathematical con artists — people who will throw figures, statistics, or complex-looking formulas at you — who want to swindle you out of money, attention or even votes. There will never in your life be a sign that says "what this person is saying relates to a mathematics problem, and you already have the tools you need to prove or disprove them!"

When in doubt, pause and think about what you're seeing, in this book or elsewhere. You may be surprised by what you'll notice!

Misunderstanding the Question: The Birthday Paradox

The surprising conclusion to the last problem provides us a neat transition to the next theme. Throughout the last half of this chapter, we will consider counterintuitive questions relating to *time* — dates, time zones, clocks, and holidays — starting with a well-loved classic.

Suppose you are taking a class. How many people would need to be in the class to guarantee at least a fifty percent chance that two people would share a birthday?

Take a guess. Do you suppose you would need 50 people? 100? 180? 250?

The answer is twenty-three!

This number may sound much too small, but we will demonstrate that it is indeed correct! To do so, we will look at the problem in two ways: first, we will demonstrate the answer directly. Then, we will justify each step to further your understanding of *why* the answer makes sense.

Imagine that there are ten classrooms each containing twenty-three students. Each student writes down his or her birthday on an index card, and those cards are put together into ten decks of twenty-three birthdays. What does a fifty percent chance of two people sharing a birthday mean? It means that out of those **ten** decks, on average **five** would contain at least two cards with matching birthdays.

To begin justifying this, you can think about how to measure the *negation* of the statement: how likely is it that *no two* people share a birthday in a given classroom?

You write your birthday down first. Your birthday can fall on one of 365 possible days.[5]

Another student writes their birthday down. If you two don't have the same birthday, theirs can occur on any one of the 364 dates that remain.

The third student's birthday must occur on one of the $(365 - 2) = 363$ remaining dates; the fourth student's has 362 possible dates for their birthday, and so on. As each additional person writes down their birthday, the number of possible dates their birthday can fall on decreases by one. Hence, the twenty-third student's birthday can occur on any one of $(365 - 22) = 343$ dates.

To calculate the total probability, we must divide each number above by 365 possible days, and then multiply their results. Thus, the probability that *no* student in a class of twenty-three shares their birthday with another student is

$$\frac{365}{365} \cdot \frac{365-1}{365} \cdot \frac{365-2}{365} \cdot \frac{365-3}{365} \cdots \frac{365-22}{365}$$

$$= 1 \cdot \frac{364}{365} \cdot \frac{363}{365} \cdot \frac{362}{365} \cdots \frac{343}{365} \approx 0.4927$$

Or, about 49.27%.

Now, we can justify this approach by saying that there are only two possibilities: either *no* two students share a birthday, or some students share a birthday. This second group encompasses all other possibilities, including that exactly two students share birthdays, or two pairs of students do, or three students share a birthday, or all twenty-three happen to have the same birthday.

---

[5] Ignoring leap years, and some other convolutions and complications, for the time being.

Table 3.14.   Given a number of students, the probability that at least two share a birthday.

Number of students	Probability of birth date match	Probability of no match
10	0.116948178	0.883051822
15	0.252901320	0.747098680
20	0.411438384	0.588561616
25	0.568699704	0.431300296
30	0.706316243	0.293683757
35	0.814383239	0.185616761
40	0.891231810	0.108768190
45	0.940975899	0.059024101
50	0.970373580	0.029626420
55	0.986262289	0.013737711
60	0.994122661	0.005877339
65	0.997683107	0.002316893
70	0.999159576	0.000840424

Since we've accounted for all possibilities, the two options must sum to 100%. So, in a class of twenty-three people, there is a 100% − 49.27% = 50.73% chance that at least two students have a birthday in common.

To develop an intuition for what's going on here, we can perform the same computation for groups of every size, from 1 (a sole student, who has a 0% chance of sharing a birthday with a second student) to 366 (a 100% chance that at least two students share one of the 365 birthdays in common). A sample of these values is shown in Table 3.14.

We can also enter these numbers in an appropriate computer program and to produce the graphs in Figures 3.2 and 3.3.

Wow! The probability that two people do not share a birthday drops to almost 0% in a room containing slightly more than 50 people! In Figure 3.3 we zoom in on the range between zero and fifty people, so you can see more clearly this crossover point.

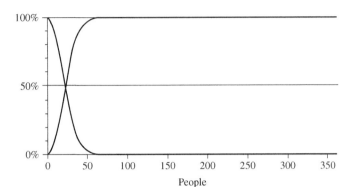

Figure 3.2. The percent probability that no two people will share a birthday (falling) versus the probability that at least two people will (rising), up to 366 people.

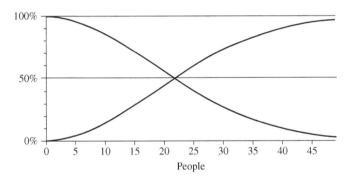

Figure 3.3. A zoomed-in view of the previous graph. The likelihood that at least two people will share a birthday surpasses the opposite around 23 people.

The point where it becomes more likely a room contains two people sharing a birthday than the inverse happens somewhere around twenty-three people, as we stated earlier.

You may find it difficult to digest these numbers, or to use them to build your intuition. We can walk you through step-by-step to demonstrate the logical basis of the solution.

- Each term in the probability computation was less than or equal to 1 because the numerator in each case was smaller than the denominator. (For instance, $\frac{364}{365}$.)

- When you multiply a number by a value between 0 and 1, the resulting product is smaller than the original number.[6]
- Since each successive term in the calculation is less than 1, the product of all terms will get smaller as each new person enters the classroom.
- At some point there will be more people than possible birthdays, so you can tell that eventually the probability that no two people share a birthday will drop to 0.
- If a series of values starts at 1 (that is, 100%), and continues downward toward 0 with each new person, there must be some number of people that produces a probability smaller than any given value. That is, if there is a downward curve that connects the points 1 and 0, at some point it must pass through any value between these.
- If you are looking for a specific value — in this case 0.50, or 50% — and know that every term is smaller than the last, then that value can only occur between the term that is slightly greater than it and the one that is slightly less than it.

This is sufficient to demonstrate that a solution *must* exist, but doesn't help justify why that solution is twenty-three *in particular.* The numbers are not easy to reason about on their own, and the statements we made (while accurate) are not easy to remember without examples or experience. So at this point, you might be able to surprise your friends with this counterintuitive answer, but might not be able to explain *how* it works!

To strengthen your memory and intuition, we could continue to take the direct approach, breaking down potential misunderstandings and re-asserting each fact in the context of the original problem until every topic was covered. In lieu of that, we're going to take a digression through two topics that will help us better explain this apparent paradox: socks and handshakes.

---

[6] The opposite is of course true as well: when you multiply a number by a value greater than 1, no matter how small the difference, the product will be greater than the original number.

The sleeping spouse, revisited

Recall in Chapter 2 that we solved a problem concerning the minimum number of socks you would need to take from a drawer whose contents you couldn't see in order to get a matching pair. In that problem, the goal was to ensure that *at least one* pair of socks could be created from the contents of a drawer containing hosiery of three different colors. We will expand on this theme in two directions.

The likelihood of a situation, and its opposite

First, let's look in greater detail at how likely it is that a pair of socks is found. Instead of removing a certain number of socks all at once, we will take each sock out of the drawer, carry it into another room — say, the kitchen — compare it to the existing socks, and, *no matter what*, leave it in the kitchen.

At the outset, a sock of some color is brought to the kitchen and laid on the table. That sock's color doesn't matter, since there is no way to form a pair with only one sock. When the next sock is retrieved, it can be of the same color as the first, or one of the two remaining colors. By now, you have seen enough examples of probabilities to agree that there is a one-third chance the second sock is of the same color as the first, and a two-thirds chance it is a different color (see Table 3.15).

In Figure 3.4, we illustrate all possible routes to making a pair of socks, in each case stopping as soon as a pair is found (as marked with a checkmark).

Starting at the left, we assume the first sock chosen is black. Then, in one case (the bottom-most branch), the second sock we retrieve is black, and we're done (check).

Table 3.15. The relationship between the number of socks and two related probabilities.

Socks	1	2	3	4
Chance of a pair	0	1/3	7/9	1/1
Chance of no pair	1/1	2/3	2/9	0

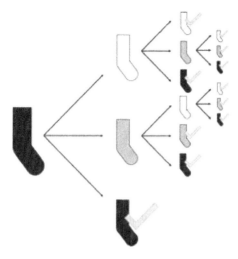

Figure 3.4.   A tree of possibilities. Assuming the first sock is black, the next sock could be black, and make a pair (one-third chance), or not (two-thirds chance).

Otherwise, at the second step, we have two different possibilities: black and white, on the top branch, and black and gray in the middle branch.

Assuming the second sock is white, then you can follow the branches between the second and third steps to see that if the third sock is white or black, we will have made a pair, and the process is complete. If, on the other hand, the third sock is gray, then we have to continue to the fourth layer, drawing a fourth sock. No matter its color, we are guaranteed a match.

In Table 3.15, we translate this diagram into numerical terms. After drawing a second sock, if it is necessary to retrieve a third sock (as will happen two-thirds of the time), there is a two-thirds chance that it will match one of the others, and a one-in-three chance that it has a different color.

We can multiply the two-thirds chance the second draw won't match the first with the one-in-three chance that the third draw will produce the third color (black–white–gray, and black–gray–white, above) to find that there is a two-in-nine chance that you will draw three socks with no two matching.

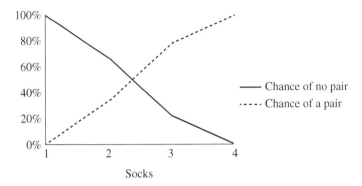

Figure 3.5.   A familiar graph. After each successive draw, the chance of making a pair rises as the chance of not being able to make a pair falls.

And, as you already know, retrieving a fourth sock will guarantee that you can make at least one pair of socks, so the chance of having no pairs of socks after four draws is 0.

There are two related questions being answered here. You are calculating how likely it is that you *can* form at least one pair of socks, and that you *cannot* form a pair, after each draw. We can graph these two complementary probabilities, as shown in Figure 3.5.

Observe here that when there is only one sock, you can never create a pair — the chance of having a pair of matching socks is 0%. While obvious, this observation undergirds the intuition that the chance of there being no pairs starts at 100%.

Further, when there are more socks on the table than possible colors (four socks and three colors), you are guaranteed to have at least one pair. If you were to extend this graph forever to the right, the probabilities wouldn't change: with five, or six, or twenty socks, there would always be a 0% chance of having no pairs, and a 100% chance of having a pair.

Finally, the probability that something happens, plus the probability that it doesn't happen, always add up to 100%. This seems sensible, perhaps, but this logic is susceptible to very subtle flaws. Sometimes it's not clear how to define complementary probabilities that should both add up to 100%.

This leads to the second major theme to be explored: how can you identify and quantify categories when answering questions of probability?

Identifying all categories; or, why are we interested in calculating the opposite?

Consider a lottery that offers a jackpot for matching all numbers, and smaller prizes for matching some numbers. The opposite of "winning the jackpot" is "*not* winning the jackpot," not "not winning anything." On the other hand, the opposite of "not winning anything" is not "winning the jackpot," but rather "winning some prize." In a six-number lottery, a ticket can match all six winning numbers, or five, or four, ... or none. These seven "buckets" — types of possible results — form the entire universe of possibility, and thus the individual probabilities sum to 100%.

If tickets with four, five, or six matching numbers are "winners" and tickets with zero, one, two or three matching numbers are "losers," then you can calculate the likelihood of a ticket having zero, one, two, three, four, five, or six matches, and gather together each of the "winning" and "losing" probabilities using addition to find out how common winning really is.

Framed in this way, it might be hard to see how the mistakes we alluded to are possible. We are being cautious, choosing our words and scenarios carefully. In general, identifying the right question to ask, and computing the answer, can be quite difficult.

Returning to socks, let's choose a specific pitfall and see how it changes our calculations.

What are the chances of finding a *single pair* of matching socks given four socks of three different colors? You might say that there is the same probability as before — 100% — but this slight change in wording adds a twist. The opposite of "a single pair" is not *no pairs*, it is "not a single pair." Therefore, you shouldn't count any case where *two pairs* of socks can be formed. (For instance, two white socks and two black socks.)

We can formalize this by enumerating all possible cases.

Table 3.16. When taking four socks from a drawer, three socks of the same color can come out in four different orderings.

1st Draw	2nd Draw	3rd Draw	4th Draw
B	B	B	—
B	B	—	B
B	—	B	B
—	B	B	B

Case 1: Four socks of the same color.

Given three colors, black (B), white (W), and gray (G), there are three ways to have four socks that are all the same color: BBBB, GGGG, or WWWW.

Case 2: Exactly three socks of the same color.

There are four ways to find three socks of the same color, and some other sock. In Table 3.16, we enumerate all possible ways of pulling three black socks out of the drawer.

In each of these cases, the empty space can take one of two values: G or W.

We can produce similar tables for the other two colors. Thus, if we chose G instead of B as our "base color," then each of the empty spaces could be filled with either B or W.

There are four rows in Table 3.16, three different "base colors" (B, W, G), and in Table 3.17, an "empty space" can be filled with one of the two values. Thus, there are $4 \cdot 3 \cdot 2 = 24$ different ways to find three socks of one color, and one sock of a different color, as you can see in Table 3.17.

The next case produces similar results.

Case 3: Exactly two socks of the same color.

In Table 3.18, we examine all situations where there are exactly two socks of one color, and two other socks *of two different colors*. If one of the other two socks were the same color as the first — black, in the case of this table — you would produce a three-sock scenario, which we just already counted in the previous case.

Table 3.17. Enumerations of all 24 possible ways to draw four socks, with three of the same color.

	Black				Gray				White			
Black					B	B	B	G	B	B	B	W
					B	B	G	B	B	B	W	B
					B	G	B	B	B	W	B	B
					G	B	B	B	W	B	B	B
Gray	G	G	G	B					G	G	G	W
	G	G	B	G					G	G	W	G
	G	B	G	G					G	W	G	G
	B	G	G	G					W	G	G	G
White	W	W	W	B	W	W	W	G				
	W	W	B	W	W	W	G	W				
	W	B	W	W	W	G	W	W				
	B	W	W	W	G	W	W	W				

Table 3.18. If you know that exactly two socks of the same color came out of the drawer, they came out in one of six possible orders.

1st Draw	2nd Draw	3rd Draw	4th Draw
B	B	—	—
B	—	B	—
B	—	—	B
—	B	B	—
—	B	—	B
—	—	B	B

You can see here that there are six rows in the table, two ways to fill in the blanks — GW or WG — and three different starting colors — B, G, or W. Using the same type of calculation as before, you can count $6 \cdot 3 \cdot 2 = 36$ different ways to produce one pair of socks, and two socks of different colors.

Table 3.19. All 12 ways to one pair of black socks, and one of each of the remaining colors.

	Gray/White				White/Gray			
	B	**B**	G	W	**B**	**B**	W	G
	B	G	**B**	W	**B**	W	**B**	G
	B	G	W	**B**	**B**	W	G	**B**
Black	G	**B**	**B**	W	W	**B**	**B**	G
	G	**B**	W	**B**	W	**B**	G	**B**
	G	W	**B**	**B**	W	G	**B**	**B**

We enumerate all of the ways to draw exactly one pair of black socks, and no other pairs, in Table 3.19. The tables for white and gray would be very similar.

Case 4: Two different pairs.

What cases remain to be enumerated? We started this section by asking how many ways there are to find two pairs of two different colors, so you must consider this case. Observe, though, that you cannot simply reuse the logic from Table 3.18, because you would end up double-counting certain pairings. For instance, **BB**_ _ with white as the other color, and _ _ **WW** with black as the other color would produce identical results: **BB**WW and BB**WW**.

A common challenge when trying to enumerate is avoiding these kinds of duplicates. To avoid double counting, you can pair the colors without repeating, by following the arrows in Figure 3.6. Thus, we pair black with gray and black with white, then gray with white.

We can use Table 3.18 as the basis for this case. There are six rows in that table, and three distinct pairings as we just saw, yielding $6 \cdot 3 = 18$ different ways to create two pairs of socks of different colors (Table 3.20).

In order to calculate probabilities, we need to know how many different combinations there are in total. Adding all four of the previous cases together, we find that there are $3 + 36 + 24 + 18 = 81$ different combinations of four socks. This makes sense, though, because

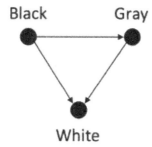

Figure 3.6. How to count pairs of colors without double counting. Black pairs with gray and white; gray pairs with white.

Table 3.20. All 18 possible combinations of two pairs of two different colors.

Black/Gray				Black/White				Gray/White			
B	B	G	G	B	B	W	W	G	G	W	W
B	G	B	G	B	W	B	W	G	W	G	W
B	G	G	B	B	W	W	B	G	W	W	G
G	B	B	G	W	B	B	W	W	G	G	W
G	B	G	B	W	B	W	B	W	G	W	G
G	G	B	B	W	W	B	B	W	W	G	G

each time you draw one sock from the dresser, it can have one of the three possible colors, yielding $3 \cdot 3 \cdot 3 \cdot 3 = 3^4 = 81$ different ways to pull out four socks. Since these two numbers are equal, you can be confident that there are no other ways to draw four socks!

Having enumerated all possible combinations, we could find any derived probability. For instance, the probability that you draw four matching socks is $\frac{3}{81} \approx 3.7\%$.

Our original question concerned the probability that you draw *a single pair* of matching socks. So, the case of four matching socks is out, as are the cases where there are two pairs of two different colored socks. "Not a single pair" happens $3 + 18 = 21$ times out of 81, so we get "a single pair" $81 - 21 = 60$ times out of 81.

Therefore, the answer is: given any four socks of three colors, the probability that you find one *and only one* pair of socks is $\frac{60}{81} \approx 74\%$.

Consider this from the perspective of the birthday paradox. When twenty-three people have written down their birthdays, there is slightly more than fifty percent chance that some pair of people share a birthday; further, there is some smaller chance that *two pairs* of people will share birthdays, or that three people will share the same birthday, or that *all twenty-three people* would have all been born on the same date! It is far easier to calculate the opposite — the chance that *no two* people share a birthday — and then subtract than it is to consider all cases. Recognizing that the probability of some *easy to calculate* thing and its opposite always add to one hundred percent saves us from the nightmare of having to quantify that complicated thing!

In this section, we showed you how and why calculating the opposite of a scenario is useful. We have not yet explained why the answer to the birthday paradox is quite so small. We will use another, similar, tool to bridge this last gap.

Counting comparisons: The handshake problem

Let's consider a different scenario. Imagine that in addition to writing down their birthdays on index cards and comparing, each person in the classroom also shakes hands as they compare. How many handshakes do you suppose take place?

The pattern begins in a straightforward fashion, as you can see in Figure 3.7. The second student shakes hands with the first; the third

Figure 3.7.   Patterns of handshakes. As each new person compares their birthday to those already present, they shake hands with each of them.

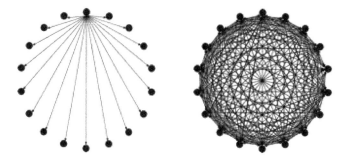

Figure 3.8. (Left) When the eighteenth student shares her birthday, she shakes hands with seventeen other students. (Right) In aggregate, many handshakes have happened!

shakes hands with the first two; the fourth shakes hands with the first three students; and on and on.

When the eighteenth student arrives, she shakes hands with the other seventeen, as illustrated in Figure 3.8. By now though, quite a few handshakes have taken place!

What an explosion! Looking at the cumulative number of hand-shakes as each new student enters the classroom in Table 3.21, you can begin to get a sense for how quickly the numbers grow.

By the time the eighteenth student enters, there will have been nearly ten times as many handshakes as people!

The graph of the ratio of people to handshakes (Figure 3.9) demonstrates the nature of this relationship. When only two people are in the classroom, there has been one handshake for two people, for a ratio of 2.0. When the eleventh person enters, there will have been a total of five times as many handshakes as people; when the twenty-first person enters, that number climbs to ten times as many hand-shakes as people!

In Chapter 2, we introduced triangular numbers in the discussion of dropping eggs from a building. As you may recall, triangular numbers are the numbers that are formed by adding successive integers: thus, the first triangular number is 1, the second is $1 + 2 = 3$, the third is $1 + 2 + 3 = 6$, and so forth.

Table 3.21. When the pth person enters the room, they shake $p-1$ hands, adding to the total. The ratio of people to handshakes drops quickly.

People, p	New handshakes	Total handshakes, h	Ratio p/h
2	1	1	2.0000
3	2	3	1.0000
4	3	6	0.6667
5	4	10	0.5000
6	5	15	0.4000
7	6	21	0.3333
8	7	28	0.2857
9	8	36	0.2500
10	9	45	0.2222
11	10	55	0.2000
12	11	66	0.1818
13	12	78	0.1667
14	13	91	0.1538
15	14	105	0.1429
16	15	120	0.1333
17	16	136	0.1250
18	17	153	0.1176

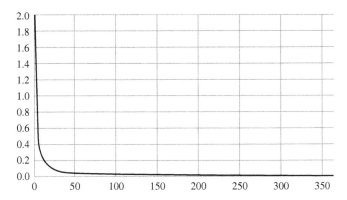

Figure 3.9. A graph of the ratio of people to handshakes. As the number of people increase from 2 to 365, the number of people per handshake drops to almost zero.

If you remember, this is the formula we presented, then

$$\frac{n(n+1)}{2}$$

The handshake problem is also related to triangular numbers, but you can't use the previous formula directly. People don't shake hands with themselves, so at each step, there is one fewer handshake added than there are people. Due to this, we can substitute $(p - 1)$ for n in the previous equation. Thus, the number of handshakes h given a certain number of people p is:

$$h = \frac{(p-1)((p-1)+1)}{2} = \frac{(p-1)(p-1+1)}{2} = \frac{(p-1)p}{2}$$

For completeness, you can calculate the ratio of people to handshakes like so:

$$\frac{p}{h} = \frac{p}{\left(\frac{p(p-1)}{2}\right)} = \frac{p}{1} \cdot \frac{2}{p(p-1)} = \frac{2}{p-1}$$

You can confirm that this produces the same values as the earlier table by inserting different values of p into this formula. You can also see that there will have been eleven handshakes per person, and 253 handshakes in total, when the twenty-third person enters the classroom!

What does this tell you, exactly? Among the most common ways people misunderstand the birthday paradox is to look at it the problem from the perspective of only one person. That is, they instinctively think that the problem is asking how likely it is that anyone else in the classroom has *a specific birthday* (for instance, their own). When you consider the classroom as a whole, though, and see how quickly the number of comparisons grows, it's much easier to understand and believe the real value!

Tools for understanding

The birthday paradox's solution seems counterintuitive at first glance, but with the right tools you can understand what the answer is and why it is so small. By the time you fully understand the lessons of the previous section, you should have a new intuition to replace the old — of course the answer is small! How could it not be?

Be careful, though — a little knowledge can be dangerous. You may begin to assume that any problem that sounds similar to the birthday problem has an answer that is similarly paradoxical. Consider this question: How many people would you have to meet to guarantee a fifty percent chance that one of them has *the same birthday as you*?

We highlighted this earlier as one exceptionally common misunderstanding of the birthday paradox. You have all the tools you need to demonstrate that the answer to this question is much larger than twenty-three. The answer to this question is closer to the first guess most people make when responding to the "classic" birthday paradox!

Assume that when you meet a new person that their birthday is decided randomly, and not related to the birthday of anyone else you've already met. Then, each time you meet someone, there is a $\frac{364}{365}$ chance they do not share your birthday. This number never decreases, because we don't care if those people have the same birthday *as each other*.

The chance that you continue to meet new people, all of whom do not have the same birthday as you, is the product of successive terms of this form. Thus, to calculate how many people you would to meet to have a 50% chance that one of them has the same birthday as you, you can continue multiplying $\frac{364}{365}$, over and over, until you reach a number less than 0.5.

Break out your calculator and find the value of $\frac{364}{365}$. Store the result. Multiply that result by itself. Keep multiplying the result by the original value of $\frac{364}{365}$, over and over, until you reach a number smaller than 0.5.[7]

---

[7] On a four-function calculator, take the following steps: type 364/365, hit "M+" to add that value to the calculator's memory, hit the multiplication key, hit "MR" to recall the value you stored, and then hit the enter/equals key. Many calculators will repeat the

After repeating this calculation for quite some time, you will find that you have to multiply $\frac{364}{365}$ by itself 253 times before reaching a number smaller than 0.5. That is to say, you would have to meet 253 people before you would have a 50% chance of meeting someone with *your* birthday.

That's quite a difference!

$$\frac{364}{365} \cdot \frac{364}{365} \cdot \frac{364}{365} \cdot \frac{364}{365} \cdot \ldots \cdot \frac{364}{365} < 0.5$$

$$\left(\frac{364}{365}\right)^{252} = 0.50089\ldots$$

$$\left(\frac{364}{365}\right)^{253} = 0.49952\ldots$$

Developing robust intuition in the field of probability takes a little skill and a lot of practice. We will set aside questions of probability for now, and instead consider counterintuitive problems with clear and guaranteed solutions, starting with a principle you may not realize you have already seen.

To get us there, we pose one final question about birthdays: How many people have to enter a room to *guarantee* that at least two people share the same birthday?

The Pigeonhole Principle

Consider a letterbox (a piece of furniture composed of small wooden boxes arranged in a grid). These boxes can be used to hold anything: knickknacks, office supplies, letters (as a kind of *ad hoc* mailbox), or as homes for carrier pigeons. If there are eight boxes, what is the minimum number of items — pens, letters, or pigeons — you would need to guarantee that at least one box has two items in it?

Unsurprisingly, the answer is nine. As simple as this observation is, it was only formalized around 1834 by the mathematician

---

last calculation upon hitting enter subsequent times, which you can feel free to make liberal use of here.

Peter Gustav Lejeune Dirichlet (1805–1859), who named it the *Schubfachprinzip* ("drawer principle"). Through a somewhat convoluted etymology, we know it today as the "pigeonhole principle."

Formally stated, the pigeonhole principle says that given m container and n objects, where $n > m$, at least one container must contain more than one object.

The principle's simplicity belies some counterintuitive results. Sometimes it shows up in surprising places, as we will see later in this chapter. Other times the applications are more clearer.

The meanings of "container" (or hole) and "object" can be quite flexible. Recall once again that if you withdraw four socks from your drawer, containing three different color socks, you are guaranteed to find at least one pair. In this case, the colors — white, gray, and black — behave as the "holes," and the socks themselves act as the pigeons. There is no way to put four socks into three holes without putting two socks together in at least one hole (as we have seen twice already).

If there are 365 holes — birthdays — then 366 pigeons (people) are surely sufficient to guarantee that at least two people share a birthday. (This presentation is much easier to understand than the last section, isn't it?)

We wouldn't remember this principle today if there weren't more surprising examples of its power. We will pose, then, a series of rapid-fire problems that will demonstrate some uses of this principle.

Question: How many letters would a mail carrier have to deliver to ten apartments to guarantee that at least one apartment receive three letters?

Answer: No more than twenty-one letters would be necessary. In this case, if each mailbox received two letters, they could hold at most twenty letters total; the twenty-first and final letter would necessarily go into a mailbox already containing at least two letters.

Note, however, that the principle doesn't answer how many letters each apartment receives; it only tells us that *at least one* mailbox has three or more.

Question: At a particular school, each student's locker is marked with the initials of their first and last name. So, David Smith's locker has the letters "DS" written on it.

What is the greatest number of students that could attend the school before you could guarantee that two students share the same locker?

Answer: 677 students.

There are twenty-six letters in the English alphabet. Each of the twenty-six possible initials for a first name can be combined with one of twenty-six possible initials of a student's last name. $26 \cdot 26 = 676$, thus one more student than that, 677, would guarantee that at least two students share initials.

Question: How many cards would you have to draw from a deck to be guaranteed a pair? How about three-of-a-kind? Four-of-a-kind?

Answer: The one pair case is straightforward enough: there are thirteen different values (Ace, 2, 3, ..., Jack, Queen, King) in a standard deck of cards, so on the fourteenth draw you would be guaranteed **some** pair. By similar logic, you would need to draw at least twenty-seven cards to guarantee a three of a kind, and forty cards to guarantee you had drawn four of *some* card.

Question: At least how many people in the City of New York have the same number of hairs on their head?

Wait, what?

Finding the answer to this question requires that you find (or know, already) two pieces of information: the population of New York and the number of hairs on a typical head.

Answer: The City of New York has a population of roughly eight million. An internet search suggests that the typical person has approximately one hundred thousand hairs on their head. Even assuming that in the worst case scenario some people have as many as two hundred thousand hairs on their head, by the same principle as we already seen, at least some forty people should have the same number of hairs on their head:

$$\frac{8,000,000}{200,000} = 40$$

Good luck finding them, though — or verifying that they share the exact number of hairs!

The pigeonhole principle is just one tool that allows us to talk about what happens at the edges of problems, no matter how large or small the numbers are. While it is not able to say anything about what those solutions are, sometimes just demonstrating that some solution must exist can help you understand otherwise difficult problems.

Many of the most interesting lessons in mathematics occur when we investigate these edges, the boundaries where our expectations of normal behavior start falling apart.

Many problems that are otherwise easy become difficult when you look at their edges, whatever those edges look like. In the following section, we will look at different kinds of edges, and the counter-intuitive problems that can be found there!

The Edges of Logic

This set of quick puzzles should remind you to heed our warning from earlier in the chapter: some mathematics problems are meant to mislead you. In these situations, the first answer that pops into your head is probably incorrect!

The day before yesterday

A man claims that the day before yesterday, he was eighteen, and next year he will turn twenty-one. How is this possible?

For most people, this would be impossible. There is, however, one way for the man's statement to be true.

If his birthday is on the 31st of December, and the man is speaking on the 1st of January, then the day before yesterday he was one year younger than he is at the moment. Every day this year, but one, he will be nineteen, and every day next year but one he will be twenty. But at one point during the next year, on the 31st of December, he will be twenty-one years old.

Concretely: if this man was born on the 31st of December, 2000, and you first talk to him on the 1st of January, 2020, he was indeed 18 the day before yesterday (the 30th of December, 2019). Next year, he will be 21. Specifically, on the 31st of December, 2021, he will be 21, despite having been 20 for the other 364 days of the year!

East and west

A couple in a long-distance relationship are up late one night talking on the phone. The man, who lives in an east coast state, remarks that it's currently 1:30 AM, and the woman, who lives in a west coast state, realizes that's the time where she is, too. How is this possible?

It helps here to realize that time zone boundaries do not always follow political boundaries. Notably, some portions of states choose to follow the time zone as their neighbors, if that part of the state is closer to a major city or other area of economic activity in the neighboring state.

Looking at a map of time zones in the United States, you can spot these exceptions: Northwest Indiana keeps the same time as Chicago; Southwestern Indiana uses Central Time alongside its neighbors Illinois and Kentucky. The parts of Michigan's Upper Peninsula which share a border with Wisconsin follow Central Time, as well. Western Texas, around El Paso, uses Mountain Time, unlike the rest of the state. The eastern halves of both Kentucky and Tennessee follow Eastern Time, whereas both these states' western halves are in the Central time zone. Northern Idaho, along the Washington state border, uses Pacific Time. And several plains states — North Dakota, South Dakota, Nebraska, and Kansas — have western counties that use Mountain Time (Figure 3.10).

The two exceptions we interested in, however, are Malheur County, in eastern Oregon, and the western panhandle of the state of Florida. All three states along the Pacific Ocean use Pacific Time, with the

Figure 3.10.   United States time zones. (*Courtesy*: US Geological Survey.)

exception of portions of Malheur County; all the states that border the Atlantic Ocean use Eastern Time, except the Florida panhandle south of Alabama. Normally, these two areas have one hour difference, but once a year at 2:00 AM on the night Daylight Savings Time ends, Florida turns its clocks back an hour. During this brief window, clocks in both western Florida and eastern Oregon will show the same time.

On this one day of the year, for one brief hour, clocks in eastern Oregon (a west coast state) and western Florida (an east coast state), will read the same time.

This problem isn't particularly fair. The solution requires the knowledge of some odd trivia, and demands that you consider two different kinds of edge cases: an edge case in space (a time zone in a state that differs from the rest of the state) and an edge case in time (a moment where 1:59 AM is followed by 1:00 AM). If you were shown this question on a test, no one could fault you for not being able to find the answer.

Nevertheless, there is a reason to study problems like this, beyond the practice it provides in thinking about counterintuitive solutions.

Many problems in the real world are messier than this. In the history of timekeeping, there are thousands of tiny stories like this; stories that arise out of the interactions between idealistic dreams, practical considerations, and human politics. These stories can remind us that the perfect world of mathematics is not always the same as the world people live in.

The next problem concerns a different kind of edge.

Back where you began

You are at a remote cabin, and walk ten miles north, ten miles west, then ten miles south, only to find yourself back where you started. Where is the cabin located?

Once again, we have a question that cannot be true except in extreme cases.[8]

What are the natural extremes on the Earth? Well, the North and South Poles are pretty extreme!

If you start at the South Pole, travel ten miles north, travel any distance east or west, and travel ten miles back south (as shown in Figure 3.11), you will end up back at the pole!

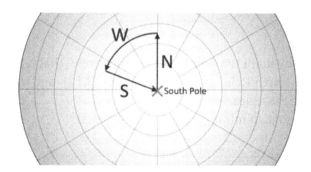

Figure 3.11.   Walk north, west, then south from the South Pole and you will end up where you started.

---

[8] Yes, cases plural. Readers who have already seen this riddle elsewhere may be surprised to learn there are multiple answers.

There are other answers, though. Consider this: around each pole, you can draw circles of any circumference, from nearly zero miles to the circumference of the globe itself. You can travel any of these circles in either direction — east or west — eventually returning to where you started.

Of particular interest to us, there is a class of circles of circumference $\frac{10}{n}$ miles (e.g., 10 miles, 5 miles, 3.33 miles, 2.5 miles, 2 miles, ...). For a given circle of circumference $\frac{10}{n}$, by walking 10 miles west, you will make exactly n laps of the pole, returning to where you started.

To be more specific: there is exactly one circle of circumference 10 miles that is centered on the North Pole. If you walk 10 miles east or west along this circle, you will end up back where you started. There is also a place closer to the pole where you can walk due east or west for 10 miles and end up circling the pole twice. And there is another circle where you can walk 10 miles and circle the pole three times... and so forth.

The largest of these circles is ten miles in circumference, running at a distance of approximately 1.59 miles from the North Pole:

$$c = 2\pi r$$
$$10 = 2\pi r$$
$$\frac{10}{2\pi} = r$$
$$r \approx 1.59$$

Start at any point ten miles south of this circle (that is, 11.59 miles south of the North Pole, as shown in Figure 3.12), and you will find an answer to this question.

This gives us an infinite number of answers![9]

---

[9] For fun, if you wish to reduce the number of answers back down to one, you could remark that there is only sea ice at the North Pole and thus, no one in their right mind would build a cabin there. This represents yet another "edge"—mathematically, you could say that there are an infinite number of answers, but at the edge where mathematics meets the real world, there is only one answer!

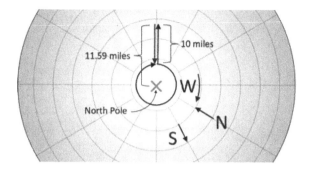

Figure 3.12. Starting south of the North Pole, go north, and walk a circle that is exactly 10 miles around, once, before turning south again.

Five points on a globe

We define a *closed hemisphere* as all points on exactly one half of the globe, including the points on line dividing the globe in half. Thus, the *closed northern hemisphere* includes the northern hemisphere and the equator, and the *closed southern hemisphere* includes the southern hemisphere and the equator.

Choose five points at random on the globe. Show that at least four of the five points must be in the same closed hemisphere.

First, we need to define an additional term. The dividing line between two hemispheres will always be defined by a *great circle*, which is a circle of the largest possible diameter that can be drawn on a sphere, or a circle on the sphere that has its center at the center of the sphere, and is formed by finding the intersection of the sphere with a plane passing through its center.[10]

In Figure 3.13 you can see how a great circle is formed. A random plane passes through the center of the sphere; cutting away one half

---

[10] The shortest path between two points on a sphere lies along the great circle between them. Planes fly routes that lie approximately along these great circle paths, although they often alter their paths to avoid weather, to benefit from tailwinds, and to account for other considerations.

The proof that the great circle path is the shortest possible is beyond the scope of this text, as it relies on calculus.

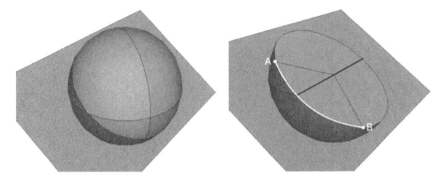

Figure 3.13.   A great circle. (Left) A plane intersects the sphere through the center of the sphere. (Right) The upper hemisphere cut away; the line of the equator is visible. The shortest path between A and B follows the white arc, lying on the great circle passing through them.

of the sphere reveals the great circle formed by the intersection of the plane and the surface of the sphere.

You can start by observing that if four or five points are on the same great circle then this is trivially true. Similarly, if three points fall on the same great circle, then the two remaining points fall in opposing hemispheres, or the same hemisphere. Either way, you can choose one closed hemisphere that encompasses at least four points.

If no great circle passes through three points, you can always draw a great circle through any two points. Three points do not lie on this great circle, and *by the pigeonhole principle*, at least two of the remaining points must lie together in the same hemisphere.

Some problems seem more difficult than they need to be, but by examining edge-cases, we can find a simple solution.

Between two hard places

Two trains are traveling towards one another from a starting distance of 100 km. One is traveling at 20 km/h; the other is traveling at 30 km/h. A bird, of a species known for its remarkable speed, sets out from the first train at 50 km/h at the precise moment both trains depart their starting locations, flying towards the second train. When it reaches the second train, it turns around to return to the first train, and so on until the trains pass one another.

How far will the bird have flown at the moment the trains pass one another?

Resist the temptation to perform extensive algebra here!

The trains' combined speed is 50 km/h, and they have a total of 100 km to cover. It should be clear that it will take 2 hours for them to cover that distance. Since the bird's flight time will also be 2 hours, then it, too, will cover 100 km (that is, 50 km/h × 2 hours) in that time.

Christmas and New Years

Christmas is celebrated on the 25th of December, and New Year's Day is on the 1st of January. The two holidays are separated by 7 days.

What is the next calendar year where Christmas and New Year's Day will fall on the same day of the week?

Your first instinct may be to say that the two holidays will fall on the same day next year. Take a second look, though: if Christmas is on a Monday, the New Year's Day 7 days later falls *in the next calendar year.*

What can you say about the relationship between these two days, then?

Imagine all the days of the year being laid along a ruler. Each day corresponds to some measure along that ruler. January 1st corresponds to day 0, January 2nd corresponds to day 1, ..., January 31st is day 30, February 1st is day 31, and so forth.

In a typical year (that is, a non-leap year), New Year's Day is day 0, and New Year's Eve (the 31st of December) is day 364. Counting backward from the end of the year, you can see that Christmas falls on day 358, 6 days before New Year's Eve.

In this model, all dates that fall on the same day of the week as New Year's Day will then have a day number that is evenly divisible by 7. Thus, January 8th, day 7, is 7 days after day 0 and must fall on the same day of the week as New Year's (as must January 15th — day 14 — January 22nd, January 29th, and so forth). If January 1st is a Monday, then we can reliably say so too will be the 8th, the 15th, the 22nd, the 29th, the 5th of February...

Christmas is day 358, and as you can see, $358 = 51 \cdot 7 + 1$, which means that in a normal calendar year Christmas happens on the day of the week *after* New Year's Day! That is to say in, say, 2015, New Year's Day fell on a Thursday, while Christmas 2015 fell on a Friday. In a normal year Christmas will *never* fall on the same day as New Years, because there are 51 weeks and *one day* between the two dates.

What about leap years? In this case, we fare no better: leap years add an additional day between New Year's and Christmas! That means, if New Year's falls on a Thursday during a leap year, Christmas will happen on a Saturday! During a leap year, there are 51 weeks and 2 days between the two holidays. There no way to win this one!

Beyond the Edge: Mathematical Loops

The previous problem leads us to a new topic. Sometimes mathematics problems have edges, but what happens when they don't? What happens when something in mathematics loops on itself, forever?

There is an old nursery rhyme that is relevant here:

Solomon Grundy,
Born on a Monday,
Christened on Tuesday,
Married on Wednesday,
Took ill on Thursday,
Grew worse on Friday,
Died on Saturday,
Buried on Sunday,
That was the end,
Of Solomon Grundy.

Nursery rhymes are often silly or absurd, so you could say that Solomon Grundy lived his whole life in a week's time. But his little biography could easily be real; nowhere does it say that he was married the Wednesday right after his Christening. One out of every seven people (on average) are born on a Monday, and one out of seven people die on a Saturday, but you would never claim *those* people only lived a week!

We could benefit from a different way of accounting for certain types of numbers. Luckily, that already exists, and it's called modular arithmetic.

Counting time

Modular arithmetic, or *clock arithmetic*, is a way of doing mathematics in contexts that involve loops. It is called clock arithmetic because of its similarity to an analog clock: 10 hours after 10:00 is 8:00, not 20:00[11] (Figure 3.14).

It is a familiar, if not always natural, way to look at math. You already know that hours loop with a period of 12 or 24 hours; for every hour, the count of minutes starts again at zero; and for every 7 days the week begins anew.

Most mathematical operations involving clocks, or modular arithmetic, behave the way you would expect from your experience with clocks and calendars. Outside the simplest examples, though, the operations you're familiar with can be quite cumbersome to use. Here, we will define terms and build tools that will make more complex calculations possible.

Figure 3.14. Clock arithmetic. (Left) An analog clock reading 10:00. (Middle) A clock that only tells hours reading 10. (Right) On a clock, $10 + 10 = 8$.

---

[11] If you are in the military or a country where a 24-hour clock is the norm, 20 o'clock exists. An equivalent formulation, then, would be: "twenty hours after 20:00 is 16:00, not 40:00."

We say two numbers are *congruent* modulo some number, called the *modulus*, if the two numbers are some multiple of the modulus apart. On a clock (which has a modulus of 12), 3 and 15 are congruent because they are exactly 12 hours apart. Similarly, 27, 39, … and –9, –21, … are all congruent to 3, because they all occur at a separation of some whole number multiple of 12 hours from 3.

That definition is more complicated than it needs to be, and likely to lead to mistakes, so we'll develop a few equivalent formulations.

Start by numbering the days of the week, starting at 0. Thus, Sunday = 0, Monday = 1, Tuesday = 2, and so on. If day 1 of some month starts on day 1 of the week, we can calculate which numbers are congruent to 3; that is, which dates fall on a Wednesday. The 3rd of the month falls on Wednesday, as does the day 7 days later (the 10th), and the day 7 days after that (the 17th), and so on.

$$3+7=10$$
$$10+7=17$$
$$17+7=24$$
$$24+7=31\ldots$$

So 3, 10, 17, 24, 31, … are all *congruent* to one another, modulo 7. You can quickly verify this by looking at a calendar and finding a month that starts on Monday, then checking the dates associated with each Wednesday in that month. You will discover that no matter the month, so long as it begins on a Monday, the 3rd, 10th, 17th, 24th, and 31st of the month will fall on Wednesday. These numbers are all congruent to one another, modulo 7.

We can also go the other direction, looking at the past. Subtract 7 days from the 3rd day in the month, and you will find that the day four days prior to the beginning of the month will be a Wednesday. If the previous month has 30 days, then the 30th will be day 0 (a Sunday), the 29th will be day –1 (a Saturday), … and the 26th will be day –4, a Wednesday.

Days –11, –18, –25 and so on will all be Wednesdays as well:

$$3-7=-4$$
$$-4-7=-11$$
$$-11-7=-18$$
$$-18-7=-25\ldots$$

..., −25, −18, −11, −4, 3, 10, 17, 24, 31, ... are thus all congruent to one another, *modulo 7*, and belong to the same *congruence class*. It should be clear to you that for a given modulus, there is no limit to the numbers that belong to a congruence class, just as you can imagine the weeks marching forward and backward in time forever.

A consequence of congruence is that the difference between any two congruent numbers will always produce some multiple of the modulus.

Consider two numbers that are congruent modulo 7, such as 3 and −4. Note that the difference between them is a multiple of 7.

$$3-(-4)=-7=7\cdot-1$$

Similarly, 31 and 10 are congruent modulo 7, and their difference is 21 — which is the product of 7 (the modulus) with 3:

$$31-10=21=7\cdot3$$

Further, dividing any number of members of the same congruence class by the modulus will produce as remainder the same value:

$$\frac{24}{7}=3R3$$

$$-\frac{11}{7}=-2R3$$

We can take an arbitrary number, say 178, and find what congruence class it belongs to by finding the remainder of its division by 7:

$$\frac{178}{7}=25R3$$

So 178 is a member of the same congruence class as all the numbers we've seen so far.

We can define the idea of congruence class more formally in terms of these divisions. The definition we suggest is this: the congruence class of a number is the *non-negative* remainder r of the integer division a/b with integer quotient q such that

$$a=qb+r$$

Note that this statement of numbers in terms of their products and remainders is equivalent to saying that *a* is congruent to *r*, modulo *b*. Substituting an earlier example,

$$31 = 4 \cdot 7 + 3$$

Stated in the terms we have defined so far, this line indicates "31 is congruent to 3, modulo 7."

Why is this definition useful? Because, under this definition, *b* always tells you how many times you have "gone around the clock." Days 3 and −11 are two weeks apart, and day −11 is two weeks before day 3; thus day −11 is the Wednesday that comes two weeks earlier than day 3.

Ensuring that any definitions you use are consistent (as opposed to full of exceptions) confers a number of advantages. Consistent rules are easier to remember, easier to apply, and less likely to result in mistakes. And, since you're making them up here, you're free to make any rules you want, as long as they seem to be useful!

One consequence of these choices may not yet be obvious. By convention,[12] when performing modular arithmetic, we refer to the congruence classes using the numbers from zero to one less than the modulus, as these correspond to all possible remainders. Thus, in arithmetic *mod 7*, we use the numbers 0, 1, 2, 3, 4, 5, and 6 to represent all possible congruence classes.

Let's explore how this works by exploring some operations between two numbers, a_1 and a_2.

$$a_1 = q_1 b + r_1$$
$$a_2 = q_2 b + r_2$$

---

[12] For now, it suffices to say "by convention;" in truth, there are good algebraic reasons to do this that are beyond the scope of this chapter. One reason for using the numbers between 0 and *n* − 1 is that 0 is the *additive identity*, no matter which modulus we use.

As you may recall, the additive identity is the number *I* that, when added to any other number *a*, does not change its value under addition. Thus,

$$I + a = a + I = a$$

Here, the values of q_1 and q_2 don't particularly matter, nor does the modulus b. What does matter is that the modulus is the same for both numbers, and that all numbers are integers. Let's see if we can formally demonstrate how addition works under modular arithmetic.

$$
\begin{aligned}
a_1 + a_2 &= \left(q_1 b + r_1\right) + \left(q_2 b + r_2\right) \\
&= q_1 b + r_1 + q_2 b + r_2 \\
&= q_1 b + q_2 b + r_1 + r_2 \\
&= \left(q_1 + q_2\right) b + r_1 + r_2
\end{aligned}
$$

The left half of the last line is divisible by b, and the two remainders in the right half correspond to the congruence classes of a_1 and a_2. We have demonstrated, then, that the sum of two numbers in modular arithmetic is equivalent to the addition of their congruence classes!

If the 24th falls on a Wednesday, and we wish to know what day of the week it will be 15 days later, we can identify the congruence classes of these two numbers, and add those instead! Therefore:

$$
\begin{aligned}
24 &= 3 \cdot 7 + 3 \\
15 &= 2 \cdot 7 + 1
\end{aligned}
$$

So, 15 days after the 24th will be $3 + 1 = 4$ which, by our system, corresponds to Thursday.

Doing the same thing for multiplication produces the following:

$$
\begin{aligned}
a_1 \cdot a_2 &= \left(q_1 b + r_1\right) \cdot \left(q_2 b + r_2\right) \\
&= q_1 q_2 b^2 + q_1 b r_2 + q_2 b r_1 + r_1 r_2 \\
&= \left(q_1 q_2 b + q_1 r_2 + q_2 r_1\right) b + r_1 r_2 \\
&= \left(...\right) b + r_1 r_2
\end{aligned}
$$

We abbreviated the entire contents of the parentheses to highlight the fact that you shouldn't particularly care about what is being multiplied by the base b as it doesn't affect the final answer. The only part that should matter, from your point of view, is the fact that multiplying two numbers produces the same result as multiplying their congruence classes!

In case you struggled to follow the last two demonstrations, what you need to know for now is:

- under modular arithmetic, addition of numbers is the same as adding their congruence classes, and
- under modular arithmetic, multiplication of numbers is the same as adding their congruence classes.

Finally, we introduce the following notation. Congruence of two numbers is indicated by a three-bar equals sign, followed by the modulus in parentheses, as follows:

$$a \equiv b \pmod{n}$$

We can rewrite some earlier findings using this notation. Here, we present examples of congruence, addition, and multiplication under modular arithmetic.

$$-11 \equiv 3 \pmod 7$$
$$2 + 10 \equiv 2 + 3 \pmod 7$$
$$10 \cdot 15 \equiv 3 \cdot 1 \pmod 7$$

You can verify each of these statements on your own. For practice, try reading each aloud. The middle line, for instance, is read as "two plus ten is congruent to two plus three, mod seven."

Using these ideas, you can re-evaluate the previous problem. As a reminder, that question asked, "what is the next calendar year where Christmas and New Year's Day will fall on the same day of the week?"

As before, we denote the 1st of January as day 0. You can easily verify that January 8th, 15th, 22nd, and 29th all happen on the same day of the week as the 1st. It becomes more difficult to reason about days in other months, so you might want to build a table like the one we have provided in Table 3.22.

In each row, we indicate the number of days in the given month, then calculate an "offset" to add to the current month's offset to generate the starting day of the week for the next month. An offset is the length of that month, modulo 7, and indicates how many days each month "pushes" the following months.

Table 3.22.   Each month shifts the start of the subsequent month by some number of days. You can add these offsets, modulo 7, to find the months' congruence classes.

Month	Start day	Length	Offset
January	0	31	3
February	3	28	0
March	3	31	3
April	6	30	2
May	1	31	3
June	4	30	2
July	6	31	3
August	2	31	3
September	5	30	2
October	0	31	3
November	3	30	2
December	5	31	3
January	1	—	—

If each month had 28 days, then all their offsets would be equal to 0, and all months would start on the same day of the week. In reality, most months are 31 days long, and thus, push the months that follow those months forward by 3 days.

By this table, if January first falls on a Thursday (day 4), then February first falls three days of the week later, on Sunday (day $3 + 4 = 7$, which is the same as day 0). March has the same start day as February, so it, too, begins on Sunday.

April's start day is 3 days later than March's, so April begins on Wednesday, which you can verify by adding 3 days to March's start day (Sunday, day 0, plus 3) or by adding 6 days to January's start day: Thursday, day 4, plus 6: $4 + 6 = 10 \equiv 3 \pmod 7$.

Christmas is 24 days after[13] December 1st, and December's start day is 5, so it falls on day $5 + 24 = 29$. Since day $29 \equiv 1 \pmod 7$, you

---

[13] If this does not make sense to you, remember that December 1st is 0 day after December 1st, the 2nd of December is 1 day later, ... and thus the 25th of December is 24 days after the first.

Table 3.23. International holidays and their offsets, and what day they would fall on assuming the year begins on a Sunday.

Holiday	Country	Date	Offset + Day	Day
New Year's Day	International	January 1	$0 + 0 = 0 \equiv 0$	Sunday
Anzac Day	Australia/NZ	April 25	$6 + 24 = 30 \equiv 2$	Tuesday
Golden Week (Begins)	Japan	April 29	$6 + 28 = 34 \equiv 6$	Saturday
Labor Day	International	May 1	$1 + 0 = 1 \equiv 1$	Monday
Independence Day	US	July 4	$6 + 3 = 9 \equiv 2$	Tuesday
Bastille Day	France	July 14	$6 + 13 = 19 \equiv 5$	Friday
German Unity Day	Germany	October 3	$0 + 2 = 2 \equiv 2$	Tuesday
Halloween	International	October 31	$0 + 30 = 30 \equiv 2$	Tuesday
Armistice Day Veteran's Day (US)	International	November 11	$3 + 10 = 13 \equiv 6$	Saturday

know that Christmas happens one day of the week after New Year's Day, confirming our earlier result!

You can also calculate the offset for other holidays. Ignoring holidays that always fall on specific days of the week (e.g., the US holidays Memorial Day and Labor Day always fall on Mondays, and Thanksgiving is always on a Thursday; internationally, Easter, Mother's Day, and Father's Day all fall on Sundays), there are nevertheless a wealth of examples to explore. Assuming the year begins on a Sunday and is not a leap year, we calculate in Table 3.23 the day of the week for a selection of international holidays.

Let look at some ways this system can help your problem solving.

Counting Friday the 13ths

Friday the 13th is considered an unlucky date in some cultures. What is the greatest number of times Friday the 13th can appear in any year? What is the smallest number?

The 13th day of the month will only occur on Friday if that month started on a Sunday. You can group months by congruence classes based on the information in Table 3.22. Leap years shift the start date

Table 3.24.   Aggregating the months of the year according to their congruence classes. The months that fall after February 29th in a leap year have their class shifted forward by 1.

Class	Months (non-leap year)	Months (leap year)
0	January, October	January, April, July
1	May	October
2	August	May
3	February, March, November	February, August
4	June	March, November
5	September, December	June
6	April, July	September, December

of each month after February forward by one day, so you should add another column to your table that accounts for this fact.

You *could* use this table (Table 3.24) to calculate which months would have a Friday the 13th based on the beginning of the year. If the year began on Sunday, then months that belong to congruence class 0 will also begin on Sunday and will contain a Friday the 13th; if New Year's Day is a Saturday, then class 1 months will contain a Friday the 13th; and so on.

However, *this is not what the question asked for*!

With a glance at Table 3.24, you can see that no congruence class has more than three members. Similarly, no congruence class has fewer than one member. So in any given year, there will be at least one, and at most three, months containing the day Friday the 13th.

This demonstration of how to apply modular arithmetic might not have been very counterintuitive, but it was simple! It does allow us to look at some counterintuitive problems in a new light, though.

Another way of looking at Faro shuffles

In Chapter 2, we discussed how to track cards as they moved through the deck during in and out shuffles, but we left some questions unresolved. Notably, we left unanswered a question of how cards behave when they are in the lower half of the deck, and how to predict the

Table 3.25: The movements of the cards of an eight-card deck after successive in shuffles.

Position	1	2	3	4	5	6	7	8
Start	1	2	3	4	5	6	7	8
Shuffle 1	5	1	6	2	7	3	8	4
Shuffle 2	7	5	3	1	8	6	4	2
Shuffle 3	8	7	6	5	4	3	2	1
Shuffle 4	4	8	3	7	2	6	1	5
Shuffle 5	2	4	6	8	1	3	5	7
Shuffle 6	1	2	3	4	5	6	7	8

position of any card after any number of shuffles. We have now provided you with the tools that will allow you to follow this discussion!

In Chapter 2, we looked at a series of problems involving a deck of eight cards. In some of those discussions, we investigated *in shuffles*, where the deck is cut exactly in half, and then the top half is shuffled *into* the bottom half.

We have reproduced the table of in shuffles from that chapter in Table 3.25. After each in shuffle, the cards moved from one location to another.

You can track these movements in a manner similar to the way we tracked the movements of the cards in the Kings and Aces problem. In that problem, the task was to reconstruct the original configuration of the deck based on the result of a particular procedure: given eight cards in a deck, deal the top card face up, then move the new top card to the bottom of the deck, and repeat until no cards remain in the deck. In Chapter 2, we saw two different ways to reconstruct the original order.

Comparing any two lines of Table 3.25, you can see that the card in the first position moves to the second position, the card in the second position migrates to the fourth position, and so on. We have illustrated these transitions in Figure 3.15.

Starting at any card, trace the arrow leaving that card until you end up back where you began. Eventually, you will find that there are two loops:

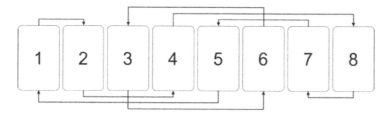

Figure 3.15.   Mapping the movement of each card after a single in shuffle defines loops.

Table 3.26.   The result of one in shuffle with a nine-card deck.

Position	1	2	3	4	5	6	7	8	9
Start	1	2	3	4	5	6	7	8	9
In shuffle	5	1	6	2	7	3	8	4	9

- The first loop, of length six, moves any card on its path back to its starting location after six shuffles. Starting at card 1, the loop proceeds **1**–2–4–8–7–5–**1**... We can say that this loop has a *period* of six.
- The second alternates the positions of cards three and six with a period of two.

You can see and confirm this by referring to Table 3.25; note that the third column alternates in value between 3 and 6.

Since the motivating example for looking at modular arithmetic is the study of loops, it should come as no surprise that the loops we see here can be analyzed using this problem!

As you might recall, cards in an eight-card in shuffle behave the same as those in a nine-card shuffle, whether that be an in shuffle or an out shuffle. For reasons that will shortly become clear, we will use the nine-card in shuffle as the motivating example.

For reference, recall how in shuffling changed the order of the cards in the nine-card deck, as seen in Table 3.26. The deck was split into a four-card top half and a five-card bottom half, and then the top half was shuffled into the bottom. The card numbered "1" was shuffled under 5, 2 was shuffled between 6 and 7, 3 between 7 and 8, and 4 between 8 and 9.

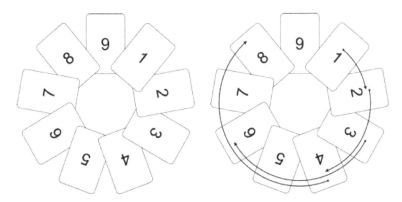

Figure 3.16. (Left) Nine cards in a circle. (Right) Doubling the position of the first four cards.

What if, instead of in shuffling a deck of cards, we shuffled the cards around a loop?

Take nine cards, Ace through 9, and arrange them in order in a loose circle, as illustrated on the left of Figure 3.16. Now, move each card around the circle in accordance with its position in the second row of Table 3.26. We will refer to *position p* as the position initially occupied by the card with the number *p* on its face (and *position 1* for the Ace). Card 1 starts in position 1 and moves to position 2, card 2 moves from position 2 to position 4, and so on.

After an in shuffle, a card located in the top half of the deck has its index, or position, doubled; you can verify this by examining the diagram on the right in Figure 3.16. Thus, the Ace goes to position 2; the 2 goes to position 4; and so on.

In Chapter 2, we noted that the positions of cards in the bottom half of the deck double as well, but in the opposite direction. In an eight-card in shuffle, card 8 moves from the first position from the back to the second position from the back, and so forth.

This worked for the eight-card case, but it's not clear how we should handle the odd case; after all, the ninth card remains fixed in place the entire time! Try an experiment, then: with the cards that would be in the bottom half of the deck (5, 6, 7, 8, and 9), perform the same procedure as just now, moving each card clockwise *p* places

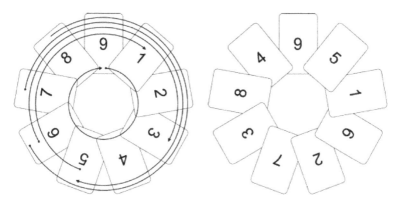

Figure 3.17. (Left) Using the same doubling procedure on the other five cards. (Right) The final ordering.

around the circle from its respective starting point, as shown in Figure 3.17.

It seems the experiment worked! By comparing the right diagram in Figure 3.17 to the result of the in shuffle in Table 3.26, you can verify that rotating each card by p places in a circle produces the same effect as shuffling.

Extending the idea of modular arithmetic to this example, then, you can see that the new position of a card with starting position p in a deck of n cards is determined by the following equation:

$$2p \ (\text{mod} \, n)$$

It will be useful to start bringing in algebraic formalism here. In this case, we want to define a shuffle as a function, f, which operates on the position of a card, x. Then to say that a single in shuffle moves card 1 to position 2 is to say:

$$f(1) = 2$$

We can provide a complete definition of this by listing all inputs and their corresponding outputs as follows:

$$f(1) = 2$$
$$f(2) = 4$$

$$f(3)=6$$
$$f(4)=8$$
$$f(5)=10\equiv 1\,(\mathrm{mod}\,9)$$
$$f(6)=12\equiv 3\,(\mathrm{mod}\,9)$$
$$f(7)=14\equiv 5\,(\mathrm{mod}\,9)$$
$$f(8)=16\equiv 7\,(\mathrm{mod}\,9)$$
$$f(9)=18\equiv 0\,(\mathrm{mod}\,9)$$

These outputs correspond exactly to the outputs of the tables and figures above.

What about multiple shuffles? It will be helpful to come up with notation that will help us describe repeated shuffles on the same deck. You can choose to denote multiple shuffles in any way that makes sense to you; in this case, we will choose to write a subscript indicating the number of shuffles next to the function name, like so: $f_2(x)$. If some other notation makes more sense to you, feel free to use that instead.

Shuffling twice moves card 1 to position 4; we can state this in our new notation as

$$f_2(1)=4\equiv 4\,(\mathrm{mod}\,9)$$

With this notation established, we can restate the facts above in purely algebraic terms as shown in the following:

$$f_0(x)\equiv x\,(\mathrm{mod}\,n)$$
$$f_1(x)\equiv 2x\,(\mathrm{mod}\,n)$$
$$f_2(x)\equiv 4x\,(\mathrm{mod}\,n)$$
$$f_3(x)\equiv 8x\,(\mathrm{mod}\,n)$$
$$\vdots$$
$$f_s(x)\equiv 2^s x\,(\mathrm{mod}\,n)$$

Follow the pattern. Does the final line make sense to you? Do you believe that after *s* shuffles, the formula we wrote is true? Try to convince yourself of what has been written so far is correct, and if you have written down something else, see if there's a way to reconcile the two statements.

In Chapter 2, you saw that a nine-card deck returns to its initial configuration after six in shuffles. Let's plug these values into the formula above.

$$f_6(x) \equiv 2^6 x \pmod 9$$
$$f_6(x) \equiv 64x \pmod 9$$

We already know that multiplication of two numbers under modular arithmetic produces the same result as multiplying their congruence classes. So it stands to reason that you must find what 64 is congruent to, modulo 9.

$$64 = 7 \cdot 9 + 1$$
$$64 \equiv 1 \pmod 9$$

In normal arithmetic, 1 is the multiplicative identity,[14] and the same holds true here! Performing six in or out shuffles to a nine-card deck will return it to its original configuration *because* 64 is congruent to 1, so no matter which position a card started in, it will be in that same position after "being multiplied by 1."

We started out asking how many shuffles of each type are required to return a deck to its original state. Well, you finally have enough information to generalize for decks of all sizes!

For odd *n*:
For both in and out shuffles, the number is the same. Find *s* such that $2^s \equiv 1 \pmod n$.

For instance, with a fifteen-card deck:

$$2^1 = 2 \equiv 2 \pmod{15}$$
$$2^2 = 4 \equiv 4 \pmod{15}$$

---

[14] Recall that the multiplicative identity is the number that, when multiplied by another number, does not change its value.

In algebraic terms, the multiplicative identity is the number *I* such that for any number *a*,

$$I \cdot a = a \cdot I = a$$

$$2^3 = 8 \equiv 8 \ (\text{mod } 15)$$
$$2^4 = 16 \equiv 1 \ (\text{mod } 15)$$

So, after four consecutive in shuffles, or four consecutive out shuffles, a fifteen-card deck will be restored to its original order.

For even n:

In shuffles with an even deck behave the same as repeated shuffles of a deck one card larger. Thus, find s such that $2^s \equiv 1 \ (\text{mod } \boldsymbol{n} + \boldsymbol{1})$.

Out shuffles with an even deck behave the same as in shuffles with a deck two cards *smaller*. Since this number will also be even, it will behave the same as an *odd* deck *one* card smaller. Thus, find s such that $2^s \equiv 1 \ (\text{mod } \boldsymbol{n} - \boldsymbol{1})$.

With a six-card deck:

In shuffles	Out shuffles
$2^1 = 2 \equiv 2 \ (\text{mod } \boldsymbol{6} + \boldsymbol{1})$	$2^1 = 2 \equiv 2 \ (\text{mod } \boldsymbol{6} - \boldsymbol{1})$
$2^2 = 4 \equiv 4 \ (\text{mod } \boldsymbol{7})$	$2^2 = 4 \equiv 4 \ (\text{mod } \boldsymbol{5})$
$2^3 = 8 \equiv \boldsymbol{1} \ (\text{mod } \boldsymbol{7})$	$2^3 = 8 \equiv 3 \ (\text{mod } \boldsymbol{5})$
	$2^4 = 16 \equiv \boldsymbol{1} \ (\text{mod } \boldsymbol{5})$

So, after three consecutive in shuffles, or four consecutive out shuffles, a six-card deck will be returned to its original order.

You can try this process with decks of any size, and check your work against the table we provide. Due to the inter-relationships between decks of adjacent sizes, you can see repeating patterns in Table 3.27.

Treating a deck of cards like a circle of cards may not have been an intuitive approach, but it has certainly proven to be powerful!

Building Your Intuition

This chapter covered many of the same types of problems as given in Chapter 2, with a twist that should be familiar to you by now.

You have seen many problems that yielded to the similar approaches as those in Chapter 2 — testing boundaries, intelligent guessing and

Table 3.27. The number of successive in shuffles and out shuffles necessary to restore decks of varying sizes to their original order.

Cards	In shuffles	Out shuffles
1	1	1
2	2	1
3	2	2
4	4	2
5	4	4
6	3	4
7	3	3
8	6	3
9	6	6
10	10	6
11	10	10
12	12	10
13	12	12
14	4	12
15	4	4
16	8	4
17	8	8
18	18	8
⋮	⋮	⋮

testing, drawing pictures, and so forth — but each of these problems, or their solutions, demanded some counterintuitive leap.

How did you fare? Did you see the trick in each puzzle, and in each solution? It is impossible to teach brilliance; nevertheless, this kind of practice can cause your brain in ways that make each new puzzle that much easier. If you've maintained the habit of thinking about your thinking from Chapter 1, you should by now have a thorough accounting of your own ability to "see through" problems, to see the kernel of an approach that each problem demands of you.

If you haven't quite gotten the knack yet, no worries! Learning to see through problems requires a great deal of practice, so simply reading and following the solutions we presented should have given you some perspective on what problem solving means. We have planted the seeds of future growth.

You may have noticed something peculiar, though: as we made our way to the end of this chapter, we introduced more and more algebraic notation, as well as several lightweight proofs. The problems in this chapter have been counterintuitive enough; if you've ever struggled with algebra, bringing more in at this stage may have challenged what intuition you had left!

We will make the case for algebra in the following chapters. This chapter and Chapter 2 have afforded you the opportunity to see how to perform mathematics without many symbols (or by inventing your own symbols). We will continue to make parallel demonstrations — showing you how to solve mathematics problems with pictures, tables, and other useful transformations — but algebra and proof will play a greater role as we make our way through the remainder of this book, as these facilitate some very Important and useful solutions to mathematical problems.

Chapter 4

Important and Useful Mathematical Solutions to Problems

The Importance of Understanding Mathematics

The chapters, thus far, have illustrated some fun mathematical curiosities and entertainments. By now, we hope that you've begun to see some of the beauty and pleasure of doing mathematics, and come to appreciate that mathematics appears in a variety of different and often unexpected forms.

However, some may as yet remain unconvinced of the utility or importance of mathematics. Recreational mathematics has a long history, and many useful mathematical applications have come out of purely theoretical explorations initially undertaken for the mere pleasure of problem solving. We might even argue that, from the perspective of the learner, "fun" mathematics is more beneficial because the motivations for learning are often self-reinforcing: if a student enjoys solving a certain kind of problem, tackling these problems will lead to deeper and deeper understanding of richer and more powerful tools, eventually carrying the student to the limits of existing knowledge.

This is not the chapter where we make that argument, although we hope you notice the thread of that idea woven throughout this text. If you find one type of problem particularly enjoyable, we strongly encourage you (once you have finished the final chapter here) to find a specialist book that covers that topic in detail, and immerse yourself in it until you understand its every nuance.

When, at the end of your deep dive into mathematics, you come up for air, you may have cause to wonder: "what is the point of all this?"

This chapter attempts to answer to *that* question.

We will investigate the history of two major problems — predicting the future, and finding out where you are — and through them, demonstrate *why* people have been interested in mathematics, and (hopefully!) how it can be useful to you.

How to Predict the Future

Making a fortune using mathematics

François-Marie Arouet (1694–1778) was born in Paris in 1694. He was the youngest of five children of François Arouet, a lawyer and minor bureaucrat, and his wife, Marie Marguerite Daumard. His father wished for the young François-Marie to follow in his footsteps and become a lawyer, but his exceedingly strong-willed son had his own ideas. By the time he graduated school, in 1711, he knew he wanted to become a writer.

His father nevertheless struggled to provide him a profession, arranging for him to work in a number of legal posts to complete his training, but the young Arouet would not be deterred. He pretended to follow his father's wishes, but wrote poetry and essays all the while, his wit earning him reputation (and scorn) from people of status.

In 1718, after being jailed for 11 months in the Bastille for insulting the then-ruler of France, Arouet decided to officially change his name to the much more poetic name we remember him by: Voltaire.

Voltaire is known to us now as one of the most influential writers of 18th century Europe. However, in 1718 his career was still young, and his fortunes far from secure. His first play after leaving prison was a financial success, but later works frequently flopped. His acerbic wit was both his talent and his curse: a play he penned in 1725 was so well-regarded it was chosen to be part of the celebration of the king's wedding. Meanwhile, in 1726 his retort to a noble's insult escalated by strides, until he faced a choice: indefinite imprisonment or banishment to England. He chose the latter.

After two and a half years abroad, he secured his return to France, and in due time, to Paris. His time abroad had been good for his safety and education, if not for his finances. Voltaire's father, who died in 1722, had left most of his fortune in trust until his son could demonstrate sufficient responsibility to merit its dispensation, and for this and other reasons, the writer's finances had been stretched thin.

In 1729, he encountered the mathematician Charles Marie de La Condamine (1701–1774) at a party, and after some discussion, the two hatched a plan. The French government, desperate to finance its debts, had come up with a scheme to make certain otherwise unattractive bonds more appealing. The general idea was this: as long as he held the bond, a bondholder could pay 1/1000 of its face value to purchase a lottery ticket. Periodically, the government would hold a drawing, and the lucky winner would receive a jackpot of 500,000 livres — quite a substantial sum of money! This would encourage creditors to buy and hold existing bonds, while theoretically generating a new stream of revenue for the government in the form of ticket sales.

What de la Condamine had figured out, and what he shared with Voltaire, was that every lottery ticket had the same chance of winning. So, the owner of a 1,000 livre bond — whose market value was well less than its face value by this point — could, for the price of 1 livre, purchase a ticket that offered as much chance of winning as a ticket purchased against a 100,000 livre bond for the cost of 100 livres!

In order to corner the market and guarantee victory, de la Condamine and Voltaire would need a great deal of capital, as well as allies to purchase huge numbers of tickets. Voltaire's wit, charisma, and social network enabled him to gather confederates, and together they formed the syndicate that would dominate the bond lottery for over a year, winning time and time again!

Voltaire's disdain for authority in general and the French government in particular would be the syndicate's undoing. Not content to quietly profit at the government's expense, Voltaire left mocking clues that pointed the government to the syndicate's existence. He and the rest of the conspirators were made to stand trial but, not having done anything strictly *illegal*, were let go and grudgingly allowed to keep their winnings. The lottery was canceled shortly thereafter.

As epilogue, Voltaire used his lottery winnings and social savvy to identify promising investment opportunities — what we would today call "insider trading" — and amassed such great wealth that he would never again need to worry about balancing commercial success against artistic purity. He continued to defend humanism and justice, engaging and enraging the greatest scholars and rulers of the day for another 48 years. Thanks to his winnings, he spent this time largely free from financial worry.

The universal motive

One reason to learn mathematics is to make money. Voltaire's story should resonate with us, if only because many people dream of the kind of financial freedom he was able to attain. The process is not necessarily as simple as "learn math, make money," though.

The idea that de la Condamine noticed, and that he and Voltaire exploited, is related to a concept called *expected value*. Expected value is a way to relate the cost of something — for instance, a lottery, or a game of chance — to the size of prizes we are likely to win. Since this chapter is about important and useful mathematical problems, we will build up the case for historical *motivations*, *explanations*, and *connections* between problems. By so doing, we will be able to explain the idea that allowed the syndicate to outsmart the government, and in time explain the foundations of modern governments.

To truly understand their scheme, then, we have to go back in history to 1654 and the roots of probability theory (see Figure 4.1).

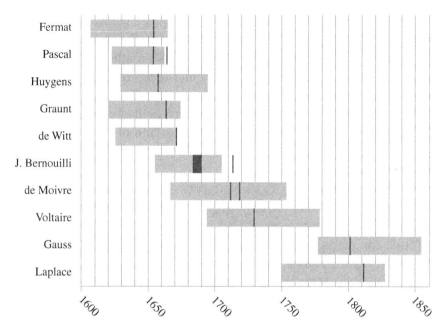

Figure 4.1. Some important names and moments in probability and statistics. Individuals are sorted by the date of their contribution, not birth. (Some contributions posthumous.)

Learning to count

The odds be with us

The study of probability began in earnest with the study of a number of gambling-related problems. Antoine Gombaud, Chevalier de Méré (1607–1684) was a minor French intellectual and unabashed gambler. Although not a nobleman, he adopted the title "Chevalier" (knight) after one of the characters in a story he wrote, and over time the name stuck.

In the early 1650s, Chevalier de Méré's gambling losses had started to get embarrassing (to him, if not the people getting rich at his expense). Seeking answers to explain his losses, he wrote letters to notable mathematicians of his day, hoping to engage their intellect and avoid future losses.

One of his letters found its way to his friend, a young but brilliant mathematician named Blaise Pascal (1623–1662). Pascal's curiosity was piqued, and in 1654 he and another notable mathematician of the age, Pierre de Fermat[1] (1607–1665), sent a series of letters back and forth discussing these problems. The two solved de Méré's problems, while laying the groundwork for the modern field of probability.

The letter posed two related questions. One of them, called by Pascal *the problem of the dice*, is similar to some problems you have already seen.

Consider this: a pair of gamblers play a game with two six-sided dice. The first one bets his opponent that after a certain number of rolls, two sixes will show up at least once; the opponent wins if at no point do the dice show two sixes. How many rolls are necessary for the first to have an advantage?

Chevalier de Méré, a capable mathematician in his own right,[2] broke the problem down like this. He correctly reasoned that a single six-sided die should show the number six one-sixth of the time. He knew from experience that, after four rolls of a single die, the probability of getting at least one six is slightly more than half.

Next, he reasoned (again, correctly) that two dice should both show sixes one-thirty sixth of the time. By proportions, he reasoned that if the odds are slightly in his favor that one die shows a six after four rolls, then after twenty-four rolls of two dice, he would have the same probability.

---

[1] If you know Fermat's name, it's most likely for his eponymous Last Theorem.

Fermat's Last Theorem stood unproven for over 300 years, until the early 1990s, when it succumbed to a mathematician named Andrew Wiles (to much fanfare in the popular press).

[2] Pascal, in one of his letters to Fermat, articulated respect for de Méré's ability, while bemoaning his intellectual blind spots:

I have no time to send you the proof of a difficult point which astonished M. [de Méré] so greatly, for he has ability but he is not a geometer (which is, as you know, a great defect) and he does not even comprehend that a mathematical line is infinitely divisible and he is firmly convinced that it is composed of a finite number of points. I have never been able to get him out of it. If you could do so, it would make him perfect.

$$\frac{4}{6} = \frac{2}{3} = \frac{24}{36}$$

Why was he wrong?

We have already seen problems like this: The Birthday Paradox (and its companion problems) from Chapter 3 specified strategies for counting discrete events like dice rolls. Given that chapter was all about counterintuitive problems, it's no wonder that de Méré was so confused! Let's look at how to correct de Méré's logic, using skills we have already practiced.

Recall that we can use the probability that an event does *not* occur to find the probability that it *does* occur. In this instance, each time the Chevalier rolls one die, there is a five-sixths chance he does *not* see a six. After four rolls, then, we can calculate the probability that he never sees a six.

$$\frac{5}{6} \cdot \frac{5}{6} \cdot \frac{5}{6} \cdot \frac{5}{6} = \left(\frac{5}{6}\right)^4 = \frac{5^4}{6^4} = \frac{625}{1296} \approx 0.4822 = 48.22\%$$

Subtracting this number from one (i.e., 100%), we can then find the chance that de Méré wins — that is, that he rolls *at least one* six.

$$1 - \frac{625}{1296} = \frac{1296}{1296} - \frac{625}{1296} = \frac{1296 - 625}{1296} = \frac{671}{1296} \approx 0.5177 = 51.77\%$$

So far, we can see that his chances of winning are better than half, confirming what Chevalier de Méré knew from experience.

We shall continue following in Pascal's footsteps by examining the game of two dice. Note again that de Méré was correct in deciding that the odds of getting two sixes was one in thirty-six, as seen in Tables 4.1 and 4.2. The former table shows all thirty-six possible rolls of two six-sided dice, and the sums of their values; the latter counts the number of ways each value can be found, and from there calculates the probability that a single roll will produce each value. One out of the thirty-six squares in Table 4.1 has a twelve; thus, we record a count of 1 and a probability of 1/36 in the row marked "12" in Table 4.2.

Table 4.1.   The sums of faces on two
six-sided dice.

	1	2	3	4	5	6
1	2	3	4	5	6	7
2	3	4	5	6	7	8
3	4	5	6	7	8	9
4	5	6	7	8	9	10
5	6	7	8	9	10	11
6	7	8	9	10	11	12

Table 4.2.   All values that can be generated by two dice,
their frequencies, and the associated probabilities.

Value	Count	Probability
2	1	1/36
3	2	2/36 = 1/18
4	3	3/36 = 1/12
5	4	4/36 = 1/9
6	5	5/36
7	6	6/36 = 1/6
8	5	5/36
9	4	4/36 = 1/9
10	3	3/36 = 1/12
11	2	2/36 = 1/18
12	1	1/36

Using the same logic as before, we can calculate the probability Chevalier de Méré does *not* roll two sixes in twenty-four rolls, then subtract from 1 to find the probability that he wins.

$$\frac{35}{36} \cdot \frac{35}{36} \cdot \frac{35}{36} \cdot \frac{35}{36} \cdots \frac{35}{36} = \left(\frac{35}{36}\right)^{24} \approx 0.5086 = 50.86\%$$

$$1 - 0.5086 = 0.4914 = 49.14\%$$

By rolling only twenty-four times, the Chevalier went from having a slight advantage over his opponent to having a slight disadvantage!

It would, in fact, take twenty-five rolls for the odds to be in de Méré's favor, and twenty-six rolls for him to enjoy the same advantage he had in the one-die game!

It's not about winning or losing, it's about how much money we make

The second problem that Pascal and Fermat discussed, and the one that took up much more of their time, requires a bit more introduction.

Most of the probability problems we have demonstrated cost nothing to play, and had only one winning position. For instance, in the Sleeping Spouse Problem (where your goal was to make a matched pair of socks), "playing" (pulling a sock out of the drawer) cost no money, and "winning" (making a matched pair) conferred no financial reward.

Most games of chance do not work this way. Gambling, whether on sporting events or games of chance, is attractive to some people precisely because of the possibility of winning big, and the risk (cost) of playing.

Thus, we propose a simple game. You and a friend have a fair coin. Each of you chooses one side of the coin, and whoever wins three tosses first wins the game. You both wager the same amount of money: if you win, you take her money, and if she wins, she takes yours.

If the game is interrupted part way through, and one of you has to leave, how much money should trade hands to "buy your way out" of the wager?

This is the question that engaged these intrepid mathematicians, and it was one that dated back centuries, called *The Problem of Points*.

They attacked it with gusto, and quickly found a solution. Pascal reasoned that if the game were tied — for instance, your two heads to your friend's two tails — each of you could take your money back with no issues.

It should be clear that this is a fair game: at the beginning of the game, there is an equal chance that either one of you will win, collecting all the money.

Assume you had each wagered thirty-two dollars.[3] If the game were tied, you could each walk away with thirty-two dollars and no hard feelings.

However, if your friend had the advantage at two tails to one, then surely she should walk away with at least thirty-two dollars. This is because if the next flip turned up heads, the two of you would be facing a tie, which we've already addressed; whereas if the next flip came up tails, she would win all sixty-four dollars. By Pascal's logic, then, she should be entitled to her original wager plus half of your wager, walking away with forty-eight dollars.

Examine the left side of Figure 4.2. We illustrate three different positions. Each position shows a number of wins in the center. On the left, you can see the position 2-1, meaning that your friend has won

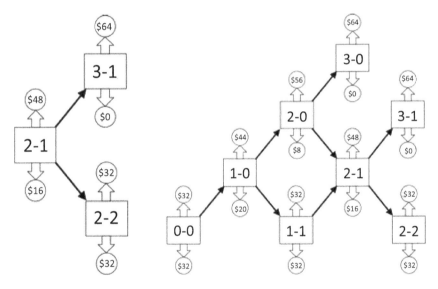

Figure 4.2. (Left) If two players are tied, their relative equity in the game is equal. If one wins, their equity is equal to the sum of the wagers. The value of a buy-out in the 2-1 case is then the average of each of these values. (Right) Continuing backward, we calculate the value of a buyout for all possible game states.

---

[3] In their letters, Pascal and Fermat mostly reference the gold currency of the time, the *pistole*, with a value that varied between ten and twenty-four *livres*. We will speak more about the value of money later in this chapter.

twice and you have won once. If you win the next round, you can follow the lower arrow to the position 2-2, whereas if your friend wins the next round, you can follow the upper arrow to the position 3-1. In this case, your friend would be entitled to all the money (sixty-four dollars, represented by the arrow pointing up), and you would earn zero dollars (shown in the arrow pointing downwards). Similarly, you can see how much money she would take inside the upwards arrows leading from each of the other positions.

Next, continuing to work backwards (as shown in the right half of Figure 4.2), Pascal reasoned that should you have to buy your way out of the game while down by two (2-0), then your friend would have a 50% chance of winning, and a 50% chance of finding herself in the situation described in the prior paragraph (2-1). So she should be entitled to forty-eight dollars, plus half your remaining money, for a total of fifty-six dollars.

Finally, if your friend were up by one (1-0), then there would be a 50% chance on the next flip that you would tie at 1-1, and a 50% chance that you would end up in the situation in the preceding paragraph (2-0). Pascal reasoned that the second player would be entitled to her original wager, plus half of what she would have taken from you in the previous paragraph, or twelve dollars. This would give her a total of forty-four dollars.

Examining Figure 4.2, we can see that the cost of "buying out" the game before its conclusion is the average of the costs of the two options that could proceed from it. By working backwards from known points — win positions and ties — Pascal was able to come up with a value for each possible buyout, one by one.

Pascal's reasoning, although sound, is not always easy to follow. We have deliberately refrained from translating his solution to modern language to give us a sense of how tentative and experimental his, and Fermat's, explorations of this subject were.

A mathematical revelation

Pascal and Fermat ended up solving the Chevalier de Méré's problems rather quickly, albeit not to their own satisfaction. The two mathematicians continued to correspond, speaking on topics that extended naturally outward from the two questions that motivated their discussion. The previous game, which we might think of as a "best of five series," led to discussions of games of other lengths, and to games with more than two players.

Pascal's explorations of these related topics led him to notice a pattern underlying all of them.

Contemporaneously with their correspondence, Pascal was studying a mathematical structure that might seem inconsequential at first glance. He started by writing the number 1 on one edge of a piece of paper. Then, in each direction, he wrote more 1s at regular intervals. Next, the value below and between any two values was computed as the sum of the values in the cells above and to either side of the former. We provide an illustration of this process in Figure 4.3.

Pascal did not discover this triangle, but his *Traité du triangle arithmetique*[4] (*Treatise on the arithmetic triangle*) was, to that point, the most thorough study of this figure that the world had seen.

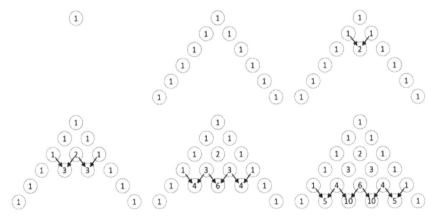

Figure 4.3. Generating Pascal's Triangle. Starting with a single 1 (top left), Pascal wrote more 1s in either direction (top center). To fill in each intermediate value, he added the values from the cells to the right and left in the previous row (next four diagrams).

---

[4] The timeline for Pascal's writing is somewhat confusing. His correspondence with Fermat concerning probability largely occurred during the latter half of 1654, but some authors sources date his *Traité* to 1653. Fermat makes reference to the *Traité*,

Table 4.3.   An alternate view on Pascal's Triangle. The columns in this table correspond to diagonals in Figure 4.3. The sum of the values in each row work out to powers of 2.

	1	2	3	4	5	6	7	8	Sum
1	1								1
2	1	1							2
3	1	2	1						4
4	1	3	3	1					8
5	1	4	6	4	1				16
6	1	5	10	10	5	1			32
7	1	6	15	20	15	6	1		64
8	1	7	21	35	35	21	7	1	128

In playing with the figure, he discovered a number of interesting properties. In Table 4.3, we highlight one in particular. The sums of the values in each row make what should by now be a familiar progression; they are exactly the powers of 2!

We can rationalize the reason this happens as follows: each cell contributes to the value of two of the cells in the following row. By the method we used to construct the triangle, then, any row should have double the sum of the previous row. Since we started with 1, which is equal to 2^0, each doubling produces a subsequent power of 2.

Still curious about the Problem of Points, Pascal looked at edge cases, wondering what the fair cost of buyout was for "best of" series of increasing sizes. The case studied in the problem of points was a "best of five," where the first player to three wins would take the prize; Pascal was interested in finding patterns for games to the best of three (two wins out of three), best of five (three wins out of five), best of seven (four wins out of seven), and so on.

He observed that, in a "best of three" game with one player up by one (i.e., with a score of 1-0), on the next throw, half the time the

---

calling it a "new work," in August 1654. The version we know addresses the Problem of Points, which Pascal's writing suggests was still novel to him that same month.

Regardless, its final publication did not occur until 1665, after Pascal's death.

second player could catch up, leading to a score of 1-1, and half the time the first player would win with a score of 2-0. Averaging the two values by the same logic as before, the first player could end the game at 1-0 and justly walk away with ¾ of the total wager.

In dollar terms, with a score of 2-0, the first player would take sixty-four dollars and the other zero; with a score of 1-1, both the first and second player could walk away from the game with their original wager of thirty-two dollars; and so at 1-0, the first player would be entitled to ¾ of the total, or forty-eight dollars. These are the same values shown in Figure 4.2.

The next game in this sequence is a "best of five," that is, a game to three points. In this case, if the first player were up two points to zero (2-0), as we have already seen, he could fairly walk away with 7/8 of the total wager, or fifty-six dollars.

In a game to four points (a best of seven series, similar to many sporting events like the World Series of Baseball or the National Basketball Association (NBA) Championships), Pascal calculated that if the first player were winning 3-0, he could take 15/16 of the total wager. We can re-trace his logic as follows.

Observe that each of the previous values is of the form $\frac{2^n - 1}{2^n}$ (or, equivalently, $1 - \frac{1}{2^n}$). In calculating these values, Pascal reasoned backward from winning states to not-yet-winning states, for instance, determining the value of a 2-1 game by averaging the values of a 3-1 win and a 2-2 tie. His approach helped him realize that it was not the number of points each player *possessed* at a given moment that determined the fair buyout price, but rather how many points they *needed to win*.

Consider this: in a best of seven series, if one player needs one head to win, but the other needs four tails, then the only way for the second player to win is for the coin to show tails four times in a row, which will only happen one out of sixteen times! Fifteen out of sixteen times, the first player will win, carrying away all the money.

But this same reasoning applies in games of other lengths: in a game to ten points, if the first player has nine points and the second player has six, then the first player *needs* one point to win, and the second needs four. Each of them has the same chance of winning as they did in the previous paragraph, and so the buyout should have the same price!

Pascal's solution, then, was to sum the numbers of needed points, and then find the row with that many cells in the triangle in Table 4.3. He then constructed a fraction with denominator equal to the sum of the entire row, and numerator the sum (starting from either side) of the number of cells equal to the number of points one player needed to win.

In the case of a 3-0 best of seven series, player one needs one point to win, and player two needs four to win. The sum of these two values is five; we can see row five of Table 4.3 highlighted in Table 4.4.

If the game ended here, then player one would be entitled to the sum of the first four cells (as his opponent needs four points to win), divided by the sum of the entire row. We can verify that this fraction, again, equals 15/16:

$$\frac{1+4+6+4}{1+4+6+4+1} = \frac{15}{16}$$

The power of this approach is that it applies to any values, without having to calculate the intermediate steps. We could, for instance, state that if player one needs three points to win and player two needs five, then player one is entitled to 99/128 of the money wagered:

$$\frac{1+7+21+35+35}{1+7+21+35+35+21+7+1} = \frac{99}{128}$$

If each player of this game had wagered sixty-four dollars, and their scores were in accordance with the values we stated (e.g., 7-5 in a game to 10 points, 2-0 in a game to 5 points), then the first player would be entitled to ninety-nine dollars. Easy!

A question remains: why is this true? What justifies this result? Pascal himself seemed satisfied with his conclusions and logic, but we may remain unconvinced by his reasoning. We, however, have the

Table 4.4.   The fifth row of Pascal's Triangle.

1	4	6	4	1

Table 4.5. All thirty-two ways to flip two coins. The point at which one of the players wins is highlighted; if player 1 wins after three flips (first four rows) we nevertheless specify all possible combinations for the other flips.

1	2	3	4	5	Winner	1	2	3	4	5	Winner
H	H	**H**	H	H	Player 1	T	H	H	**H**	H	Player 1
H	H	**H**	H	T	Player 1	T	H	H	**H**	T	Player 1
H	H	**H**	T	H	Player 1	T	H	H	T	**H**	Player 1
H	H	**H**	T	T	Player 1	T	H	H	T	**T**	Player 2
H	H	T	**H**	H	Player 1	T	H	T	H	**H**	Player 1
H	H	T	**H**	T	Player 1	T	H	T	H	**T**	Player 2
H	H	T	T	**H**	Player 1	T	H	T	**T**	H	Player 2
H	H	T	T	**T**	Player 2	T	H	T	**T**	T	Player 2
H	T	H	**H**	H	Player 1	T	T	H	H	**H**	Player 1
H	T	H	**H**	T	Player 1	T	T	H	H	**T**	Player 2
H	T	H	T	**H**	Player 1	T	T	H	**T**	H	Player 2
H	T	H	T	**T**	Player 2	T	T	H	**T**	T	Player 2
H	T	T	H	**H**	Player 1	T	T	**T**	H	H	Player 2
H	T	T	H	**T**	Player 2	T	T	**T**	H	T	Player 2
H	T	T	**T**	H	Player 2	T	T	**T**	T	H	Player 2
H	T	T	**T**	T	Player 2	T	T	**T**	T	T	Player 2

benefit of hindsight, and can explain this problem in ways that are much more agreeable.

In the interest of space, in Table 4.5 we consider a best of five series, enumerating all possible series of throws, *even those that would have ended the game earlier!*

Observe that there are a number of symmetries in this table: for instance, the row five rows down from the top lists the winner as player 1, while the row five rows up from the bottom shows player 2 winning.

We can answer the Problem of Points by counting, here. That method bears a strong resemblance to methods we used in the first two chapters. We wish to develop it more fully, though, and to do so it will be instructive to view this data from a different perspective.

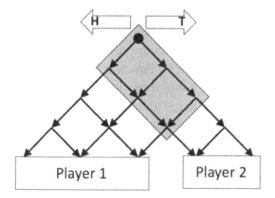

Figure 4.4. Mapping coin flips to directions. In a game where player 1 needs two heads to win and player 2 needs three tails to win, each player's equity in the game can be determined by the number of paths in this graph.

Select a possible game position at random. Say for instance that player one needs two more heads to win, and player two needs three more tails. Choose at random directions to assign to the sides of a coin; thus, we could say "left is heads, right is tails." Then count how many different ways there are to travel from the starting position to all possible end positions, and how many of each result in a win for each player.

Figure 4.4 illustrates these possibilities. At each intersection, starting at the top, we can trace either the arrow to the left (heads) or the one to the right (tails). One possible sequence here would be left, right, left, right, or LRLR. We can trace this path down and see that it results in a win for player 1.

Inside the gray block, the outcome of the game is still uncertain, but once past it, one of the players has already effectively won. Nevertheless, we should account for every path.

Again, we could enumerate and count each path, but as we have seen, there's an easier solution. Observe that the sums we would find correspond exactly to the values of Pascal's Triangle in the row with five entries. Thus, there are $(1 + 4 + 6) = 11$ ways for player 1 to win and $(4 + 1) = 5$ ways for player 2 to win, justifying Pascal's approach. (Note, too, that these numbers correspond to the odds of winning listed in the left half of Table 4.5.)

This transformation — between heads and tails, wins and losses, left and right — is the major revelation of Pascal's Triangle. In any problem that considers the enumeration of things that can go one way or another, with equal chances, Pascal's Triangle is likely to make an appearance. We provide some examples of these in Table 4.6. All of these questions have the same answer: 128.

The final row of Table 4.6 reveals the most versatile power of Pascal's Triangle: it provides a way to find the number of combinations of k objects out of a set of n total by reading a number off the triangle, rather than by performing the actual count.

Given a set of size n, there is one way to choose zero objects, and n ways to choose one object. We can use these two values to identify the row that corresponds to the number of combinations for a set, and then count upward to find the desired number of combinations.

For instance, if n is equal to 7, we want to find the row beginning 1, 7, ... Now, suppose the problem we wish to solve is "how many ways are there to choose three different scoops of ice cream from a selection of seven flavors?" Starting at 0, we can count to the third position: 1 is in the zeroth position; 7 is in the first position; 21 is in the second position; and finally, 35 is in the third position. So, there are 35 ways

Table 4.6. Pascal's Triangle can help answer questions where there are choices between two alternatives. In each example here, the answer is the same — 128.

Question	Example
How many different ways are there to flip a coin seven times?	HTHTTHH
Two teams are playing in a best-of-seven championship. What are all possible final outcomes?	WLWLLWW
If we examine any seven people in a line each person is either a man or a woman, at random, how many different combinations are there?	WMWMMWW
How many different ways are there to travel seven blocks north and east in a city laid out in a grid?	NENEENN
How many binary numbers are there that are seven digits in length?	1010011
We have seven letters: A, B, C, D, E, F, G. How many ways are there to take combinations of these letters?	A,_,C,_,_,F,G

to choose three flavors of ice cream out of seven — and we can know this without having to enumerate them!

The number of combinations of k objects from a set of n (denoted variously as C_k^n, $_nC_K$, or $\binom{n}{k}$) can be calculated as shown in the following equation:

(The factorial function, $n!$, represents the multiplication of all numbers between n and 1.)

$$\binom{n}{k} = \frac{n!}{k!(n-k)!}$$
$$= \frac{n\cdot(n-1)\cdot\ldots\cdot(1)}{k\cdot(k-1)\cdot\ldots\cdot(1)\cdot(n-k)\cdot(n-k-1)\cdot\ldots\cdot(1)}$$
$$= \frac{n\cdot(n-1)\cdot\ldots\cdot(n-k+1)}{k\cdot(k-1)\cdot\ldots\cdot(1)}$$

For demonstration, here is how you can apply this formula to compute the number of ways to choose five objects from a set of eight:

$$\binom{8}{5} = \frac{8!}{5!\cdot(8-5)!} = \frac{8\cdot7\cdot\ldots\cdot2\cdot1}{(5\cdot4\cdot3\cdot2\cdot1)\cdot(3\cdot2\cdot1)} = \frac{8\cdot7\cdot6}{3\cdot2\cdot1} = 56$$

Referring to the triangle in Figure 4.5, counting rightward from the zeroth cell to the fifth in the bottom row, you can verify that the entry corresponding to this calculation indeed has a value of 56!

Pascal's mathematical output dried up shortly after he produced these works. He was in chronically ill health for most of his adult life. After experiencing a religious vision in a dream in late 1654, he devoted the last eight years of his life to religion and philosophy. His most well-known argument, Pascal's Wager, leveraged a probabilistic approach to argue for the value of belief in God — so it's clear that he wasn't quite done with probability!

The Arithmetic Triangle that bears Pascal's name is clearly a thing of immense beauty and power. The world of mathematics is richer for Pascal having brought attention to it! One might wonder, though, whether spending so much time studying such a beautiful creation might make a person prone to visions of higher truths.

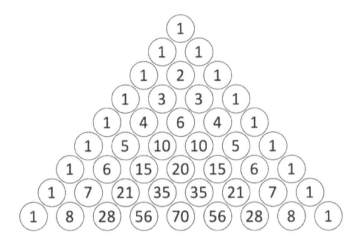

Figure 4.5.   Nine rows of Pascal's triangle.

Determining value

The correspondences of Pascal and Fermat marked the beginning of the study of probability in Enlightenment Europe. Word of their communications spread to other mathematicians. Christiaan Huygens (1629–1695), a young Dutch scholar living in Paris in the 1650s, heard about these problems and wrote up a study of the two previous problems in a short work called *De Ratiociniis in Ludo Aleae* (*Reasoning in Games of Chance*, 1657). He independently expanded on the ideas discussed by Pascal and Fermat, thereby bringing these ideas to a larger audience.

The Problem of the Points, in particular, opened the door to the study of expected value. We have already seen how to determine the equity conferred to players when leaving a game early; by putting to words the idea of "expectation," Huygens opened the door to the mathematical analysis of other probabilistic enterprises. At first, these seemed limited to games of chance, but in short order scholars realized that this power extends well beyond gambling.

Try, try, again

In this section, we are going to cheat. To explain expected value, we will be using tools that 17th century mathematicians did not have available. This is in no way to belittle their discoveries, only to aid your understanding!

We propose a simple game. Your friend has a fair coin (that is, one that has equal probability of coming up heads and tails). You can pay her one dollar to flip the coin. If it comes up heads, you lose your money; tails, you get your money back, and win an extra dollar.

The expected value of a single play in this game is zero. Half the time, you spend one dollar to play and win an additional dollar; half the time, you lose your dollar. If you keep playing this game forever, you might have long winning or losing streaks along the way, but the ratio of your total winnings to the number of times you play the game will trend toward zero (even as your net winnings stay above, or below, zero).

This example might be too simple to see the consequences clearly, so let's up the ante. We propose a different game: you and your friend roll a single (fair) six-sided die. Whatever number shows on the die's face, you win that amount of money in dollars. You friend offers to let you play each round of this game for three dollars. Should you take her up on this offer?

In this game, you can win six different amounts, each with equal probability. If you roll a 1, you win $1; if you roll a 6, you win $6. There are six options and six different amounts you can win. The sum of the six possible winnings, divided by six (that is, the arithmetic mean), will tell you how much you can expect to win each time you play:

$$\frac{\$1+\$2+\$3+\$4+\$5+\$6}{6} = \frac{\$21}{6} = \$3.50$$

A cost of three dollars per play is a good deal, then!

We can simulate many simultaneous games, keeping track of the winnings for a simulated player to demonstrate this. In this case, we will simulate five separate players, and see how each of them performs over one hundred rolls.

In Figure 4.6, we can see the result of this simulation. Some players dip briefly below zero, briefly losing money; over time, though, all the players show a net profit. The dashed line illustrates the expected value over time. If each roll costs three dollars, and expected winnings

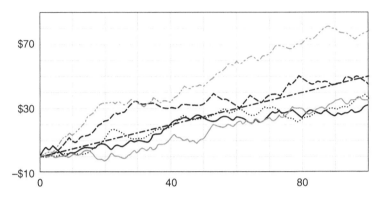

Figure 4.6. Five players play one hundred rounds of a dice game that costs $3 per play. The expected profit of $0.50 per game is represented by the dashed line.

are fifty cents more than the cost to play, then after one hundred rolls we can expect most players to have won fifty dollars. Most of them end up pretty close to this value!

Your friend realizes this game is a bad deal for her, so she proposes to change the cost of each game to $3.50. Should you continue to play?

Well, we have already computed the expected value of this game, but you might be convinced that you can still get ahead. After all, looking at the graph in Figure 4.6, almost every player spends a little time above the line of expected value; maybe you'll get lucky!

In Figure 4.7, we display the same data from the previous five simulations, but this time using a cost per game of $3.50. We can compare this to the previous figure and see the same peaks and valleys, but more of the players spend more time in the red — that is, losing money.

Still, there's an outlier on top — maybe that could be you! Let's continue the playing games, then. Figure 4.8 extends the exact same simulation to one thousand iterations, with some surprising consequences. One player is winning big — but that player is the one who had lost the most money at the one hundred mark! Meanwhile, the player who was winning the most money in Figure 4.7 has regressed, breaking about even after ten times as many games as before. And the biggest loser at this stage is a player who had been winning pretty regularly at the one-hundred-game mark.

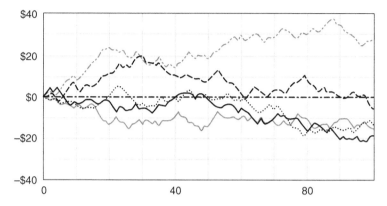

Figure 4.7. The same data as Figure 4.6, but this time each game costs $3.50. Each player's accumulated winnings — or losses — hovers around the expected value of $0, represented by a dashed line.

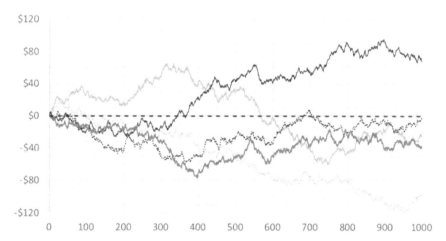

Figure 4.8. All five players continue to play for one thousand games. Each player has streaks of good and bad luck.

It bears emphasizing that these are *utterly unremarkable* simulations. While building this simulation, we chose the first set of data that came out of the computer, with no tricks involved.

Tables 4.7 and 4.8 highlight some interesting patterns from the data. Table 4.7 shows each of the five simulated players' winnings (or, losses), and you can trace how fortune favors each one. For instance,

Table 4.7.   Sampling each player's accumulated winnings every one hundred games. Negative numbers are indicated in highlighted cells and with parentheses.

Games	Player 1	Player 2	Player 3	Player 4	Player 5
100	$28.00	$(13.00)	$(5.00)	$(18.00)	$(15.00)
200	$14.00	$(42.00)	$(12.00)	$(20.00)	$(18.00)
300	$49.00	$(50.00)	$(19.00)	$(12.00)	$(30.00)
400	$42.00	$(49.00)	$(37.00)	$23.00	$(72.00)
500	$32.00	$(27.00)	$(49.00)	$45.00	$(51.00)
600	$(4.00)	$(35.00)	$(77.00)	$43.00	$(49.00)
700	$(39.00)	$7.00	$(77.00)	$54.00	$(37.00)
800	$(46.00)	$(20.00)	$(83.00)	$89.00	$(27.00)
900	$(18.00)	$(9.00)	$(108.00)	$92.00	$(42.00)
1000	$(23.00)	$(1.00)	$(99.00)	$68.00	$(40.00)

Table 4.8.   Average winnings (or losses) per game, over one thousand games. Each of these values gets closer and closer to zero.

Games	Player 1	Player 2	Player 3	Player 4	Player 5
100	$0.28	$(0.13)	$(0.05)	$(0.18)	$(0.15)
200	$0.07	$(0.21)	$(0.06)	$(0.10)	$(0.09)
300	$0.16	$(0.17)	$(0.06)	$(0.04)	$(0.10)
400	$0.11	$(0.12)	$(0.09)	$0.06	$(0.18)
500	$0.06	$(0.05)	$(0.10)	$0.09	$(0.10)
600	$(0.01)	$(0.06)	$(0.13)	$0.07	$(0.08)
700	$(0.06)	$0.01	$(0.11)	$0.08	$(0.05)
800	$(0.06)	$(0.03)	$(0.10)	$0.11	$(0.03)
900	$(0.02)	$(0.01)	$(0.12)	$0.10	$(0.05)
1000	$(0.02)	$(0.00)	$(0.10)	$0.07	$(0.04)

note that Player 3, despite losing the most overall, *neither wins nor loses money* between the 600th and 700th rounds.

Table 4.8 shows the amount, on average, that each player has won (or lost) per game. Despite some substantial losses, even our big loser — Player 3 — has spent only ten cents, on average, to play each

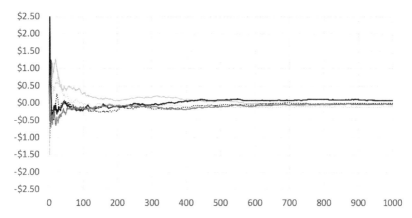

Figure 4.9.   Each player's average winnings per game, over one thousand games. The lines all get arbitrarily close to zero.

game. One player (Player 2) has effectively broken even after one thousand games; and the big winner so far, Player 4, has only won seven cents per play!

We can graph these averages to see how little each player deviates from expectation. On a long enough time scale, the amount each player wins or loses gets closer and closer to zero, as we can see in Figure 4.9. It makes sense to use the term *expected* value, since not one of our players defies expectation!

Consider this: no matter how good we are at some game (say, tennis), if we play against the top pro in the world for the length of an entire match, we're likely to lose. On the other hand, we might get lucky in the short term, and win a few points, and maybe even a few points in a row. Our "expected value" for the match is a *loss*, even if we manage to score a few lucky points along the way.

Testing the waters

The principle we're quietly benefitting from here rightfully has a name: the *law of large numbers*. First observed by Jacob Bernoulli (1655–1704) during his studies of probability in the late 1680s, it appeared in his posthumous work *Ars Conjectandi* (*The Art of Conjecturing*, 1713). In essence, the law states that if you perform some experiment — in this case, a game — a large number of times, the average result will approach closer and closer to the expected value.

Bernoulli's work was notable for three reasons. First, he performed his research with only a superficial understanding of those who had come before him — he was able to read a copy of Huygens' *Reasoning in Games of Chance* — but it's clear from his work that he had not read the letters of Pascal and Fermat, nor Pascal's *Treatise*, nor was he able to acquire several other texts that would have informed his research.

Second, he progressed much further in the study of probability than those who came before him, despite not having access to their writings.

Finally, he undertook to explain not only games of chance, where the outcomes are assumed to be fair and the sole output of cards, dice, or coins; he also sought to explain more human problems. Bernoulli realized that probabilities could be used to make sense of any situation that occurs on a large scale, such as problems concerning public health, or economics. We will touch on these problems in due time; for now, we will continue to explore games of chance.

Let's look again at Table 4.7. An extremely lucky player could have won up to twenty-five hundred dollars after one thousand games;[5] meanwhile, our luckiest player has only won sixty-eight dollars. The law of large numbers tells us that after a significant length of time, more central values — more "typical" results, closer to the expected value — are much more likely than extreme outliers.

The games of chance we have explored thus far all have similar outcomes, based on fair outcomes with six-sided dice or two-sided coins. What, though, can these tools say about other games of chance?

By this point, we have considered some general strategies for handling games of this sort. The Problem of Dice illustrated how to calculate the chance of winning based on random processes. The Problem of Points demonstrated how to calculate equity in an incomplete game; while the most recent dice problem showed what

---

[5] A player can win at most $6 per game, at a cost of $3.50; the greatest possible winnings per game is thus $2.50.

expected value means when the values are close together and winning each prize is equally probable.

In truth, we currently have enough information at our disposal to understand the story we used to open the chapter, concerning Voltaire's lottery winnings. But it may be difficult to understand some of the more important ramifications of these principles without further demonstration. This demonstration should take the form of an asymmetrical game, with prizes of vastly different values and chances of winning — and a lottery is just the ticket.

Imagine a hypothetical lottery game where players attempt to guess five numbers out of fifty. At the time of the drawing, fifty balls, numbered 1 through 50, are spun around in a container, and five are drawn out one at a time without replacing them. Players are rewarded for having some matches, regardless of order. Thus, if the drawing produces the numbers 2, 7, 8, 19, and 33, a player who has the numbers 19 and 8 on their ticket, and none of the other winning numbers, would win a small prize. Matching all five numbers rewards the player with a grand prize.

The value of each prize is listed in Table 4.9.

In this game, it is impossible to see the same number twice, because the balls are not placed back into the container after they are drawn. We can use the formula for combinations to count the number of different tickets in this lottery: with fifty different numbers, drawing five balls per game, there are over two million possible tickets:

$$\binom{50}{5} = \frac{50!}{5! \cdot (50-5)!} = \frac{50 \cdot 49 \cdot 48 \cdot 47 \cdot 46}{5 \cdot 4 \cdot 3 \cdot 2 \cdot 1} = 2,118,760$$

Table 4.9. Proposed prize values for a hypothetical lottery.

Matches	Prize
1	$1
2	$10
3	$100
4	$1,000
5	$100,000

There will only ever be one set of numbers that can match the final drawing. So, the odds of winning the grand prize are 1 in 2,118,760.

There are five ways to be "missing" the fifth number. Assuming the same drawing as above, we could have a ticket with some number other than 2, yielding _, 7, 8, 19, 33; some number other than 7, yielding 2, _, 8, 19, 33, and so on. The empty slot can take on one of forty-five different values — any of the "non-winning" numbers 1, 3, 4, 5, 6, 9, ... Thus there are $5 \cdot 45 = 225$ different ways to choose a ticket that matches four of the winning numbers.

Leveraging again the formula for combinations, we can see in Table 4.10 that the number of tickets matching each prize is equal to the number of ways to choose k numbers out of the five winning numbers, multiplied by the number of ways to choose the remaining numbers out of $(50 - 5) = 45$. We can also calculate the number of losing tickets in the same way.

Table 4.10.   Using the formula for combinations to compute the number of winning (or losing) tickets of each type, and their associated probabilities.

Matches	Formula	Probability	Odds
5	$\binom{5}{5} \cdot \binom{45}{0} = 1$	0.000047%	1 in 2,118,760
4	$\binom{5}{4} \cdot \binom{45}{1} = 225$	0.0106%	1 in 9,416
3	$\binom{5}{3} \cdot \binom{45}{2} = 9{,}900$	0.467%	1 in 214
2	$\binom{5}{2} \cdot \binom{45}{3} = 141{,}900$	6.69%	1 in 14.9
1	$\binom{5}{1} \cdot \binom{45}{4} = 744{,}975$	35.1%	1 in 2.84
0 (lose)	$\binom{5}{0} \cdot \binom{45}{5} = 1{,}221{,}759$	57.6%	1 in 1.73

The prizes don't match up with the odds of winning very well, do they? Matching two numbers is 5.25 times more rare than matching one, yet the prize is ten times larger; matching four numbers is 44 times more rare than matching three, yet the prize is *only* ten times larger! These numbers work well enough for our demonstration, though.

Imagine that we are lottery fanatics. We purchase a huge number of tickets for every drawing. With these odds, and the prizes listed in Table 4.9, how much can we expect to win for each ticket purchased?

In the dice game from earlier, each possibility had the same chance: one sixth. In that case, we were able to easily justify taking the mean of all possible values. What if we looked at that process in a different way? What if, for instance, we were to enumerate all possible results, compute the probability of finding each result, and multiply that number by the expected prize? By this method, we would produce Table 4.11. Taking the sum of all the values produces the expected value of a single play of the dice game, re-affirming what we learned in the previous section.

In the lottery we described, you can use the same method to compute the expected value per ticket. Multiply the probability that you win each prize by the value of the prize, then add the values together, as we have done in Table 4.12.

The sum of the individual values produces the expected value of a ticket. Assuming that the game is fair, each ticket purchased should return $1.64 in winnings average, over a large number of tickets.

Table 4.11.   Computing expected value for the dice game in a different way.

Roll	Probability	Prize	Value
6	1/6	$6	$1.000
5	1/6	$5	$0.833
4	1/6	$4	$0.667
3	1/6	$3	$0.500
2	1/6	$2	$0.333
1	1/6	$1	$0.167
		Total:	$3.50

Table 4.12. By computing the probability of winning each prize, and multiplying that number by the dollar value of the prize, you can determine the value of a lottery ticket.

Matches	Probability	Prize	Value
5	0.000047%	$100,000	$0.047
4	0.0106%	$1,000	$0.106
3	0.467%	$100	$0.467
2	6.69%	$10	$0.669
1	35.1%	$1	$0.351
		Total:	$1.64

Again, this doesn't account for extremely good nor extremely bad luck; it only says, that if you buy a *truly large* number of tickets, your winnings per ticket should be about equal to this value.

Note that we have said nothing thus far about the cost to play. The symmetrical two player games we discussed earlier were all fair: each player wagered the same amount, and each one had (at the beginning of the game) even odds of winning. The person running the lottery can set the price of a ticket as high or as low as they want — but there's no guarantee that the price will be good! If the price is too high, no one will want to play; if it's too low, the lottery will go bankrupt over time.

Here, the expected value tells you two things: again, your average winnings over time should approach $1.64 per ticket. Also, and just as important, the amount that the lottery should budget for payouts approaches $1.64 per ticket *sold*. To maintain their business, lottery officials would be well advised to make sure they charges *more* than this value as the price per ticket. Two or three dollars, in this case, would be enough to charge to leave the lottery with a healthy profit.

As a matter of curiosity and intuition-building, we can illustrate the likelihood of winning each prize in lotteries using different numbers of balls (Figure 4.10), and the expected value of a ticket in each case (Figure 4.11). In the first figure, we can see that the chance of matching five numbers starts very high in a drawing with very few balls, but drops to almost zero very quickly.

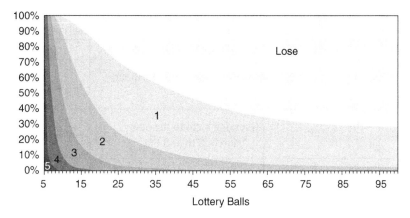

Figure 4.10.   As more balls are used in a lottery, the chance of winning each prize drops. The chance of winning biggest prize, for matching five balls, becomes almost invisible once ten balls are in play. In most lotteries the chance of winning all but the smallest prizes is vanishingly small.

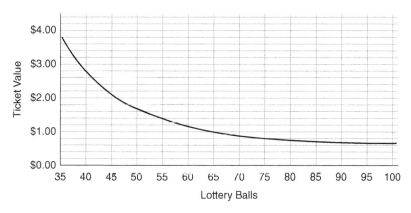

Figure 4.11.   As the number of lottery balls grows, the chance of winning each prize drops, causing a drop in the expected value of a ticket.

The value of a ticket starts very high to cover the high likelihood that any one of the tickets would be a winner: it's not even possible to buy a losing ticket in a lottery with fewer than ten balls! Therefore, Figure 4.11 does not use the same *x*-axis as the previous figure, since the value of a single ticket would have to be so high in smaller lotteries.

Table 4.13. Comparing the odds of winning the grand prize in lotteries of different sizes with the odds of guessing someone's birthday to increasing levels of precision.

Balls	Tickets	Event with similar probability	Odds
11	$\binom{11}{5} = 462$	Guessing someone's birthdate "March 27th"	1 in 365
18	$\binom{18}{5} = 8{,}568$	Guessing birth to the hour "March 27th at 8 PM"	1 in 8,760
39	$\binom{39}{5} = 575{,}757$	Guessing birth to the minute "March 27th at 8:32 PM"	1 in 525,600
94	$\binom{94}{5} = 31{,}598{,}138$	Guessing birth to the second "March 27th at 8:32:07 PM"	1 in 31,536,000

The expected value of a ticket in each of these lotteries is pretty high (certainly higher than the value of real-world lottery tickets!), but we put these probabilities into perspective in Table 4.13 by comparing the odds of winning each prize to the chance of guessing the date and time of someone's birth to increasing levels of precision. The odds of winning a 94-ball pick-five game are about the same as guessing a random stranger's birthday to the *second*!

So, while we can calculate (and be confident in) the worth of each ticket, we shouldn't count on ever winning the grand prize unless we can purchase all possible tickets!

Taking the guesswork out of winning

What would happen if we could buy all the tickets?

Most popular lotteries use jackpots. Starting at some modest value, if no one wins the grand prize in a given drawing, some fraction of ticket sales "roll over" into the next jackpot. This has the effect of building excitement for the game (the lottery jackpot hitting a new high becomes the top story on TV news and newspapers), driving up sales.

In the previous example, we calculated the expected value of lotteries with fixed grand prizes. Logically, lotteries using a jackpot should change the expected value of a ticket as the jackpot increases in value, and indeed they do!

We can consider, in this case, the real world PowerBall® lottery. As of 2018, this lottery is played in forty-four states, Washington DC, and several US territories.

For two dollars per play, players can pick tickets consisting of five numbers in the range from 1 to 69, plus an additional sixth number, the eponymous PowerBall™, which can have a number between 1 and 26. Based on these numbers, there are almost three hundred million possible combinations of numbers (that is, tickets).

$$\binom{69}{5} \cdot \binom{26}{1} = 292,201,338$$

We won't demonstrate the calculation of all the odds for this game; review Table 4.10 to recall the process. We have computed all the relevant probabilities in Table 4.14, and copied the prizes from the lottery website in order to determine the real-world expected value of a ticket.

Table 4.14. Expected values in a real-world lottery. Vendors, states, and the lottery itself each take a cut of ticket sales, leaving thirty-eight cents per dollar to be paid out in the form of prizes.

Matches	Odds	Probability	Prize	Value
0 + PB	1 in 38.32	2.609%	$4	$0.1043
1 + PB	1 in 91.98	1.087%	$4	$0.04349
2 + PB	1 in 701.33	0.1426%	$7	$0.009981
3	1 in 579.76	0.1725%	$7	$0.01207
3 + PB	1 in 14,494	0.006899%	$100	$0.006899
4	1 in 36,525	0.002738%	$100	$0.002738
4 + PB	1 in 913,129	0.0001095%	$50,000	$0.05476
5	1 in 11,688,053	0.000008556%	$1,000,000	$0.008556
5 + PB	1 in 292,201,338	0.0000003422%	$40,000,000*	$0.1369
			Total:	$0.3798

Note, here, that this value is roughly $0.38, while a ticket costs $2, meaning that if you purchase many tickets your average *losses* should approach $1.62 per ticket.

The prize in the final line indicates the minimum jackpot, but we have marked this value with an asterisk. This is because every time someone wins the jackpot, it starts fresh at forty million dollars, so the jackpot can never be lower than this value. But remember, as long as there is no winner, the jackpot can grow without bound, even as the other prizes stay fixed in value!

We can repeat the calculations for expected value as the jackpot rises in value. Only one value varies — the money in the jackpot — while all other values (probabilities, small prizes, costs) remain constant. Thus, we can illustrate the jackpot size versus expected value as a simple line graph (Figure 4.12).

From appearances, it seems that once the jackpot hits five hundred million dollars, assuming we can find the money to buy every single ticket, making a profit is a sure thing!

We should be wary of these numbers. First off, there's the logistical implications to consider: under normal circumstances, point-of-sale lottery processors are not capable of handling purchases of five hundred million dollars. To purchase every possible ticket, we would

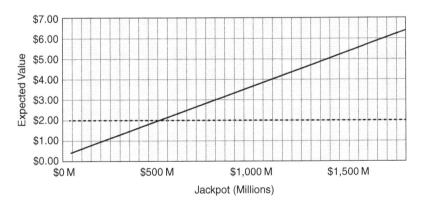

Figure 4.12. The expected value of a lottery ticket as the jackpot grows, compared to the cost ($2) of a single ticket. When the jackpot hits $500,000,000, the expected value of a ticket crosses the cost line.

have to enlist the aid of thousands of confederates, and odds are that they wouldn't work for free.

Second, when the jackpot hits record values, more people buy tickets, increasing the likelihood that we would have to split it equally with other winners.

Third, the displayed jackpot is calculated as the value of a thirty-year annuity. Most winners choose to receive their winnings in a single lump-sum payment, which cuts the value of the jackpot to approximately 63% of the publicized total.

Finally, the government always gets its share. If you purchased the winning ticket in New York City, you would pay 25% Federal tax, 8.82% State tax, and 3.876% City tax, for a total of 37.696% tax. In Figure 4.13, we add to the previous graph some of the biggest jackpots in history, and compare them to the lump-sum, after-tax value of the jackpot. Note, too, that the biggest ever jackpot was split three ways, meaning that while the expected value of a ticket may have briefly crossed the threshold of profitability, the *actual* value of a ticket has *never* exceeded the price of the tickets themselves!

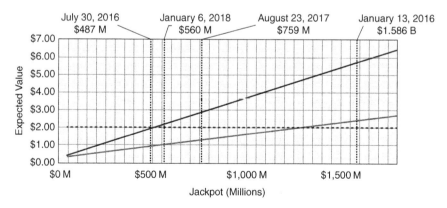

Figure 4.13. The expected value from the previous graph (black) is reduced in value when the prize is taken as a lump-sum payment, and taxes are accounted for. If the winning ticket is purchased in New York City, the expected value of a ticket only exceeds the price when the jackpot hits $1.31 billion. The all-time record jackpot, of $1.586 billion, was split between three winners.

Case studies in beating the odds

All this work isn't to say that there has never been a time that some-one has beaten the odds in a game of chance. Most established games of chance — lotteries, casinos, and sports betting — are run by people who are more educated and sophisticated than this book can possibly make you, and thus almost never lose money. (Keep this in mind if you try to gamble to make a profit!)

Sometimes, though, gaps appear. Earlier, we looked at what would happen to expected value if the jackpot varies, while all other values (probabilities, small prizes, and costs) remain the same. Here, we will look at real-world examples of how clever players have taken advantage of each of these three factors to make money, despite the odds.

Case 1: A Guaranteed Win. Jai Alai is a sport that is sometimes called "the fastest sport in the world." Played on a closed court somewhat similar to a racquetball court, a ball (*pelota*) is hurled at the walls at high speeds with a woven implement called a *cesta* (or *xistera*), which is mechanically similar to a Lacrosse stick.

In the format practiced in Palm Beach, Florida, two programs run per day, one in the afternoon and one in the evening. During each program, eight teams compete in six matches. Predictably, many small bets take place — on the results of individual matches, on the standings at the end of a single game, and so on — but the big prize is reserved for the person who successfully selects the winner of all six games in the program.

Since there are eight teams competing in each match, we can fill out a ticket with one of eight options per match. And since there are six matches per program, we can multiply eight by itself six times to find out how many ways there are to fill out a ticket:

$$8 \cdot 8 \cdot 8 \cdot 8 \cdot 8 \cdot 8 = 8^6 = 262{,}144$$

On a typical night in the early 1980s, ten-thousand bets would be placed on this result, at the cost of two dollars per ticket. Normally, some lucky bettor would pick the winners of each of the six games every couple of days, turning two dollars into thousands. But, in early

1983, the jackpot went unclaimed, over and over... for 146 programs! By then, the jackpot had grown to over half a million dollars — quite a substantial sum!

An anonymous group of bettors, seeing that there were 262,144 ways that one of eight teams could win each of the six matches, gathered together over a half million dollars of capital ($2 \cdot 8^6 = $524,288$, to be exact) and bought every possible ticket! Given that they bought up over twenty-six times the typical number of tickets sold in a night, the odds of them having to split the jackpot with some other lucky winner were small.

Sure enough, by the end of the night, the anonymous bettors had doubled their money, proving the value of expected value.

The story of the Jai Alai bettors illustrates one way to manipulate probabilities (without cheating!). Their strategy didn't affect the expected value of a ticket, but it did guarantee a winning outcome.

What happens if the value of a small prize changes?

Case 2: Double Prizes. Keno is a lottery game played in some US states. In one format, commonly seen in bars, a television screen shows twenty new numbers drawn (or computer generated) every few minutes from the set of numbers between 1 and 80. Players buy tickets with anywhere from one to twenty numbers on them (increasing in cost the more numbers purchased), and prizes are determined based on the number of matches on each ticket.

For instance, a player might buy a pick-six ticket and a pick-ten ticket for the same drawing. Within minutes, that drawing's twenty numbers are broadcast, and the player's odds of winning with each ticket can be determined in accordance with Table 4.15. (Each of these figures was calculated using the method shown in Table 4.10.)

While matching all twenty numbers is essentially impossible (such that many states don't even sell a pick-twenty ticket), the game's popularity is due to the prevalence of small prizes, and the way that players can casually play many games while out on the town.

In the mid 2000s, two students who frequented a bar in Massachusetts noticed that the state was running a promotion on a certain kind of Keno game. Every Wednesday for a month, the prize

Table 4.15. Odds of matching 0, 1, 2, ... numbers in a Keno drawing given tickets with different numbers of picks. Highlighted cells indicate numbers of matches that yield prizes.

	Pick 12	Pick 10	Pick 8	Pick 6	Pick 4
0	1 in 43.05	1 in 21.84[6]	1 in 11.33	1 in 6	1 in 3.24
1	1 in 8.79	1 in 5.57	1 in 3.75	1 in 2.75	1 in 2.31
2	1 in 4.21	1 in 3.39	1 in 3.05	1 in 3.24	1 in 4.70
3	1 in 3.57	1 in 3.74	1 in 4.66	1 in 7.70	1 in 23.12
4	1 in 4.86	1 in 6.79	1 in 12.27	1 in 35.04	1 in 326.44
5	1 in 10.06	1 in 19.44	1 in 54.64	1 in 323.04	
6	1 in 31.05	1 in 87.11	1 in 422.53	1 in 7,752.8	
7	1 in 142.30	1 in 620.68	1 in 6,232.3		
8	1 in 980.78	1 in 7,384.5	1 in 230, 114		
9	1 in 10,482	1 in 163,381			
10	1 in 184,230	1 in 8,911,711			
11	1 in 5,978,272				
12	1 in 478,261,833				

payouts for pick-four tickets would be doubled. Pulling out their calculators, the students realized that while the normal expected value of a pick-four ticket was $0.60 per dollar played, under this promotion, the expected value of a ticket rose to $1.20 per dollar!

During the dates of the promotions, the students bought as many pick-four tickets as the machines could print, hardly looking at them as they played. They trusted in their math, buying tickets and redeeming prizes as fast as they could. Sure enough, over the course of the month they managed to earn a hefty profit — earning within $100 of their original estimate — demonstrating once again the power of expected value!

---

[6] Most Keno games reward players who pick ten or more numbers and have zero matches in the drawing. Why is this? Think back to the birthday paradox — it becomes increasing unlikely, as a player chooses more numbers, that they won't match any of those drawn!

Table 4.16. A hypothetical version of the 1729 French lottery. There are many different kinds of bonds in circulation, and bondholders can purchase tickets for 1/1000 of the face value of the bond.

Bond Value	Circulation	Total Value	Ticket Price	Revenue
1,000,000	500	500,000,000	1,000	500,000
100,000	1,000	100,000,000	100	100,000
10,000	5,000	50,000,000	10	50,000
1,000	10,000	10,000,000	1	10,000
Totals:	16,500	660,000,000	Total:	660,000

The final of our three parameters to consider is cost. This example will take us back to the 18th Century and Voltaire, providing the conclusion to that story.

Case 3: Cheap Tickets. Recall, in Voltaire and de la Condamine's scheme, that each bond held by a creditor entitled that bondholder to purchase a single lottery ticket for 1/1000th of the face value. The government bonds not only gave the bondholder the permission to buy a lottery ticket; they also set the price for that ticket.

Regardless of their price, each ticket had the same chance of winning. And now you can see the French government's fatal error!

We have generated hypothetical values for the number of bonds in circulation in Table 4.16. While these numbers aren't based on history, they make a sort of intuitive sense: generally, small denominations of bonds or currency are printed in greater numbers than larger ones.

Reading across the top, you can see that the government issued 500 bonds with a face value of one million livres, whose initial sale amounted to a total of 500 million livres revenue. Each bond entitled its holder periodically to purchase a lottery ticket for the cost of 1,000 livres, which would translate to a recurring revenue of 500,000 livres for the government. Adding up the numbers for recurring revenue (in the right column) shows that the government could expect to earn 660,000 livres every drawing, prior to paying out the prize.

The grand prize for each drawing was set at 500,000 livres. If we pretend the numbers in this table are accurate, even after paying the winners, the French government would be making a tidy profit of 160,000 livres every drawing.

In this model, there are 16,500 tickets and a jackpot of 500,000 livres. The expected value of each ticket is 500,000/16,500 ≈ 30.3 livres per ticket. However, the number of tickets that can be bought per unit of currency varies wildly. For 10 million livres, a savvy investor could buy ten, one million livre bonds, entitling them to purchase ten lottery tickets, or they could buy *ten thousand* one thousand livre bonds!

Imagine that your local lottery sold two different kinds of tickets: one that cost one dollar, and one that cost one thousand dollars, and both had the same chance of winning the jackpot. You would be a fool to buy the more expensive ticket!

This principle is, of course, what allowed the syndicate to corner the lottery market and ensure their repeated wins.

From the standpoint of probabilities, Voltaire's syndicate could hardly lose. Under this hypothetical model, for a small investment — purchasing 1.51% of all the bonds (by face value) — they would be entitled to purchase 60.6% of all the lottery tickets!

Measuring reality

Most of us will never encounter "sure thing" lotteries in our lives, and even if we did, we might not have the capital to take advantage of them!

Most money is made the old-fashioned way: a little bit at a time. Mathematics can help you foresee income or plan a budget; or, if you're starting a business, calculate how much you need to earn to avoid bankruptcy, and maybe hire a few people over time.

These aren't the only financial opportunities available, though. A savvy investor might be interested in any of a huge variety of financial instruments to earn income — or insure against loss.

The length of a life

The question of how to price financial instruments was unresolved for hundreds of years. One of the oldest instruments, called a life annuity, works like this: an organization offers to sell annuities for a large amount (as principal), and then pay out money periodically over time to a recipient (often, the purchaser, or the purchaser's child).

The purchaser's goal is to ensure a steady stream of income over time, should they lose their job or become infirm. The seller's desire is to raise capital in the short term (for other investments, or for immediate expenses). The seller profits either when their other investments pay off over the cost of payments; or, when the recipient dies before the cost of payments exceeds the principal.

In a basic case, a man might buy a life annuity for his daughter at the cost of one million dollars. After she turns eighteen, the daughter will receive forty thousand dollars per year for the rest of her life.

Clearly, if the daughter lives past the age of forty-three, assuming the seller has not invested the principal in the meantime (by purchasing land, funding trade, etc.), the annuity would have paid out more than was put into it:

$$\frac{\$1,000,000}{\$40,000 \, per \, year} = 25 \, years$$

But if the daughter dies young — of disease, during childbirth, or because of war or catastrophe — the seller would keep the remaining principal.

Astonishingly, some of the early sellers of annuities — often monasteries and cities — didn't take into consideration the age of the recipient when pricing these!

Clearly, there were parties interested in the problem of how to quantify the length of a life.

This story takes us back to the age of Pascal. An English haberdasher,[7] John Graunt (1620–1674), was the first person to compile tables of mortality. By compiling death reports from *bills of mortality* (listings of burials) in his native London, Graunt was able to construct tables that predicted how frequently people died at different ages.

His interest was primarily in epidemiology, so the purpose of his *Natural and Political Observations Made upon the Bills of Mortality*

---

[7] That is, a dealer in men's clothing.

(1663) was first and foremost to help predict the spread of epidemic diseases through the country, as a kind of advanced warning system for the plague. Given that London was nevertheless afflicted by the plague in 1665–1666, one might say that Graunt was unsuccessful. However, in analyzing the bills of mortality, Graunt set the stage for a different kind of study — the field of statistics.

The next major triumph in statistics came within 10 years. Johan de Witt (1625–1672) was a great Dutch statesman who oversaw a period of relative peace and prosperity in the Netherlands. Elected as leader of the States of Holland in 1653, he was the *de facto* leader of the Dutch Republic for nearly 20 years.

Dutch power in the 17th century mainly came from the water: the Dutch navy was strong, and shipping and trade provided the lifeblood of the economy. With so much money flowing through the heart of the Republic, it is no wonder that it became a hub of the early financial industry. The government took part as well, selling both bonds and annuities to citizens to supplement tax revenues.

De Witt wondered, as head of the government, whether the prices of the latter were set fairly. In 1671, he compiled the *Waardije van Lyf-renten naer Proportie van Los-renten* (*The Worth of Life Annuities Compared to Redemption Bonds*), which examined the question in detail. He saw the problem as one that could be answered through the examination of mortality data: by looking at the odds that a person died at each age, he was able to calculate the expected payout of government annuities.

We can attempt to demonstrate the import of his work through a toy example. The United States Centers for Disease Control (CDC) collects and publishes data relating to mortality in America. While they collect data related to gender, race, cause of death, etc., we will only concern ourselves with the most coarse statistic: raw mortality.

Pretend that the US government was preparing to sell life annuities to its citizens. What would be an appropriate price to set on these instruments?

If you, in the role of the US government, collect data on the population (through the census, tax records, etc.) and the age of death of each citizen, you can look back at the previous year and determine, for

Table 4.17. CDC statistics for mortality in the United States (2008). Each ten-year cohort lists the number alive (out of 100,000 live births), the number that will die in the subsequent ten years, and what percent that number constitutes of the remaining cohort.

Age	No. of Alive	No. of Dying	Dying (%)
0	100,000	833	0.83
10	99,167	362	0.37
20	98,804	941	0.95
30	97,863	1,224	1.25
40	96,639	2,641	2.73
50	93,999	5,643	6.00
60	88,356	11,203	12.68
70	77,153	21,591	27.98
80	55,562	33,215	59.78
90	22,347	20,667	92.48
100+	1,680	1,680	100.00

instance, how many citizens of age 30 there were, and how many citizens died at that age. Dividing the latter by the former yields the percent mortality for 30-year-olds.

Repeating this process for all ages would reveal the expected mortality for the entire population. Assuming the general prevalence of causes of death don't significantly vary over time, you could choose a random, hypothetical person from the population, and determine their life expectancy.

In Table 4.17, we adopt a crude version of the 2008 CDC mortality data for the US. Each line represents a ten-year range. So, for instance, for every 100,000 children born, 99,167 will make it to the age of ten, meaning that 833 or 0.83% will die between birth and their tenth birthday.

For the sake of this illustration, let's look at a cohort of thirty-year-olds all buying life annuities in 2008. We will also assume that all people who die between one group and the next die exactly half-way between those dates, receiving five annual installments from the annuity in that decade (thus, we assume that all people who die

Table 4.18.   Raw mortality data per ten-year group starting at 30. Here, assume that anyone who dies between the ages of e.g., 30–40 dies halfway between these ages, and use that average value to determine how many total person-years are lived in each decade.

Age	Number	Payments	Value (months)
35	1,224	5	6,120
45	2,641	15	39,615
55	5,643	25	141,075
65	11,203	35	392,105
75	21,591	45	971,595
85	33,215	55	1,826,825
95	20,667	65	1,343,355
105	1,680	75	126,000
		Total:	4,846,690

between forty and fifty die on their forty-fifth birthday). Then, we can calculate the total number of annuity payments made (Table 4.18).

Out of 100,000 persons born in the US, 97,863 make it to their thirtieth birthday. These people should, collectively, live to receive just under five million total annual annuity payments.

There are two curious facts here: first, we haven't actually calculated what that dollar value of those payments is, although this procedure looks startlingly familiar to another process we've used to calculate dollar values in the past. Second, we can determine an average — that is, how many annual payments a typical thirty-year old would receive before dying.

$$\frac{4,846,690}{97,863} \approx 49.71$$

If the typical thirty-year old will live to receive over forty-nine payments, then that person will live (according to our data) to the age of 79.71!

The "expected value" we calculated here corresponds to *life expectancy*, as opposed to a monetary value as seen in many previous examples. De Witt was able to combine a calculation like this with information of annuity yields to determine that the government was overpaying on annuities relative to bonds, and proposed reducing interest rates to rectify the difference. So even though ours wasn't a financial calculation, it certainly has financial implications!

De Witt was a competent administrator, but his mathematical work did not get the recognition it deserved, in part because of a miscalculation on his part. Faced with the choice between a strong army and a strong navy, de Witt had strengthened the latter at the expense of the former. In 1672, the *rampjaar* ("disaster year") happened — the Netherlands were invaded simultaneously (by land) by the French and English, as well as several states of the Holy Roman Empire.[8] The citizenry blamed de Witt for failing to defend the nation, forcing his resignation. That was not enough to mollify them, though: de Witt was brutally murdered by an angry mob two weeks later.

Although his work was continued by his successor, Johannes Hudde (1628–1704), it went otherwise underappreciated for decades.

Quantifying risk

Nearly simultaneous with and parallel to the dawn of statistics was the dawn of insurance. The 1660s were not a good decade for England: after the plague year of 1665–1666, over thirteen thousand homes burned to the ground in the Great London Fire of 1666. Demand for property insurance (already present in the 1650s) reached a fever pitch after the fire, but it wasn't until 1681 that the Insurance Office for Houses was founded.

In 1686, a man by the name of Edward Lloyd endeavored to open a coffeehouse in the City of London. It quickly became well-known among the merchant and shipping class as a place to "talk shop," discussing, (among other subjects) trade, shipbroking,[9] and insurance.

---

[8] The Holy Roman Empire was a loose collection of states, the bulk of which were located in present-day Germany.

[9] That is, acting as an intermediary between people seeking to move goods and the owners of ships.

The famous Lloyd's of London, the original name in shipping insurance, began here.

The problems faced by this growing industry were similar to the problems faced by the sellers of bonds and annuities: how do you price something that is designed to protect against the unpredictable?

Recall Jacob Bernoulli from earlier in this chapter. In this same timeframe, the mid-1680s, he was beginning his seminal work on probability. There is good reason to believe Bernoulli delayed publishing *The Art of Conjecturing* only because he was looking to improve the portions relating to these kinds of real-world problems. He spends the majority of his time in the final printed chapter of *The Art of Conjecturing* motivating the law of large numbers (before finally proving it), but there's an almost palpable frustration in his quest for applicable examples, as if he is searching for some perfect example to convince the reader of the merit of his work, one that is just on the tip of his tongue.

Bernoulli sought out the work of both Graunt and de Witt. He makes explicit mention of the former, but was never able to acquire a copy of the latter. Nevertheless, it is clear that it was his explicit aim to discuss not primarily games of chance, but problems governing human existence.

His motivation for proving the law of large numbers, then, can be viewed in the context of these concerns. We can perfectly predict fair games of chance (a fair die will show each of its six faces with equal probability), but his law provides a method for quantifying, in the same terms as those games, the direction of the winds of fate.

Recording the ages and causes of death of the entire population is primarily a bookkeeping, rather than a mathematical, problem. Estimating the number of ships that should be lost at sea in a given year, or the number of homes that might burn down, presents much different challenges. The methods discovered to address these challenges, though, turned out to share many features in common.

Advanced probabilities

We must necessarily speed through this topic, as it is a far deeper and richer topic than we have room to pay adequate service to. But we hope that in describing some of the historical circumstances motivating the study of these topics, you appreciate *why* we consider these especially important and useful problems!

Abraham de Moivre (1667–1758) was born in France, but fled to England around 1687 to escape religious persecution. He spent the rest of his life there, working primarily as a tutor of mathematics, never achieving any particular degree of financial success despite his contributions to the world of mathematics.

In 1711, he published the first Latin edition of what would become the first textbook on probability theory, *The Doctrine of Chances*. The first English edition, published in 1718, was late enough to benefit from Bernoulli's 1713 text, but it wasn't until the second edition of 1738 that he left his most important mark on probability, one that shows the fingerprints of Bernoulli's work.

De Moivre was interested in finding a method to compute, in essence, arbitrary values in Pascal's Triangle to facilitate the calculation of advanced probabilities. In doing so, he first shed eyes on a function that would forever change the face of applied mathematics.

We can explain his discovery like this: take the values of Pascal's Triangle and divide the value in each row by the largest value in the row (a process called *normalization*). In each case, the largest value, divided by itself, will yield one, and all other values will produce a result that is less than one (Figure 4.14).

Graphing these rows produces a series of curves, each with a spike in the center that widens as more of the "mass" of the row spreads out (Figure 4.15).

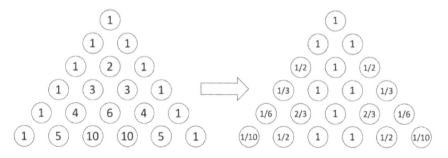

Figure 4.14.   The values in Pascal's Triangle, normalized by row. The entries in each row are divided by the maximum value of the row, allowing us to compare the shape of multiple graphs on a single axis.

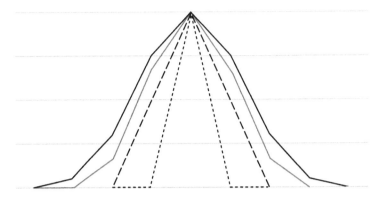

Figure 4.15. Curves for rows 1, 3, 5, and 7 of Pascal's Triangle. The values have each been normalized with respect to the central value.

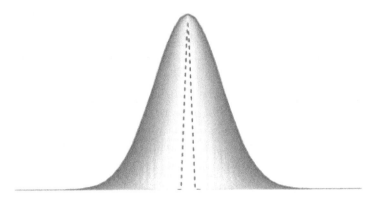

Figure 4.16. The first fifty odd rows of Pascal's Triangle: a smooth curve begins to form.

Repeating this process often enough, a smooth curve will begin to form, as seen in Figure 4.16.

De Moivre found a method to approximate individual values on this curve. More importantly though, his formula could be used as a first step to define the curve itself. However, he failed to see, or at least anticipate, this curve would end up being central to so many fields.

In the next century two mathematicians, Carl Friedrich Gauss (1777–1855), and Pierre-Simon, marquis de Laplace (1749–1827), would open the door to applying this curve (which had started with an innocent game of dice) to almost every field of human endeavor.

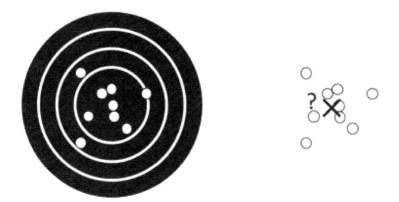

Figure 4.17.   (Left) Bullets fired at a target accumulate near the bullseye. (Right) Remove the target, and you can use the locations of the bullets to determine the location of the bullseye.

Gauss observed that if you take repeated measurements of a phenomenon — in his case, the orbit of an asteroid — the errors in your measurements can be treated as aimed more or less at the center of the target, and falling off in accuracy in accordance with the curve de Moivre described. Imagine this like multiple people shooting at a target (Figure 4.17): the holes in the target will cluster around the bullseye, with a probability that fades to zero the further you get away from the center.

If you don't know where the bullseye is, but you do know each of the measurements, then you can use a probability distribution to guess where the bullseye was!

Due to his contributions to the study of this curve, it is often called the *Gaussian distribution.*

La Place, in turn, observed that phenomenon that aren't normally distributed nevertheless often have measurements that fall along this curve. His contribution is known today as the *Central Value Theorem.*

Combine these two, and you get the foundations of most of the statistics you see reported in the newspaper today.

If you want to determine the average life expectancy of a population without first acquiring comprehensive census data, you can estimate it with a normal curve, due to the central value theorem.

If you want to estimate the risk that any given ship will fail to return to shore, the central value theorem can help you measure that.

If you want to know how many people support a candidate for office, if you want to see what the likelihood that a medicine treats a disease, if you want to determine the degree of error in a scientific experiment... all these questions, and more, can be answered with results stemming from the central value theorem.

Laplace's contributions to probability and statistics cemented the position of the normal curve as *the* normal curve — this time, meaning "typical" — for measurements of all types. And as you can see, through this work, the stories of probability, statistics, finance, insurance, and science, all became intertwined.

Unrolling the story of probability

There is no field more important in the modern age than statistics. In this half of the chapter, we covered 150 years of history, and barely scratched the surface of its effects on your daily life! Yet, statistical results define all aspects of modern existence: from surveys that appear on the news and drive our politics, to developments in science and technology, to the operations of business, finance, and government.

Recall the graph in Figure 4.9. You saw there how the average of many random processes approaches an expected value. But a method that applies to one random process can apply to others. If you are faced with any phenomenon that happens at the scale of a nation, with millions of people all living and dying, buying and selling, playing games, sometimes winning and sometimes losing, those results can be modeled and predicted using tools like Pascal's Triangle, the law of large numbers, the least squares method, and the central value theorem. This is the power of probability and statistics.

If there is one lesson you should take away from this story, it is this one: statistical (and by extension, probabilistic) methods appear everywhere and affect *everything*. If you make a little extra money in life, you may choose to invest it in stocks or bonds. Or, you may choose to purchase insurance to protect your car, or property, or to provide for your family if you die unexpectedly. Or, you may choose to start a business. In each of these cases, experts apply mathematics to estimate risk, determine likely rates of return, and price their products and services appropriately. Having an understanding of these problems can, if nothing else, help you determine when you're in over your head and can benefit from a second opinion — or, maybe this story will motivate you to learn more on your own!

Whatever your choice, you have at this point learned many new tools and skills. For instance, you have…

- … seen how to determine odds of winning in a variety of different games.
- … learned how to calculate each player's stake in an unfinished game.
- … computed the expected value of dice games, lotteries — and lifespans.
- … witnessed how difficult it is to beat the odds when dealing with the law of large numbers.
- … yet also, discovered that the law of large numbers sometimes works in your favor!
- … gotten a taste of how the least squares method and the Gaussian Distribution can be used to analyze data and help predict future events.

Winding throughout this narrative, though, is the story of mathematicians — people — who explored novel ideas for profit, or to build a reputation, or to run a government, or just for fun. All of these play into an overall narrative of "why are people motivated to study mathematics?" This is not a history book, but to understand people's motivations it does help to understand their stories.

Think about your own story. What drives you? What questions would you want answered — for fun, education, or profit? Do you think problem solving can be useful or important in your own life? If this chapter is any indication, it already does!

Our next section looks at how the humble triangle set the stage for understanding the universe around us.

Where in the Universe?

The full version of this story covers thousands of years of human existence. The bulk of it occurs in the same timeframe as our last story — the Renaissance through the Enlightenment, the 14th through the 18th centuries — but we will only be able to scratch the surface of this fascinating history.

It is possible for a modern human to learn more about the shape of the universe from their computer than ancient humans could have learned in a lifetime of observation. It is perhaps ironic that a modern human can be expected to know much less about practical astronomy or the empirical changes in the heavens than an observant human living in the darkness of ages past. To begin, then, we will sketch out what the world would have looked like to our ancestors.

Setting the stage

The two topics most relevant to our discussion are why ancient peoples observed the skies, and how they measured the earth. As you could guess, these topics are related to one another, and to our mathematical thesis.

Ancient astronomy

Early hunting and agricultural societies needed to keep track of time in order to secure food sources, either by predicting animal migrations, or to determine when to plant and harvest.

The length of the day was marked by the movement of the sun and the month by the moon. But the motions of the moon are notoriously complicated, and do not align well with the turning of the seasons.[10]

---

[10] Many religious and traditional calendar still in use today — including the Christian Ecclesiastical calendar (used to determine the date of Easter), the Jewish calendar, the Islamic calendar, and East Asian lunar calendars — use varying methods to reconcile these differences, either by adding a thirteenth lunar month periodically, or

While counting lunar months proved inadequate for measuring the year, early humans noticed that the stars, rotating through the heavens, rose and set at different times depending on the time of the year, and soon found ways to exploit these regularities.

The Ancient Egyptians in particular learned that when the brightest star in the sky, Sirius, rose with the sun, so too would the waters of the Nile rise. By learning to predict when the silt-heavy waters would nourish the land each year, the Egyptians were able to develop an agricultural base that would feed their civilization for thousands of years.

Two of the compass directions were marked out by the rising and setting of celestial objects. Due east corresponded to the sun's position at dawn on the equinoxes (the two days per year when there is equal amount of day and night), and thus, due west corresponded to the location of the sunset on these same days.

Not all heavenly bodies rose and set, though. If you were to stand in an open field as the sun set to your left, and watch the sky all night, you might notice something peculiar. All the stars travel in arcs similar to the sun and the moon, except for those directly in front of you. These stars spin in place, neither rising nor setting; their center provides the third significant compass direction, pointing to the north.

These three directions — west, north, and east — describe right angles on the earth, and imply a fourth direction — south — at right angles to the existing compass directions. Until the discovery of the magnetic compass,[11] important directions could only be determined by observing the heavens themselves.

Finally, consider how early humans must have stood in awe of their world. Each day, the sun, giver of life, would seemingly set the edge of the world on fire as it descended below the earth, only to return from the opposite direction the next morning. Sometimes, the

---

allowing the calendar to shift backwards through time. All derive in some way from these ancient methods of timekeeping.

[11] Primitive compasses were used in China as early as the 3rd century BCE, but the magnetic compass was not known in the Islamic world or Europe until the 12th century CE.

moon would cover the sun in a solar eclipse; other times, comets would appear, trailing celestial fire across the sky for weeks at a time. Mars, bloody and red, stared down balefully at the earth, its appearance seeming to promise violence and bloodshed. In a world of such marvels and terrors, where a pantheon of gods travels across the sky and stars foretell the rising of the floodwaters, it is only natural that some people would look to the heavens to try to divine the future.

Together, these factors — telling time, finding directions, and trying to divine the future — provided more than enough motivation to drive ancient peoples to observe and measure the sky.

We can see these effects in archaeological findings. Ancient documents from cultures around the world — including Egypt, Babylon, India, China, and Mesoamerica — record the locations of stars and planets, and describe methods for predicting events such as eclipses. Many festival dates, whether ancient or modern, secular or religious, align with astronomically significant events whose roots can be traced back to ancient observations. And the orientation of many archaeological findings — from burial sites, to temples, to cities — were laid out in cosmologically significant directions.

Measuring the earth

For all practical purposes, the ancient earth was flat. Prior to the Age of Exploration, much less the advent of commercial aviation, ancient man's entire world consisted of a flat disc, stretching from one horizon to the other.

To layout land, buildings, and cities, several steps were necessary. When building a city in a grid pattern, some cosmological basis direction could be marked based on observations of the sun and stars. Next, the foundations of buildings and the paths of roads could be measured and drawn using appropriate tools — for instance, chains or ropes, marked in constant increments derived from the human body, such as paces or spans. Finally, angles and distances could be combined to determine the locations, sizes, and shapes of successive buildings.

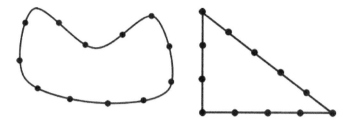

Figure 4.18.   A loop of rope with twelve knots at regular intervals (left) will, when pulled taught, draw a 3-4-5 right triangle (right).

Our knowledge of early earth measuring — geometry — suggests that it was more practical than theoretical. Each year after the flooding of the Nile River, the Egyptians faced the daunting task of redrawing all the fields and roads of the river valley. They pioneered a variety of practical applications of geometry to accomplish this. For instance, they probably used a rope with twelve equally spaced knots in it to construct right angles to redraw the corners of farmers' fields. By stretching the rope into a 3-4-5 triangle, as seen in Figure 4.18, a right angle could be reliably traced at one corner. This method was a forerunner to the discovery of the Pythagorean Theorem.

These general principles of surveying and practical measurements applied just as well to their monumental endeavors. To construct the pyramids, the Egyptians used a measure called *seked*, which is similar to the modern idea of slope. Seked is defined as the relationship between the height of a rod of fixed length (one cubit, correlating to the length of the arm from the elbow to the fingertips) to some horizontal distance, measured in palms and fingers. (In the ancient Egyptian system of measures, there were seven royal palms to a cubit, and — sensibly! — four fingers to a palm.) Common sense tells us that, by measuring the shape of each stone with tools built to a common seked, each face of a pyramid would have uniform and equal slope, resulting in the imposing structures that still stand today.

Modern archaeologists have determined that the Great Pyramid at Giza was constructed with a seked of five palms, two fingers, as you can see in Figure 4.19.

All this is, understandably, a very sketchy overview of the methods the ancient Egyptians would have used to build their empire. It is

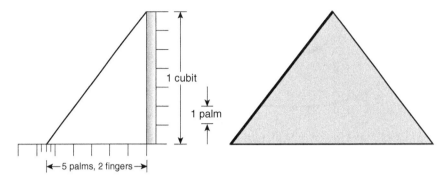

Figure 4.19. A rod of known height, and a string drawn to a point a certain distance from its base defines a slope of given seked. Tools built to this benchmark were used to ensure the pyramids had a regular slope.

sufficient here to recognize that building an empire did not require complicated tools or the understanding of underlying mathematical principles; one can draw a straight line, or a right angle, or the slope of a pyramid via eminently practical means. To advance beyond this point, though, demands a different kind of understanding, and it took the genius of the Greek geometers to connect these idea, and others, into a systematic and logical whole.

Hellenistic geometry[12]

Thales of Miletus (c. 624 BCE–c. 546 BCE) was an almost mythological personality in ancient Greece. He was known in ancient times as the first philosopher, the first person to engage in scientific philosophy,

---

[12] Almost everything we know about the Greeks comes from fragmentary texts, commentaries, and translations of translations. Prior to the modern age, it was common for authors to attribute written works and discoveries to famous people, whether real or fictional (a phenomenon called *pseudoepigraphia*, meaning "false attribution"). We have done our best to identify the correct authors for each attribution in this section.

So many texts have been lost to time that it is impossible to know for sure when and who made each discovery described here. Our story should therefore be considered an entertaining mathematical journey first and foremost.

and is the first person we know of today who was credited with a mathematical theorem.

According to legend, Thales traveled to Egypt, where he bore witness to the practical uses of geometry. While watching Egyptian geometers and builders work, he realized that some of the specific principles they applied could be abstracted into general formulations.

Consider this: a tool used to measure, for instance, the seked of a pyramid, could also tell you the size of the pyramid. For, if you were to repeatedly lay the tool on the pyramid's slope, counting the number of lengths required to reach the top, then you could determine the horizontal and vertical distances traveled by multiplying the size of the tool by the number of measurements, thus measuring the pyramid itself (Figure 4.20).

But you need not climb the pyramid! Because, if you know the pyramid's seked, you can find the length of the base (in paces, or using an appropriate ruler), and use that to calculate the number of times you would have to measure the face, and thereby calculate the height.

What if you didn't know the seked, though? The Great Pyramid was already ancient by Thales' time — nearly two thousand years old — and thus anyone involved in its construction was long dead. Thales, however, reasoned that there was nothing special about the seked used to build the pyramid, and that in fact any two triangles consisting of three equal angles would have sides in proportion to one another (Figure 4.21).

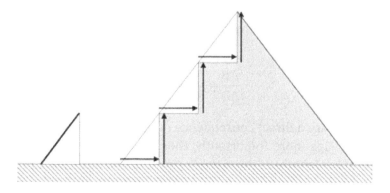

Figure 4.20.   Measuring the height of a pyramid by placing a tool up its slope.

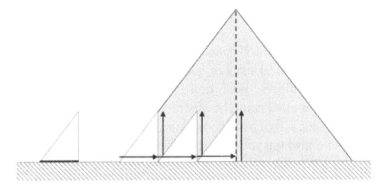

Figure 4.21.   An equivalent way of measuring the height of the pyramid.

By his reckoning, it would be possible to measure the height of the Great Pyramid using nothing more than its shadow. In Figure 4.22 you can see how this could be done. By measuring the ratio of his own height and shadow, and then measuring the length of the shadow of the pyramid and adding it to half the length of the pyramid's base, Thales could compute the height of the pyramid itself without climbing it, or even knowing how it was built!

Thales was able to generalize this property into what we now know as the Intercept Theorem. If two triangles are similar — that is, if they have three equal angles — then their corresponding sides must be proportional to one another. In Figure 4.23, you can see some examples of this.

In each case, the ratio of the length of the side marked AB to the side AD is the same as the ratio of AC to AE, as well as the ratio of BC to DE:

$$\frac{AB}{AD} = \frac{AC}{AE} = \frac{BC}{DE}$$

These ratios are a direct consequence of the definition of similar triangles. Perhaps more importantly, then, Thales' Intercept Theorem states that the lengths of the exterior segments BD and CE are proportional to the interior segments AB and AC, respectively:

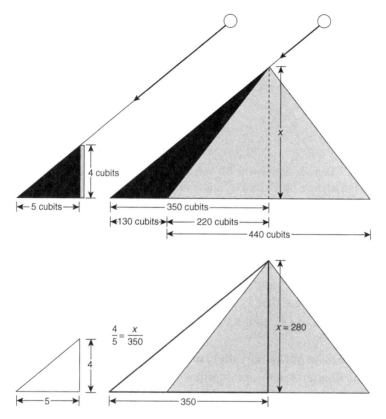

Figure 4.22. Thales' method of measuring the height of a pyramid. If the sun's rays are parallel to each other, the length of the pyramid's shadow can be used to construct a triangle that is similar to a triangle of known size.

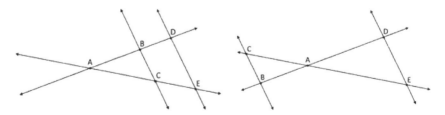

Figure 4.23. Extending the main idea: The Intercept Theorem. Given that BC and DE are parallel, many of the segments above are in useful ratios.

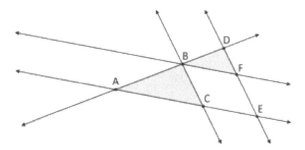

Figure 4.24. Demonstrating why BD and CE are proportional to AB and AC by constructing the triangle BDF, which is similar to ABC.

$$\frac{BD}{AB} = \frac{CE}{AC}$$

By drawing line BF parallel to CE, as in Figure 4.24, you can see why this is true. Triangle BDF is similar to triangle ABC, and segment BF is equal in length to CE, so it's only sensible that these segments are in proportion!

For the sake of the narrative, we are omitting the full proof of this theorem. Readers who are familiar with geometric proofs may be uncomfortable with this appeal to intuition, but rest assured that we will attend to your concerns in time! For now, understand that this theorem *can* be proved — but, moreover, that this real-life, intuitive application motivated Thales' logical approach to geometry, and that this theorem would in time lead to endless practical applications.

One of these applications is attributed to Thales himself. It is said that, using nothing more than his knowledge of triangles, Thales was able to find the distance to ships anchored at sea.[13] How could he have done this?

A number of different strategies have been proposed to accomplish just that. For instance, suppose you are Thales, standing at point

---

[13] Proclus (412–485), summarizing Eudemus of Rhodes' (c. 370 BCE–c. 300 BCE) *History of Geometry*, tells us that Thales used what we now call the Angle-Side-Angle Congruence Theorem to find this distance, but doesn't go into any more detail than that. Eudemus' work has, sadly, been lost.

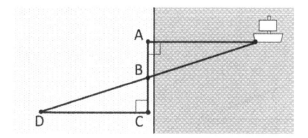

Figure 4.25.   Finding the distance to a ship at sea.

A in Figure 4.25. You can see the ship directly in front of you. Turning at a right angle, you walk along the shore, counting out a random number of paces — say, one hundred — before pushing a stake into the sand at point B. Continuing on another hundred paces to point C, you again turn at a right angle, this time headed inland.

Starting fresh at zero, you count your steps heading away from the sea. Periodically, you look toward the stake. As you walk farther and farther along CD, the stake appears to move closer and closer to the ship. Eventually, you, the stake and the ship are all in line: when closing one eye and staring directly at the stake, the ship is no longer visible. At this point, point D, you are as far from the shore as the ship was from your starting point — and you have thus measured the distance between the shore and the ship.

Some authors suggested Thales might have had a tower or cliff from which to make his observations. One way hc could have measured the distance from a tower is by sighting along a rod, recording the angle, and then turning toward the shore until he sighted another object at the same declination, but on land.

In Figure 4.26 you can see how this might work. In the picture on the left, the observer at point A sights the ship, and fixes the angle of observation. In the second, he rotates his apparatus, searching for a convenient visual marker — such as a rock lying close to the water — that is at the same angle below the tower as the ship. Finally, he climbs down the tower and counts the number of steps between the tower and the rock; that distance must be the same as the distance between the tower and the ship.

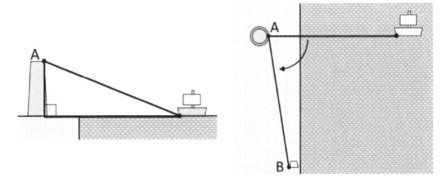

Figure 4.26. Finding the distance to a ship at sea. (Left) First, sight the ship using a tool. (Right) Next, being careful not to ruin the angle measurement, swing your tool toward the shore, looking to find some object at the same angle as the ship. The distance from the tower, A, to the object, B, must be the same as the distance from the tower to the ship.

Why should anyone walk so much, though! There has to be some way to use the Intercept Theorem to accomplish this task without walking nearly so much!

You could use the same method as in Figure 4.25, but instead of walking the same number of steps between B and C as between A and B, you could walk some fraction of that number of steps. Say, once again, that you counted one hundred steps before setting down a stake at point B. Then you could walk some fraction of that number — say, one fifth, or twenty steps — and from there follow the same procedure as the one accompanying Figure 4.25.

In this case, the triangle BCD would be similar to the previous triangle BCD, but each side would be scaled by the ratio you chose earlier (Figure 4.27). In our example, when using a ratio of one fifth, whatever distance you measured walking from C to D would have to be multiplied by five to reflect the distance between the ship and the shore. Thus, if AB is one hundred paces in length, BC is twenty paces, and CD is eighty paces, then the ship must be anchored four hundred ($5 \cdot 80 = 400$) paces from point A.

But if we allow the possibility that Thales used his understanding of similar triangles to measure this distance, then he need not have

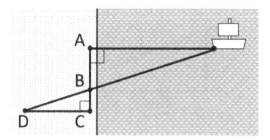

Figure 4.27.   Finding the distance to a ship at sea by walking along the beach, using similar triangles.

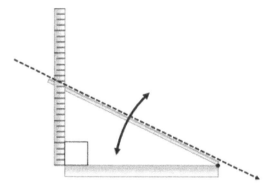

Figure 4.28.   Turning a seked tool on its side in order to create similar triangles.

walked at all! Assuming that he knew the height of his tower or cliff (from sea level) then it's quite possible he was able to replicate the discovery he made in Egypt.

Suppose that Thales was inspired by his knowledge of Egyptian building techniques, and constructed a tool like the one in Figure 4.28. You should recognize it as similar to the diagram in Figure 4.19: two rods are fixed at a right angle, and one is marked at regular intervals. By attaching a third rod that is free to pivot at the end of the horizontal rod, he would have been able to measure the seked of the ship's position from the tower. That is, by moving the third rod up or down until it was pointed directly at the ship, he could have then read the height of the vertical leg at the intersection of the two rods.

The triangle formed by this instrument would necessarily be similar to the (much larger) triangle formed by the height of the tower and

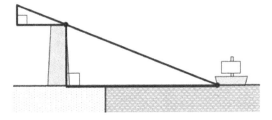

Figure 4.29.   Finding the distance to a faraway ship from a tower without moving. (The upper triangle's size is exaggerated for clarity.)

the distance between the tower and the ship (Figure 4.29). By multiplying the length of his tool by the ratio of the height of the tower to the length of the vertical leg, he could calculate the distance to the ship!

To understand this method, it may help to use real numbers. Suppose the tool is one meter in length and the tower is thirty meters tall. Were Thales to read a measurement of, say, twelve centimeters, then the distance to the ship could be calculated as follows:

$$\frac{0.12\,m}{1\,m} = \frac{30\,m}{x}$$

$$x = \frac{30\,m}{0.12\,m}$$

$$x = 250\,m$$

As you have already seen, Thales was able to use a similar method to compute the height of the Great Pyramid, so it's certainly possible that he used this method to accomplish his other famous application of practical geometry!

Knowledge of knowledge

Thales brought the Egyptian geometry with him back to Greece. Just as importantly, he brought into being a belief system: that the world might be understood through logic and observation, without recourse to mythology. His theory of the universe did not attribute natural phenomena to actions of ineffable gods, but instead suggested for the first time that the physical world is the source of both causes and effects.

Despite this triumph, this philosophy was still very tentative. The revelation that the world might be understood through reason and logic was an important one, but this new perspective on the universe opened up a nearly insurmountable mountain of potential topics to be explored. At the risk of this section becoming a litany of Greek philosophical trivia, looking at specific examples will help us to contextualize the state of understanding during this period. These will help demonstrate a key point: the Greeks were not only trying to understand the world for the first time, they were also trying to figure out *how they could figure out* if a proposed model were correct.

How can you determine whether a claim is correct? Once you have decided a claim is correct, how can you convince others? What tools do you use to support a statement that you believe to be true, and how did you come to believe those tools are effective at demonstrating truth?

For the early Greeks, the answers to these questions were one and the same: logic and rhetoric. Rhetoric took many forms — spoken arguments, dialogue, and debates — and the historical record we have from the classical period demonstrates that they were masters of the form. You have no doubt participated in rhetorical arguments of your own: perhaps as a participant in a late night discussion in a college dorm, or via Socratic questioning by a teacher in a classroom, or as a juror in a court case, or while watching a political speech or debate. Here and now, almost 3,000 years after the Greek philosophical golden age, we have retained many of the forms of classical rhetoric, but we have enriched these with other methods for determining truth. For the Greeks, though, no matter the question faced — such as "what is goodness?" or "what is the universe made of?" or "what is the relationship between the lengths of the sides of a right triangle?" — the approach was the same. Having discovered the hammer of rhetoric, the Greeks eagerly sought out nails to pound down, and they did so masterfully.

Thus began a period of almost unparalleled intellectual ambition. Every aspect of reality was open to exploration and debate, none more so than the nature of reality itself. Almost every thinker of the era threw his hat into the ring; Thales and his successors put forward countless models to explain the composition of the universe, and the earth's position in it.

As you recall from earlier, to a naïve observer in Egypt, the world appeared to be a flat disc under the dome-shaped heavens, below which the gods carried the heavenly bodies in their mythical chariots. Thales accepted that the earth was a flat disc, but his cosmology placed the world as an island, floating in an endless sea. Others agreed that the earth was a disc, but claimed it was floating in empty space. Later thinkers argued that it was a sphere; or that it moved, and there was another earth floating at the opposite side of the cosmos to balance the weight of this one.

The models varied as much as the thinkers that produced them, and later models often built on the those that had come before. From the limited perspective of human beings standing on the edge of the Mediterranean Sea and looking into the heavens, each model had its own appeal, and a certain logic. But all of them had essentially the same weakness at the core: they made no specific predictions, and they stood on nothing more than a small number of qualitative observations and rhetorical edifices constructed out of thin air.

Seeking to resolve these tensions, the Greeks looked to the best tool they knew. In their minds, the most accurate model of the universe would necessarily be the one that is the most straightforward, requiring few or no exceptional explanations, and leveraging one uniform principle to explain everything. In short, they chose the best model of the universe based on *aesthetic* principles.

In their final model, proposed by Plato and refined by Aristotle, the sphere of the earth stood at the center of the universe. It was in turn surrounded by dozens of spheres of increasing size, each of which contained a heavenly body, all moving in perfect circles.[14]

---

[14] For reasons well beyond the scope of this book, it is possible to accurately describe the motions of the heavens from the frame of reference of an observer standing on a

The great irony here is that the Greeks weren't exactly *wrong*. To look ahead a bit, modern scientific explanations are frequently aesthetically pleasing, and when given the choice between two or more competing models, scientists and mathematicians today generally prefer the one that is more simple and requires fewer distinct motivating principles. You may have heard this referred to as "Occam's Razor," often stated as the principle that "the simplest solution to a problem is often the correct one." The main difference between the Greek approach and the modern one is that the latter is often a *consequence* of other factors, and not a goal in itself.

In some cases, the purely logical and aesthetic approach to truth can yield fruit. A geometric statement that seems true based on experience can be definitively proven by running arguments backwards, in essence asking "but *why*?" again and again until there remain no more whys to ask. At the base of this tower of questioning lies nothing more than a residue of definitions and axioms — primitive statements that are taken to be self-evidently true — which nonetheless provide the foundation upon which the entire structure can rest, and from which further proofs can be built. Following threads of logic up and down repeatedly will, over time, produce a *system* of mathematics that can be applied to novel problems whose existence was first noticed as gaps in the structure. In time, the system becomes more comprehensive and, in a particular way, more beautiful.

The best way to demonstrate the interplay of these concerns — logic, rhetoric, and aesthetics — is to examine a practical example. We can use the Intercept Theorem from earlier in this chapter, after taking into account some caveats.

For one, that theorem actually encompasses several distinct claims, only one of which we will be exploring in detail. Further,

---

spinning planet orbiting a star using nothing more than a (potentially infinite) set of circular motions. This is the principle and key insight of Fourier Series, which are fundamental to electrical engineering, acoustics, signal processing, and many other fields.

So while the nested spheres of Aristotelian cosmology are considered outdated today, they were temporarily useful, *almost correct*, and that correctness manifested in a way that anticipated a much deeper mathematical model.

there are many ways to prove each portion of this theorem, many relying on more modern tools. However, to truly grasp the essence of Greek geometry, we shall restrict ourselves to the methods they used.

The elements of geometry

Euclid (mid 4th–mid 3rd century BCE) consolidated Greek geometry into the formative textbook of the field, called *The Elements*, around 300 BCE. From the surface, Euclid's *Elements* starts from simple principles to paint, step-by-step across thirteen books, a rich portrait of basic geometry, and in so doing lays out the fundamentals of mathematical writing that continue to be used to this day. A different reading of the *Elements*, however, reveals a markedly different perspective of how mathematical systems are developed.

Remember, the geometry Thales brought to ancient Greece was motivated by his observation that certain Egyptian practical applications were related to more fundamental ideas. It was his *observation* of the practical uses of geometry that motivated his investigations into pure abstract geometry. He emphatically did *not* start from basic definitions and build up abstract geometry, only to look outside and realize the Egyptians were using special cases of the same principles in their daily lives.

In due consideration of this tension, then, we can begin to appreciate the Greek approach to knowledge.

In Book VI of the *Elements*, Euclid attacks each piece of the Intercept Theorem in turn. Proposition 2 of this book (henceforth referred to as Proposition 6.2) makes the claim that, given an arbitrary triangle, a line drawn through the triangle and parallel to one of the sides cuts those sides in equal proportions. You can see a diagram of this in Figure 4.30. Here, triangle ABC is cut by the line running through DE, parallel to the base BC. Proposition 6.2 says, then, that the length of AD is to DB as AE is to EC.

How can you prove this? If you accept, *a priori*, that similar triangles have sides that are in proportion to one another, then the proof falls out quite easily. However, Euclid does not choose to go this route.

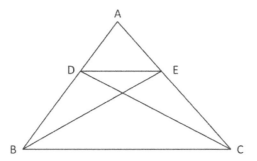

Figure 4.30.   The triangle used to demonstrate Euclid's Proposition 6.2.

In fact, he uses Proposition 6.2 to prove later, in Proposition 6.4, that the sides of similar triangles are proportional to one another.

Instead, Euclid draws two additional segments, BE and DC, thus constructing the triangles BDE and CDE. He then shows that these two triangles must be equal — or, to use more modern phrasing, they must have equal areas.

Since they have equal areas, the ratio of the area of triangle BDE to the area of triangle ADE must be the same as the ratio of the area of triangle CDE to the area of triangle ADE.

Triangles ADE and BDE have the same height — the line descending from point E, perpendicular to line AB — and thus, the ratio of their areas is the same as the ratio of their bases, AD and DB. The same rationale can be applied to show that the ratio of the areas of ADE and CDE is equal to the ratio of AE to EC.

Finally, since the ratio of ADE to BDE is equal to the ratio of ADE to CDE, the ratio of AD to BD must be equal to the ratio of AE to EC.

However, notice how this proof glosses over some facts. You might be wondering how we can make the claim that the areas BDE is equal to the area of CDE, or whether it's reasonable to conclude that the ratio of the bases of triangles of the same height are equal to the ratios of their areas. Further, there are some assertions about the equality of proportions that, while not particularly controversial, still demand some form of proof.

A well-annotated version of the *Elements*, or a close reading, reveals that Proposition 6.2 relies on several other proofs. Before you

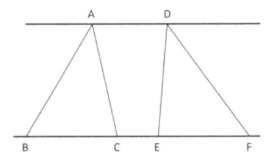

Figure 4.31. Euclid's Proposition 1.38: Given that lines AD and BF are parallel, and BC is equal to EF. Then, triangles ABC and DEF have equal areas.

can definitively state that the two triangles BDE and CDE are equal, you must find some rationale for this, which Euclid gives in Proposition 1.38. This proposition states that triangles that have equal bases and lie between parallel lines (Figure 4.31) are equal. But this proof, itself, relies on three other propositions — namely, Propositions 1.31, 1.34, and 1.36.

Continuing on in this manner, then, we can construct a list of all the ancestors of each proposition, until we arrive at the basis of all Euclidean Geometry. In Table 4.19 we summarize what each proposition demonstrates, and in Figure 4.32 we show how these propositions depend on one another.

There is a lot to take in here, so let's consider all this one piece at a time. First, notice that the dependencies in Figure 4.32 only go one direction, from top to bottom. This is a key characteristic of any system of mathematics: a tool can only be used after it has been constructed. A proof that assumes some fact A to prove another fact B, and then uses the proof of B to prove A in turn is circular, and thus not valid. As silly as it might seem to mention this, students often fall victim to this pitfall when learning to construct proofs, and we as authors struggle with this too! After all, when relating a story about the history of mathematics, it's tempting to use mathematics that had not been invented yet to explain discoveries to a modern audience — mathematics that was developed as a result of those very discoveries!

Table 4.19.   Proposition numbers from Euclid's *Elements* along with a summary of what each proposition demonstrates.

Number	Summary
1.1	A method of constructing an equilateral triangle on a given line.
1.2	A method of constructing a line of a given length at a point.
1.3	A method of cutting a line of a given length from another line.
1.4	Two triangles are equal if two of their sides, and the angle between those sides, are equal. (SAS congruence)
1.5	The angles at the base of an isosceles triangle are congruent to one another, and the angles exterior to the bases are congruent to one another.
1.7	There is only one triangle with given side lengths in a given order.
1.8	Triangles with equal sides have equal angles.
1.9	A method of constructing a line that cuts a given angle in equal halves.
1.10	A method of constructing a line that cuts a given line in equal halves.
1.11	A method of constructing a line perpendicular to a given line.
1.13	Supplementary angles are either both right angles, or they sum to two right angles.
1.15	Vertical angles are congruent.
1.16	An external angle of a triangle is larger than either of the other two interior angles.
1.18	The larger side of a triangle is opposite the larger angle.
1.19	The larger angle of a triangle is opposite the larger side.
1.20	The sum of any two sides of a triangle is greater than the remaining side. (Triangle inequality)
1.22	A method of constructing a triangle from lines of any lengths that satisfy Prop 1.20.
1.23	A method of constructing an angle equal to another angle at a given point on a given line.
1.26	Two triangles are equal if two of their angles and any one of the sides are all equal to each other. (ASA/SAA congruence)
1.27	If a straight line crosses two lines such that alternating angles are equal, then the two lines are parallel.
1.29	If a straight line crosses two parallel lines, then alternate angles are equal, and interior angles sum to two right angles.

(Continued)

Table 4.19. (*Continued*)

Number	Summary
1.31	A method of constructing a line through a point, parallel to a given line.
1.33	Straight lines joining equal and parallel straight lines are equal and parallel. (Definition of a parallelogram)
1.34	Opposite sides and angles of parallelograms are equal, and diagonals of parallelograms divide them in half.
1.35	Parallelograms that have a common base and lie between two parallel lines are equal.
1.36	Parallelograms with equal bases that lie between two parallel lines are equal.
1.38	Triangles with equal bases that lie between two parallel lines are equal.
5.1	Sums of equal multiples of magnitudes are divisible by the same multiple of sums of the original magnitudes. (Distributive property)
5.7	If two magnitudes are equal to one another, then they are in equal ratio to some other magnitude.
5.11	If one ratio is equal to second, and the second is equal to a third, then the first ratio is equal to the third. (Transitive property)
5.12	If there are several magnitudes that are in the same proportion to several other magnitudes, then the sum of the first group is in the same proportion to the sum of the second group.
5.15	If two magnitudes are in a certain ratio, then equal multiples of those magnitudes are in the same ratio.
6.1	Triangles and parallelograms which are of the same height have areas in the same ratio as their bases.
6.2	A straight line cutting a triangle parallel to one of its bases will cut the other two sides in equal proportion.

Next, even if you don't understand all the vocabulary in Table 4.19, the propositions generally have simple structure. Selecting one at random, Proposition 1.19 states that "the larger angle of a triangle is opposite the larger side." This seems sensible on the surface, and does not seek to prove too much at once. Note that the converse of this fact,

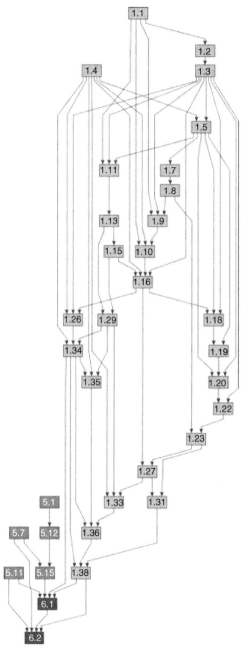

Figure 4.32.   Dependencies between propositions in Euclid's Elements.

that the larger *side* of a triangle is opposite the larger *angle*, is proved earlier, in Proposition 1.18.

Also note that the rhetorical approach here does not take into account specific details about the figures in question. It is quite reasonable to approach a novel problem that you have reason to believe is true by choosing numbers at random and testing your beliefs. For instance, Proposition 1.20 states that the sum of any two sides of a triangle exceeds the remaining side. If you were to select three lines, of lengths 1, 2 and 3, and try to construct a triangle from them, you could at best construct two overlapping straight lines. But vary the length of even one of the sides to satisfy the conditions of Proposition 1.20 — using lengths 1.0001, 2, and 3, for instance — then, by necessity, you would be able to construct a triangle.

These proofs can tell us what would happen if we substituted specific numbers, but the proofs Euclid collected did not use numbers. What did he use, then?

Euclidean geometry is built up from five postulates. A postulate is a fact, or an action, that he asks us to take for granted — one that cannot and does not rely on any other proof. First, Euclid stated that a straight line can be drawn between any two points. Second, he said that a line can be extended infinitely in either direction. Third, he stated that a circle can be drawn about any point, having any radius. Fourth, he asserted that all right angles are equal to each other. Finally, he stated that all the pairs of lines in a plane are either parallel, or meet at one point.[15]

The first three postulates describe the tools that are usually used to define this geometry. When reading certain sources, authors will make note of propositions that can be proved "using only a straightedge and a compass." The first postulate describes the use of a straightedge (e.g., a ruler, albeit one without markings), while the

---

[15] This is an alternate formulation of the fifth postulate, or *Parallel Postulate*. In the original text, Euclid states that if a line intersects two other lines such that the interior angles sum to less than two right angles, then those two lines, if extended indefinitely, would meet at a point on the side where the sum is less than two right angles.

This formulation can be somewhat difficult to wrap one's head around, so we have chosen an equivalent expression that nevertheless produces the same geometry.

second suggests either that it should be infinite in length, or that you can use it repeatedly to extend a line indefinitely. The third postulate, of course, describes the use of a compass.

He supplements these postulates with a number of definitions and common notions. For instance, Definition 1.10 states that when two lines intersect in such a way that adjacent angles are equal to one another, those angles are called "right angles," and the lines are called "perpendicular."

That any of this might seem unremarkable is a testament to the aesthetic beauty of Greek geometry. Why, you might ask, do we need to define a right angle in this way?

Every definition provided to us by Euclid allows us to talk about countless arrangements of geometric figures without making reference to their specific measurements. Indeed, if you think back to the terminology we generated in previous chapters, our motivation was always to subsume some collection of traits into a term that we could apply in a useful way. Rigorously defining a triangle as a collection of lines in a particular arrangement allowed Euclid to talk about triangles in a useful way, and to prove properties that apply to all triangles regardless of measure.

It is hard to overstate the impact of Euclid's *Elements*. As we will see throughout the remainder of this chapter, the forms that we trace back to him continue to influence the structure of mathematics to this day. Students in a modern classroom can easily feel overwhelmed by a seemingly endless stream of definitions and vocabulary to be memorized. The *Elements*, however, demonstrates mathematical terminology at its best; each new term expands the scope and power of Greek geometry.

One last point bears mentioning. The importance and aesthetic beauty of Euclid's *Elements* were, in part, a response to a significant controversy about the nature of numbers, and this controversy would split the mathematical world for thousands of years.

Speaking of numbers

To adequately explain the divide between arithmetic and geometry, we need to sketch out some additional historical background.

Imagine you are an ancient scribe tasked with keeping track of the king's property. For you, the most natural way to count objects would be to draw the objects themselves. Thus, to represent "five cows," you could draw five cows. In short order, you would grow tired of drawing the same symbols over and over.

Quickly, you should realize that five objects of one type are equivalent to five of another. Thus, to record "five cows," you could draw a small symbol representing a cow, and then five tally marks, whereas for five measures of grain you could draw a symbol for grain and five tally marks. It would still be quite challenging to record larger numbers, though. Looking at your hands and feet, you might decide that once you hit the number ten, you can replace ten "one" marks with a different symbol, and, extending this logic to hundreds, thousands, and so on, you might arrive at a system similar to the Aegean numerals.

Prior to the 7th century BCE, the Greeks used the Aegean numerals, a system consisting of five primary symbols representing ones, tens, hundreds, thousands, and ten thousands. In order to construct a given number, these symbols could be repeated as necessary to sum up to the appropriate number. So, for instance, 31,407 was written as shown in Figure 4.33. This system was cumbersome in its own way, but adequate for the task of running a business or state in early Greece.

Now, being a clever scribe, you might realize that there is a way to further reduce the amount of writing necessary to keep track of numbers. Each number you use — ten (Greek δέκα, pronounced "deka"), hundred (ἑκατόν, "hekaton"), thousand (χίλιοι, "k^{h}ilioi"), and ten

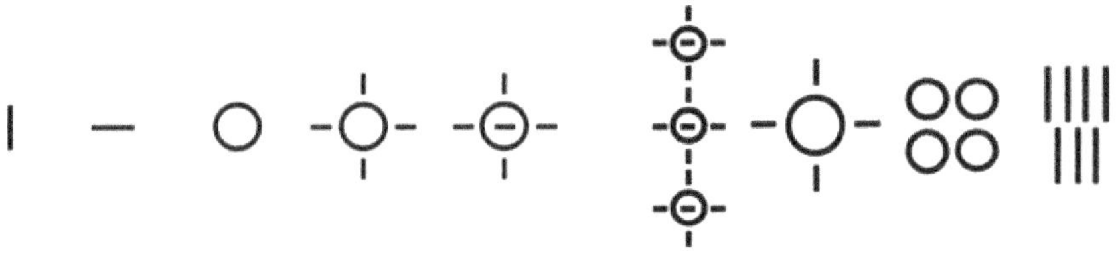

Figure 4.33. Aegean numerals. (Left) 1, 10, 100, 1000, and 10,000. (Right) The number 31,407.

thousand (μύριον, "myrion") — has a different starting letter, and, supplementing these with a vertical mark for one and a letter to represent five (πεντε, "pente"), you can simplify the process of writing any number.

Translated into English, we might write the numbers 1, 5, 10, 100, 1000, and 10,000 as I, P, D, H, X, and M[16] respectively, and thus in this system the number 31,407 could be written as MMMXHHHHPII. Here, you can count the repetitions of each letter and add them up to yield the result; thus, three M's signifies three ten thousands, one X indicates one thousand, and so on. If this seems similar to the system of Roman numerals, that is no coincidence — this system, called the Attic numerals, was a precursor to the Roman system.

There is another aspect of this system that is worth highlighting. Earlier, we demonstrated that building an appropriate vocabulary to talk about geometric figures — lines, triangles, perpendicular lines, and so forth — simplifies the process of working with them. Similarly, by leveraging the initials of already existing number words as the basis for their number system, the Greeks were able to abstract the idea of numbers even further, from a count of marks on a tablet to simple symbols.[17] This meta-skill — recognizing when some operation or object can be simplified by replacing it with a more abstract symbol or term — comes up again and again and again in mathematics.

---

[16] Here, we are eliding certain numbers that are difficult to represent typographically. The Greeks had characters for the numbers 50, 500, 5,000, and 50,000, which were written as a large pi (Π) with the right leg shorter than the left, almost like a capital gamma (Γ). Then, a small version of the appropriate letter was written in the top right corner. Thus, 500 could be written by placing a small H at the top right of the Γ.

[17] By now, you should notice that we keep returning to this idea—that it is permissible to create and define mathematical terminology and symbology to satisfy particular needs. It bears repeating, if only because too often students see mathematics as something that is handed down rather than created. This fact is both a strength and a weakness of mathematics. The strength comes from the knowledge that we can benefit from the work of everyone who has come before us; the weakness is that all too often students lose sight of why symbols, terms, and abstractions were useful, and come instead to view them as burdensome facts to be memorized.

The Attic numerals existed during the early phase of Greek mathematics covered previously, from around Thales' time to roughly the lifetime of Euclid. They were replaced with a third system of numbers around the 3rd century BCE, but in the intervening years we can imagine that whenever the Greeks performed arithmetic computations, these were the numbers they used.

Why does this matter? According to some accounts, Thales passed the mantle of leadership to Anaximander (c. 610–546 BCE), who in turn taught a student named Pythagoras (c. 570–495 BCE), whose name you are almost certainly familiar with. Pythagoras loved numbers but, in time, found that not all numbers are alike.

Impossible numbers

Pythagoras founded a group called the Pythagoreans, whose beliefs were somewhere between a school of philosophy and a cult. Their contributions to ancient mathematics were motivated by a mix of academic interest and religious fervor. The Pythagoreans believed that the universe was numerical, and that numbers and ratios were the essence of all things.

One of the central ideas of Pythagoras' school was the importance of harmony. Pythagoras himself was said to be obsessed with musical harmony; he identified certain ratios between musical notes (most notably, the 2:1 ratio between octaves), believed that harmonious music could heal the soul, and that the planets' moved in harmonious ratios that produced heavenly music that was silent to mortal ears.

Suffice it to say, to the Pythagoreans, numbers were serious business. All numbers could be spoken of — as whole numbers, or as the ratios of whole numbers — and their cosmology did not admit the existence of any other numbers. So when the Pythagorean, Hippasus (ca. 5th century BCE), proved that there are other, irrational numbers, you can understand why (according to legend) they threw him into the sea! To the Pythagoreans, the irrationality of the square root of 2 was tantamount to heresy.

Hippasus' argument is a simple example of proof by contradiction, and valuable in its own right as a demonstration of that method.

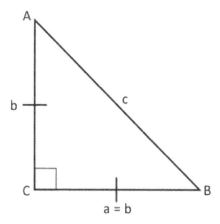

Figure 4.34. An isosceles right triangle. (The lines through the sides AC and BC indicate that they have the same length.)

Proof. Begin with an isosceles right triangle, shown in Figure 4.34. Since it is isosceles, sides a and b are equal in length. Assume that the lengths a, b, and c are whole numbers, and share no common factors (because, if they did, you could divide each side's length by the common factor to yield a smaller triangle for which this assumption holds). By this assumption, we can express the ratio as the ratio of integers.

By the Pythagorean theorem, $a^2 + b^2 = c^2$; further, since and are the same length, $a^2 + b^2 = b^2 + b^2 = 2b^2$, producing the equation $2b^2 = c^2$.

Since we assumed that c is an integer, and we know that $b^2 = \frac{c^2}{2}$, we can say that c^2 is even. Numbers are either even or odd, and as you saw in Chapter 3, the square of an odd number is odd (being the product of two odd numbers), while the square of an even number is even. Thus, c itself must be even.

Any even number can be expressed as two times some other number; let us call this number d. Then we have the equations $c = 2d$ and $c^2 = (2d)^2 = 4d^2$.

We can substitute this expression into an earlier equation to find $2b^2 = 4d^2$. Dividing both sides by 4, we find that $d^2 = \frac{b^2}{2}$, thus b must be even, by the same logic as before.

But if both b and c are even, then they share a common factor — namely, 2. As we assumed at the beginning, c and b do *not* share any common factors; thus we have a contradiction, and *b*: *c* cannot be expressed as a ratio of integers!　　　　　　　　　　　　■

This discovery caused quite a stir in the world of mathematics. Until this point, arithmetic and geometry had been treated as equals. Even ignoring the spiritual ramifications for the Pythagoreans, the knowledge that there exists a class of numbers that can be drawn geometrically, but that arithmetic had no way to describe, was deeply unsettling.

You may be wondering why a single irrational number could cause this much stress, but that misses the significance of proof by contradiction (and proof in general). It only takes a single counterexample to demonstrate that a definition or theorem does not hold. If you assume that all sheep are white, finding a single black sheep is enough to either break your definition, or to force you to create a new definition.[18] In the world of logic, discovering one black sheep is sufficient to show that not all sheep are white, and the further discovery of a flock of black sheep is technically unnecessary. So, while the Greeks would continue to prove that other numbers are irrational (mostly other square roots, from 3 to 17), the religious elements of the Pythagorean system died the instant Hippasus spoke his proof.

A way forward

Returning to more earthly matters, Eudoxus of Cnidus (c. 390–337 BCE) found a way to resolve the tension between arithmetic and geometry. Imagine that you have a rubber band which, unstretched, is 1 inch long. In stretching it to 2 inches, it must certainly pass through every intermediate length (including the square root of 2). Even if you don't know how to put into words the rubber band's length at any given moment, you could still refer to its length, geometrically, at any point in time simply by saying "the length of the rubber band

---

[18] According to one story, Plato described a man as "a featherless biped." Diogenes the Cynic, never missing an opportunity to make fun of more serious philosophers, plucked a chicken and brought it to Plato's Academy saying, "Behold! I've brought you a man!"

right now." This was the essence of Eudoxus' innovation. He found a set of principles that allowed mathematicians to talk about principles such as magnitude, addition, subtraction, and proportionality without needing to make reference to any particular numbers, by instead referring to comparative geometrical facts: greater than, less than, equal to, and so forth.

Look again at Table 4.19. The propositions from Book V of the *Elements* are, to a modern reader who is aware of algebra, quite difficult to follow. For instance, our interpretation of Proposition 5.15 reads: "If two magnitudes are in a certain ratio, then equal multiples of those magnitudes are in the same ratio." Translated into modern terms, this can be stated: $\frac{a}{b} = \frac{ma}{mb}$.

Most modern arguments about proportion, area, length, and so forth make reference to algebraic arguments and terms. Even our earlier demonstration of the irrationality of the square root of 2 uses algebraic notation! This is an anachronism: the Greeks were not aware of algebra, so the idea of representing unknown terms using symbols never quite occurred to them. (If, in algebra class, you ever wondered "what does x mean?" — take heart! You are in good company among the ancient Greeks.)

What did they do instead? The most important works of Greek mathematics until the time of the Roman Empire were highly abstract, and almost always used a geometric approach. This was, in one sense, intentional. Once Eudoxus demonstrated that geometry could stand on its own as a pure axiomatic system, most Greek mathematicians were content to live in a heady academic world of pure geometry. For hundreds of years thereafter, any particular results derived from geometric proofs were seen as a kind of afterthought, contrary to the subject's humble origins.

Practically speaking, this produced some results that can be difficult for a modern reader to comprehend. When a Greek referred to (what we would call) the square of a number, he meant a literal square, one that could be drawn with sides of some length whose value he wasn't particularly interested in knowing. When he wanted to multiply numbers whose values may or may not have been equal, he would have spoken of the "rectangle" of two lines, rather than the product of two numbers. What you might think of as addition and subtraction, the Greek geometer would think of as putting two lines (or squares, or rectangles) in a certain arrangement such that a third line (or square...) would be produced.

There are some strengths to this approach. A Greek geometer would never mistakenly try to add a square and a line, as some students in algebra classes are wont to do — clearly, this would produce a nonsensical result! Further, the proofs that this system produced were applicable to a wide variety of problems. While some particular demonstration of a proof about magnitudes might use lines, practically, that proof could be applied to any magnitudes whatsoever, whether the lengths of lines; or the areas of squares, rectangles or circles; or the volumes of three-dimensional shapes. Finally, to bring this digression full circle, this approach yielded a pure, self-contained, aesthetically pleasing, and completely logical mathematical system — the first of its kind in the history of the world.

(Im)practical geometry

Despite the separation between arithmetic and geometry, some Greeks had not forgotten the lessons of Thales and others in applying arithmetic to the results of geometric proofs. Around 332 BCE, Alexander the Great conquered Egypt and founded the city of Alexandria. After his death, his general Ptolemy the First[19] worked to make the city a great center of trade and learning, drawing some of the greatest minds in the ancient world to study and work there.

While Greeks around the Aegean Sea continued to focus on purely abstract mathematics, Alexandrian mathematicians and astronomers leveraged the principles of geometry to derive numerical results from real-world observations. Using Euclid's geometry and careful measurements, they were able to apply arithmetic to gradually expand the understanding of the cosmos. It is quite poetic that in returning to Egypt, the Greeks would return to the kind of applied geometry that inspired the founder of the field.

---

[19] A different person from the Ptolemy we will come to discuss shortly.

Recall that, in response to the profusion of cosmological models proposed by various philosophers, the Greeks decided to choose the simplest and most aesthetically pleasing explanation of the universe. From practical observation, they knew that the sun's rays hit the earth at different angles. A sundial located in Miletus, in modern-day Turkey, casts a longer shadow on the summer solstice (the longest day of the year, when the sun is highest in the sky), than a similar sundial located in Alexandria, Egypt, south across the Mediterranean Sea. The easiest, most aesthetic way to explain this phenomenon is to assume that the earth is a sphere (rather than a flat disc, or some other shape). From observations like this, combined with observations of the motions of the stars and the belief that the earth itself does not move, the Greeks decided that the universe must be a series of concentric spheres with the earth at the center.

This model, however crude, enabled the Alexandrian Greeks to begin computing rough solutions for astronomical problems, including developing the first measurements of the size of the earth, and the distance to the moon.

Building a globe

Given all that we have said about geometry, similar triangles, and ratios so far in this narrative, it should come as no surprise that these played major roles in the discoveries to follow.

Pause for a moment, and consider: how would *you* measure the size of the earth? What tools could you use that pre-industrial civilizations would have had on hand to construct a geometrical solution? What assumptions could you make to simplify your calculations? And what data would you need to gather to find a numerical answer?

You have already seen hints that could help you assemble an answer, but they're scattered throughout the text. You would have to be a very astute reader (or one who already knows the answer) to put them together!

Consider this: earlier, we mentioned that sundials in different locations cast shadows at different angles at different times of year. What conclusion can you draw from this? Or, to phrase the question differently, if the earth *is* spherical, what are the most extreme angles that could be produced between the style (that is, the tip) of a sundial, its shadow, and the ground?

First of all, this approach should remind you of a tool you have been developing since Chapter 2. By considering extreme cases, you can see how the sun must cast shadows directly below a sundial, parallel to the horizon, or at any angle in between, as illustrated in Figure 4.35.

Next, you might recall that in Egypt, Thales used the assumption that the rays of the sun are parallel to one another to measure the height of the Great Pyramid using a rod. If indeed the sun's rays are parallel everywhere, then at any given moment the line connecting the style of a sundial to the tip of its shadow is parallel to any other such line. By this assumption, you can produce any number of useful parallel lines to construct a diagram.

What else would you need? Well, in a perfect world, the two points you choose would lie on a great circle — which is a circle on the sphere whose center is at the center of the sphere — that runs through the poles, and you would know the distance between the points. (Pause to consider why.) Finally, in the best possible circumstances, at one of these points, on some specific day of the year, the sundial's shadow would lie directly below the style.

Eratosthenes of Cyrene (c. 276–195 BCE) used a conceptual model much like this to conduct the first measurement of the earth's

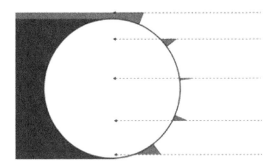

Figure 4.35. Shadows cast by straight rods at the same time of day at different points on the earth.

diameter. As the chief librarian at the Library of Alexandria, he had access to countless texts, from which he learned two facts that enabled him to perform his measurement: the location of such a point, and the distance between that point and Alexandria.

The town of Syene (modern day Aswan) lay at the cataracts of the Nile. As this dangerous, rocky portion of the river was nearly impassible by Egyptian boats, it marked the traditional edge of the Egyptian kingdom. Its importance as a frontier town meant that the roads between it and the rest of the kingdom were well-traveled, and thus well-measured; the Egyptians regularly surveyed the distance between Syene and the heart of the kingdom.

In addition to its position at the historical edge of the kingdom, the city had another claim to fame. Once a year on the summer solstice, the sun was said to rise to the center of the sky. According to legend, if you were to look down a tall, straight well on this day, your head would perfectly block the reflection of the sun on the well's waters. This was proof that the sun was directly overhead, and not at some slight angle above or below it.

Extending the observation we made in Figure 4.35, we can draw a diagram of the two cities, as seen in Figure 4.36. The angles are exaggerated for effect, but the general principle is accurate: if we assume that the two cities lie on a common great circle running through both the North and South Poles, then the distance between them describes

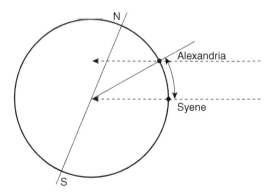

Figure 4.36. Two cities. At the summer solstice in Syene, the sun's rays shine straight down from overhead; in Alexandria they strike at an angle.

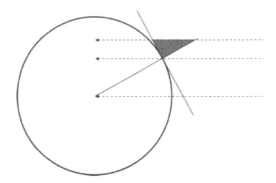

Figure 4.37. The earth is large enough that at a given point the ground can be treated as though it is flat. A rod casts a shadow of a certain length, and the rod and shadow describe a right triangle.

some portion of the earth's total circumference. On the summer solstice, parallel rays from the sun pass through both cities, but in Syene, shadows point straight down, toward the center of the earth.

The earth is much larger than this, so at any given point we can treat the ground as flat, and draw a triangle much as Thales did in Egypt. In Figure 4.37, the length of the shadow cast across the flat ground, and the height of the rod sticking up out of the earth describe two legs of a right triangle whose hypotenuse is described by the sun's rays.

Since the shadow of a rod at Alexandria is defined by the sun's rays, and these are assumed to be parallel to the line through the center of the earth, by Proposition 1.29 (from Table 4.19), we have drawn two alternate interior angles that must be equal in magnitude, marked by the Greek letter theta Θ in Figure 4.38.

Whatever angle Eratosthenes measured at noon on the solstice would have been equal to the angle described at the center of the earth between the two cities. Some modern retellings of this story state that he measured this angle as 7.2 degrees, but once again this is an anachronism: in the 3rd century BCE the Greeks did not measure angles in degrees. Instead, Eratosthenes calculated that the angle was equal to 1/50 of a complete circle, and thus, by multiplying the distance between Syene and Alexandria by 50, he found the

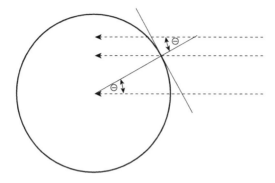

Figure 4.38. Alternating interior angles are equal, assuming the sun's rays are parallel.

circumference of the earth to be 252,000 *stadia* — within 15% of what we now know to be the true value![20]

Eratosthenes used this value to invent geography, the formal study of lands and countries around the world. On his spherical model of the world, he was able to draw evenly-spaced great circles through the poles, called *meridians*, thus inventing longitude; and by drawing parallel lines between the equator and the poles, he formally described latitude. Now equipped with this grid, Eratosthenes was able to use known measurements between places to plot a map of the known world. He was also able to make predictions about the unknown world, for instance splitting the earth into five climate zones: frozen polar zones at the extremes, a hot equatorial region in the middle, and temperate zones in between.

---

[20] There is a great deal of debate about what this value was. Eratosthenes' original writings are lost, and all that remains to us are summaries provided by later commentators. According to these accounts, Eratosthenes computed the circumference of the earth at 252,000 stadia — but it is not clear how long a single stadion was at this time!

Like many pre-modern units of measure, the length of the stadion changed over time.

Depending on the length used, Eratosthenes calculations were at most 15% off, but it is possible he was much more accurate than that — perhaps within 2% of the true value.

Heavenly angles

Normally, a discussion of mathematics wouldn't demand detailed knowledge of the tools used to find mathematical results. However, in a chapter on "Important and Useful Solutions," it's worth examining how mathematics influenced the development of practical tools.

There are two types of angle measurements that are particularly important for astronomy: angular size and parallax.

Methods of measuring angular size actually arise quite naturally out of the Intercept Theorem. You can demonstrate this in a simple manner with little more than a ruler and two coins of different sizes. Grasp the larger coin between the thumb and forefinger on one hand, and place it on a mark at a reasonable distance from the end. (In the United States, you might use a 12-inch ruler, and hold a quarter at the 12-inch mark.) Being careful not to injure your eye, place the ruler up against the bony ridge beneath your eyeball. With your other hand, grab the smaller coin and move it around until it just barely occludes the entirety of the larger coin. Place the smaller coin on the ruler, and carefully move both the ruler and both coins away from your face until you can read the two measurements. Congratulations! You have just constructed a basic *diopter*, and used it to measure the relative angular size of two objects.

Looking at Figure 4.39 you can see that this setup re-creates the basic layout used in the Intercept Theorem. Your eye acts as the vertex of a triangle, and from it you can draw two lines that touch the edges of both coins. Two isosceles triangles are formed — one with a base equal to the diameter of the smaller coin, and the one whose base is equal to the width of the larger coin. As you well know by now, the

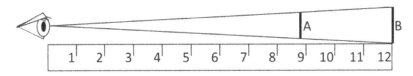

Figure 4.39.   Relative sizes and distances of U.S. coins. A dime at position A will just barely occlude a quarter at position B.

ruler measures the heights of the two triangles, and the two measures along the ruler are proportional to the respective bases.

Using American coins, the official measurement of a dime is 0.705 inches, while a quarter is 0.955 inches. If you held the quarter exactly 12 inches from your eye, then the point where the dime was the exact same size as the quarter should have occurred at the distance we compute with the following equation:

$$\frac{0.705}{0.955} = \frac{x}{12}$$

$$12\left(\frac{0.705}{0.955}\right) = x$$

$$x \approx 8.86$$

To perform astronomical measurements, you would want to have a slightly different tool. Since you (and the Greeks) don't have the distance to the moon, or its diameter, readily available, performing astronomical measurements would be simplified by using a ruler with different markings. Instead of one marked in inches, you could use a ruler marked in degrees — or one marked in fractions of a circle Figure 4.40. Then, by sliding the coin up and down the ruler, you could quickly measure apparent sizes of objects in the sky.

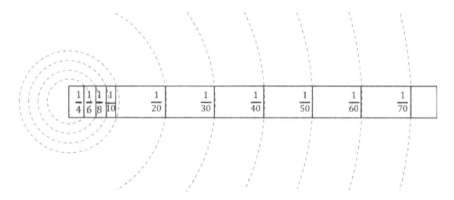

Figure 4.40.   The function of a diopter. Each solid marking represents what fraction of a complete circle (dashed) would be hidden behind a "coin" positioned at that location.

Figure 4.41. (Left) To an observer at I_1, coin A completely eclipses coin B. The observer at I_2 however sees the coins (right) as circles of equal diameter kissing at one edge.

This is a cute application of the Intercept Theorem, and one that is historically relevant, but it doesn't yet allow you to understand how to calculate the distance to the moon. To achieve that, you need to understand parallax.

Using the same two coins as before, but this time foregoing the ruler, line them up at any distance so that one covers the other completely (Figure 4.41). Now, close the eye you've been using and open the other eye; the rear coin will seem to have moved! This is the essence of parallax. Objects at different distances appear to move different amounts as you, the observer, change position.

On its own, this observation doesn't tell you much. But consider how parallax would work with multiple objects, all of the same angular size, positioned at different distances (Figure 4.42). In this diagram, the first observer sees only the first object, A, because it perfectly covers the others. Depending on the actual sizes of the objects and the position of the second observer, she either doesn't see the object at all (A), can see multiple objects at the same time, or sees one covering another (C and D).

If you imagine D as the sun, think of A as a small coin that perfectly covers the sun, B as a distant building, and C as the moon. During an eclipse, if an observer at position I_1 positioned a coin in front of his eye so as to cover the building, the moon, and the sun, a distant second observer at I_2 would not even see the coin, and would witness the moon and the building obscuring different portions of the sun.

There are many different ways to measure parallax. The simplest is the one we started with: looking at objects with one eye at a time.

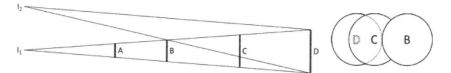

Figure 4.42.   Objects of the same angular size at different distances occlude different portions of faraway objects, depending on the location of the observer.

If the object whose parallax you wish to measure is a great distance away, like the moon, your measurements must be performed at much greater distance!

From the earth to the moon

The concept of parallax allowed the Greeks measure distances by reducing astronomical observations to geometrical figures, with which they were intimately familiar. But to use the measurements they made, they would need an absolute measure. Remember, earlier, that to find the distance from your eye to the smaller coin required knowing the sizes of the two coins and the distance to the larger one. If you have no way to measure those values directly, you would need some other way to construct your diagrams!

Every valid triangle is uniquely determined. That is, if you know three sides, you can find all three angles. And if you know the measure of two angles, you can find the third; if you can then find the measure of one side, you can describe the triangle completely!

We glossed over how the Greeks decided to believe that the stars are fixed on a great dome in the sky, but in this phase of discussion we have both the means and motive to explain this. No matter when they performed their observations, the stars themselves did not appear to change positions relative to one another, or relative to the earth. Although some philosophers theorized that the stars might be very large, and very, very far away, they were not able to measure the stars' positions with suitable accuracy to determine their distances. For all intents and purposes, the stars formed an immutable backdrop for the cosmos.

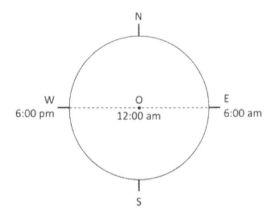

Figure 4.43.   Five observers. Observer O looks straight up at midnight; observers E and W are leading and trailing O by six hours, respectively.

Fortunately for them, and for us, this provides a convenient reference point for making astronomical measurements!

Once again, let's consider an extreme case. Imagine that there are five observers standing in a cross pattern on the earth, with one at each of the poles, and three on the equator (Figure 4.43). One of the observers on the equator, observer O, is at an equal distance from the other four: each of the others stands exactly a quarter of the way around the globe from that point.

When local time at observer O's position hits midnight, the time at observer E's position, to the east, is 6:00 am; naturally, at position W the time would be 6:00 pm.

Now, assume there's a very particular cosmological alignment. Directly overhead at midnight, observer O sees the moon move right in front of a star, such that both are in a straight line through the zenith of the sky. What will the other observers see?

In Figure 4.44, we have drawn the subjective view of each of the observers. All five would agree that the star is in the same position in the fixed reference system of the heavens, but each of them would have a different perspective on the position of the moon itself (Figure 4.45).

Here, the observer at the North Pole would say that the star is above the moon, while his counterpart at the opposite pole would say

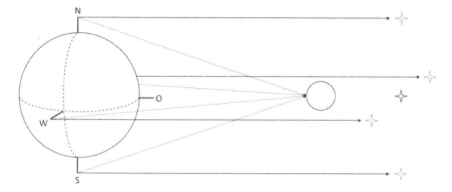

Figure 4.44. Five observers on the earth. At midnight at point O, the moon crosses directly in front of a certain star, directly overhead. At the other four positions, the observers agree on the position of the star, but say the moon is in a different relative position.

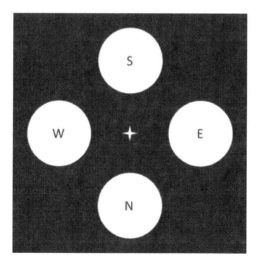

Figure 4.45. Four moons: The night sky as described by four observers located at extreme positions on the earth. The observer at, e.g., the North Pole would say the moon is in position N relative to our star, at an angle which defines the object's parallax.

that the reverse is true — and by the same amount! While, thus far, we have been referring to the concept of parallax in general, when astronomers refer to the parallax of an object, this is the precise measurement they mean. If an object covers a distant star at the equator, the angle which it appears to have moved at any point at the far edge of the earth — north, south, east, west, or anywhere in between — is its parallax.

Now, in order to find the distances to heavenly objects, we would need to draw a suitable diagram and describe some way of measuring angles between the fixed background — the stars — and the objects in question.

In Figure 4.46, we show what such an arrangement might look like. On top, you can see the same observers as before, one at the

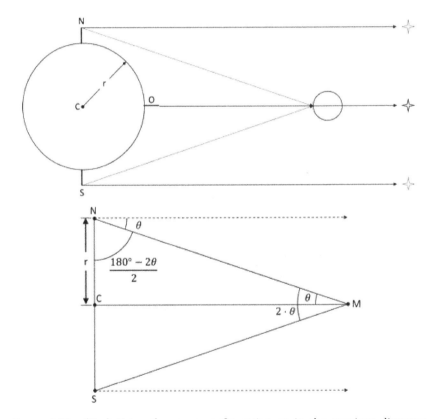

Figure 4.46. (Top) Using the same configuration as in the previous diagrams, observers at the North and South Poles measure the angle between a fixed star and the moon. (Bottom) This produces pair of triangles: the right triangle NCM with base r, the radius of the earth, or the isosceles triangle NSM with base $2r$, the diameter of the earth.

North Pole and another at the South Pole. (We have again marked observer O for reference.) N and S agree on the position of the reference star, and, at the same moment, measure the angle between the star and the center of the moon. These measurements together describe a unique triangle, one of whose sides is equal in length to the diameter of the earth. Using the properties of similar triangles, it is possible to calculate the ratio of the height of the triangle, CM, to the diameter of the earth, thus measuring the distance between the earth and the moon!

This is a greatly simplified version of the process used by Hipparchus (c. 190–c. 120 BCE) to measure the distance between the earth and the moon. In truth, the process we describe is impossible, for three reasons. First, our example assumes a heavenly configuration — with the moon directly over the equator, in front of a star in the same position — that is incredibly unlikely to occur in reality, and unlikely to be visible from all the positions we chose. Second, the positions of the observers are unrealistic. The Greek world was large, but not large enough to place observers at the three chosen points on the equator, much less the poles. Third and most importantly, it presumes the ability to tell time perfectly at multiple points on the earth, which is a much harder problem than it would seem.

How did Hipparchus measure the distance to the moon, then? The moon moves quickly, and its position (and apparent size) vary throughout the course of the month. If he had tried to make one measurement, traveled far enough across the earth to detect the moon's parallax, then made a second measurement, his calculations would have been riddled with errors. The most reliable way to make two measurements at the same time in different parts of the world is to make those measurements during an event which is indisputably simultaneous. The best candidate for this type of event is, of course, a solar eclipse.

In Figure 4.47 we lay out the most likely reconstruction of the method used by Hipparchus for measuring the distance to the moon. Point H represents Hellespont, part of modern day Turkey. Point A once again represents Alexandria, and the moon can be seen in the sky of either city at point M. The goal here is to solve the triangle

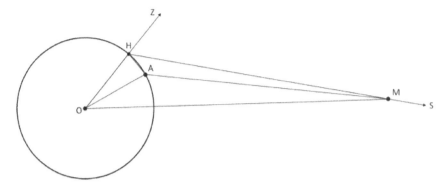

Figure 4.47.   Hipparchus' method for measuring the distance to the moon during a solar eclipse. Hellespont (H), Alexandria (A), the sky's zenith (Z), the moon (M), and the distant sun (S) are marked.

HAM — that is, to find the measure of all three angles and all three sides — and thereby measure the distance HM (or AM, which should be almost the same length).

Hipparchus' exact calculations are lost, and it is somewhat beyond the scope of this text to describe the precise methods available to him to solve the triangle. Nevertheless, to sate your curiosity, we can sketch out the beginning of his method.

Observers in both Hellespont and Alexandria recorded the extent of this eclipse. In Hellespont, the eclipse was total, represented with the ray S, connecting that city, the moon, and the sun. According to records, in Alexandria the moon covered the sun $\frac{4}{5}$ of totality. Conveniently for Hipparchus, and for us, both locations lie on the same meridian, and the latitudes of both cities are well-known.

In order to solve triangle HAM, you must find the measures of two angles and the length of one side. Angle AMH corresponds to the observed displacement of the moon at Alexandria, which we can calculate using the ratio in the previous paragraph. The length of the circular chord HA, the distance through the earth between the two cities, can be calculated almost directly from their latitudes. One more angle remains to be found, and an excellent candidate can be found in MHA, the angle between the moon and Alexandria, at Hellespont.

We will tackle these in reverse order. Thus, a line drawn from the center of the earth to the middle of the sky (the zenith, represented by the ray marked Z) can be broken down into three angles that sum to 180 degrees. The first, the angle ZHM between the zenith and the moon, would have been measured and recorded by astronomers. The third, the angle OHA between the center of the earth and Alexandria at Hellespont, can be computed using the cities' latitudes, below represented as λ_A and λ_H respectively. Note that the angle HOA at the center of the earth is equal to the difference of these latitudes. Further, recall (or convince yourself[21]) that all triangles drawn from the center of a circle to a chord are isosceles, and thus the measures of the remaining two angles, OHA and OAH, are equal. Then, the measure of angle OHA can be computed as follows:

$$\angle HOA = \lambda_H - \lambda_A$$
$$\frac{180° - \angle HOA}{2} = \angle OHA = \angle OAH$$

and

$$\angle ZHO = 180° = \angle ZHM + \angle MHA + \angle OHA$$
$$\angle MHA = 180° - \angle ZHM - \angle OHA$$

Having solved the triangle OHA, the second fact — the length of the distance HA — could be easily computed in terms of the earth's radius using a trigonometric table, one of Hipparchus' innovations.

Finally, the magnitude of the remaining angle, AMH, can be found with the tools and methods mentioned in the previous section. During this eclipse, the moon was the same size as the sun, and Hipparchus knew that the sun covers its own width 650 times per day. Or as we would say in more modern words, the apparent width of the moon is $\frac{1}{650}$ of 360°, or about 0.554°.

Since the moon eclipses the sun perfectly, they both have the same visual width, $\frac{1}{650}$ of a full circle. At Alexandria, the moon covered the

---

[21] The definition of an isosceles triangle is one where two legs are equal in length. A triangle drawn from the center of the circle has two legs of length r, the radius of that circle; thus all such triangles are isosceles.

sun $\frac{4}{5}$ of the way. Treating the sun as infinitely distant, you can say that the moon appeared to have moved $\frac{1}{5}$ of its diameter, or $\frac{1}{650} \cdot \frac{1}{5}$ of a full circle. To put this into more modern terms, the apparent parallax of the moon at Alexandria could be measured in degrees as follows:

$$\angle AMH = 360° \cdot \frac{1}{650} \cdot \frac{1}{5} = \frac{360°}{650 \cdot 5} \approx 0.11°$$

Having thus computed two angles, the third could be found by subtraction; with the length of but one side and much computation and use of his trigonometric tables, Hipparchus calculated the lengths of the remaining sides, and thus placed the moon at a distance of 71 times the earth's radius.[22]

If you know both the angular size of a heavenly object and its distance, it is a relatively straightforward process to then compute its size, assuming you have one more thing: a trigonometric table. As we will soon see, trigonometric tables relate the measures of angles and sides of known triangles, and by the properties of similar triangles, we can multiply a known triangle by a scaling factor to find the actual measure of a real-world triangle.

To bring this story to a close, then, assuming the moon is 71 earth radii away:

$$\sin\left(\frac{360°}{650 \cdot 5}\right) = \frac{r_{moon}}{71 \cdot r_{Earth}}$$

$$r_{moon} \approx 0.137 r_{Earth}$$

A universe of triangles

We hope by now that you are convinced that triangles, especially similar triangles, are useful tools for analyzing and solving problems.

---

[22] Hipparchus actually used two different methods to measure the distance to the moon. The latter is very finnicky and hard to demonstrate in the space we have available; using this method he found that the moon was between 59 and $67\frac{1}{3}$ earth radii away. (Today we know the value to be just over 60 earth radii.)

They can only take us so far on their own. Euclid tells us the rules by which triangles can be drawn, and how to relate their lengths and areas using ratios, but he is silent regarding specific applications.

Untethered from the prejudices against numbers that marked the Greeks of antiquity, the Alexandrian mathematicians made inexorable progress. But it took over 400 years from the founding of Alexandria for someone to come along and synthesize everything known at the time about geometry, arithmetic, astronomy, and cosmology, to build what would become the most significant work of practical astronomy in the history of the world.

Claudius Ptolemaeus, more commonly known as Ptolemy (c. 100– c. 170 CE), was a Greco-Roman astronomer and mathematician. He is not known for what he created, but rather how he refined and synthesized what came before him. He used Euclid's geometry, Aristotle's cosmology, Hipparchus' trigonometry and astronomy, Menelaus of Alexandria's star chart, Chaldean[23] numbers, and much, much more — not always as adeptly as those who came before him, but so singularly that his final product stood above all others.

Ptolemy's most well-known work was titled the *Mathematical Treatise*, or, later and more appropriately, *The Great Treatise*. That title was shortened in time to simply *The Greatest*, "megiste," which would in time and transmission through the Islamic world become the *Almagest*.

As flawed as we now know Ptolemy's cosmology to be, his inspiration and motivations are recognizable as ancestors of our own thinking about the nature of science. In the preface to his great work, he recognized and acknowledged the contributions of those thinkers that came before, while in turn admitting that the role of his mathematics was to link the real (numbers, measures, concrete observations) with the ideal (the unchanging cosmos, abstract geometry, and so on). His success — and failure — was in doing this so deftly that it would take centuries to correct his errors and re-shape the universe.

All this started, though, with a better understanding of numbers.

---

[23] That is, Babylonian.

Better numbers

You have probably wondered why, when so many other numerical systems are base-10, clocks and circles are based around the number 60.[24] The short answer is that this was the number system used by the Babylonians. As some of the earliest astronomers, their work influenced Ptolemy, who then influenced the world.

Why 60, though? One explanation that makes a good deal of sense is that 60 is highly divisible, which is an attractive quality when most problems would have been handled with pure fractions. Consider this: 60 can be divided by 10 (a natural human base), as well as 2, 3, 4, 5, 6, 12, 15, 20, and 30, and other common fractions (such as $\frac{2}{3}$ and $\frac{3}{4}$) can be treated quite easily.

The 360 degrees in a circle take a bit more explaining, but one theory suggests that the Babylonians saw that six equilateral triangles fit perfectly inside a circle, and if each triangle contributed 60 degrees then the circle would be divided cleanly, and into a nice number.

Whatever the reasons for their choices, Ptolemy recognized it as superior to the Greek decimal system[25] for performing the kinds of measurements and computations he would need to pin down the cosmos. The other feature that made this system superior was in the use of small parts — minutes — and second order small parts — seconds — that would enable Ptolemy to build his mathematics without recourse to clumsy fractions.

Here, we offer a brief primer on how this base-60 system — the sexagesimal system — is used to represent numbers. Whole numbers are relatively straightforward; we can write them in decimal without any conversion. Numbers smaller than one are written in base-60, separated from the whole-number part (in the modern style) by

---

[24] Especially if you live in a country that uses metric! Other than a brief period during the French Revolution, hours have always been divided into 60 minutes. And, unless you are a surveyor or a French artilleryman, you have probably never used the *gradian*, a unit of measure that divides the circle into four 100 gradian quadrants, for a total of 400 gradians in a revolution.

[25] A third system of Greek numbers, used after the Aegean and Attic numerals.

a semicolon, with further parts separated by commas. Thus, ten degrees, plus half a degree, plus a quarter of a minute, would be written as: 10;30,15. Higher precision numbers can be written by increasing the number of commas. Thus, you could write any number you needed with a string of base-60 "digits," such as 3;8,29,44,0,47... indefinitely. You might recognize this number more easily when written in decimal notation: 3.1415926535... It is the number pi!

We won't belabor this, but there are two connections you can easily make here. First, since minutes and seconds are used in the sexagesimal system in almost exactly the same way they are on a clock, if you imagine a circle as a 360-hour[26] clock, then the time since such a clock started running could be read off its face directly (although you might need a magnifying glass to make out the numbers).[27] Second, you've already seen how to convert numbers between bases in earlier chapters, and the numbers here are no exception! To convert between degrees, minutes, and seconds, and decimal degrees (as you might want to do when trying to find a location on an internet map, where decimal notation is common) you can use the procedures shown in those chapters.

In the modern age, we use sexagesimal numbers only for measuring circles, and only occasionally at that. But the sexagesimal system is *just* a number system; nothing prevents you from using base-60 numbers to measure straight lines, or to perform arbitrary calculations just as you currently perform these tasks in base-10, or as computers do in base-2. Ptolemy recognized this fact using base-60 numerals to measure both angles and lines throughout the *Almagest*.

---

[26] That is, 15 days.

[27] In our hypothetical 360 hour clock, each degree would correspond to one hour, each minute to one minute of arc, and each second to one second of arc.

To think about it in a different way, you could divide each of these numbers by fifteen to get an idea of how large an angle each unit of time is on a typical 12-hour clock. In this case, the hour hand sweeps $360/12 = 30$ degrees per hour, and thus 30 minutes of arc per minute, and 30 seconds of arc per second.

A note on sizes

Reading the rest of this discussion, readers may become confused as we make reference to certain numbers. "Here, say the line measures 60." Lacking units, you might wonder what "60" represents here. The answer is — it can represent any unit you wish!

A confusing consequence of similarity is that all measures remain proportional no matter which units are used. So, to use a common example, a 3-4-5 triangle is a right triangle no matter if the sides are measured in inches, feet, miles, or light-years — as long as the units are consistent!

We can, and do, circumvent the question of what units are being used in each of the following steps because the mathematics works no matter the units chosen. Indeed, there is a certain freedom in using so-called dimensionless units to perform calculations, since we can use whatever scale is convenient in each problem, or indeed at each *step* of a problem, as long as we ensure the ratios are expressed in common terms at the end.

Solving triangles

Here, we must pause for a moment to check your understanding. Explaining the next portion of Ptolemy's work requires that you understand, *truly understand*, how useful similar triangles can be. If there is any mathematical field more misunderstood and maligned than algebra, that field is trigonometry. In this section we intend to show you exactly why trigonometry is so useful. We won't be restricting this explanation to the methods and tools available in Ptolemy's time, but by the end of this section we hope to have shown you the idea at the heart of trigonometry, thereby demonstrating Ptolemy's motivation, and how it supports problem solving.

Earlier, we mentioned that *to solve a triangle* means finding the measures of all three sides and angles of a particular triangle. Thus, a process for solving triangles should allow you to construct any valid triangle and, using minimal information, discover each of the remaining values.

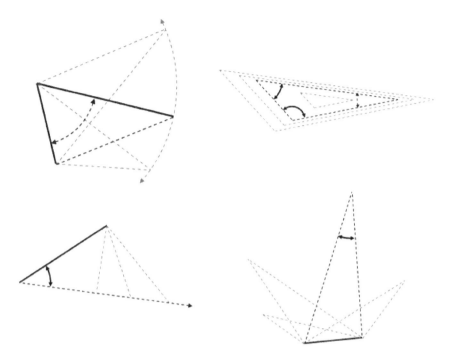

Figure 4.48.   Four cases with two pieces of given information in bold. (Top left) Two given sides can have any angle between them. (Top right) Two given angles yields the third angle, but the triangle can have any size. (Bottom left) A given angle and adjacent side can have a third point anywhere on an infinite line. (Bottom right) A given angle and opposite side can belong to many triangles.

What information do you need to determine a valid triangle? One piece of information — a point, an angle, or the measure of a side — is certainly not enough. Two pieces of information, whatever they are, are also not enough to produce a unique triangle. As you can see in Figure 4.48, two given sides can have any angle between them; two given angles defines a family of similar triangles; a side and an adjacent angle define two vertices, while the third vertex can be positioned anywhere on the line defined by these two; and given an angle and the side that subtends it, the angle can be placed in countless positions.

In general you need three independent pieces of information to completely solve a triangle, but that information could be the

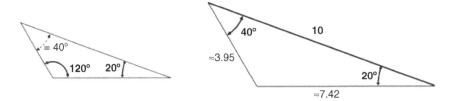

Figure 4.49.   (Left) Given two angles, the size of the third angle can be calculated, but this is not enough to specify one unique triangle. (Right) Given a side length, and a way to measure angles, a specific triangle can be drawn. The lengths of the remaining sides can then be measured.

measure of three sides, the measure of two angles and one side, or the measure of two sides and an angle.[28] The measures of the three angles aren't independent, since the values of the first two define the other. That is, if we were to provide the measure of two of a triangle's angles, you could find the third by applying the property that the sum of the angles must equal 180 degrees. However, if we were to subsequently tell you the length of any one of the three sides, you would be able to draw one specific triangle, and thereby measure the remaining two sides (Figure 4.49).

Remember, this is no mere theoretical question. Solving a triangle from two angles and a side is precisely what we demonstrated in Figure 4.47, when measuring the distance to the moon! When trying to model a real-world problem, it's often impractical to measure specified lengths and angles directly (e.g., the distance to the moon). In most cases, by measuring three other pieces of data related to the problem, it is possible to construct and then solve a triangle, thereby producing the desired value. So how can we simplify the process of solving triangles?

We can start with an example. In Figure 4.50 we provide an arbitrary triangle with three sides of specified length, and two triangles similar to the first one, with only one side marked on each. Can you find the lengths of the marked sides?

---

[28] There is an exception to this: two sides and an angle not between the two sides do not *uniquely* determine a triangle in certain circumstances, but they do define *one of two* triangles.

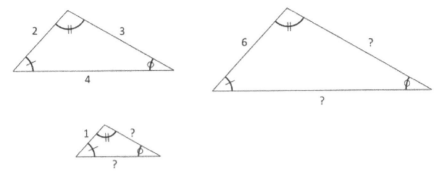

Figure 4.50.   Testing your belief. Given any number of similar triangles, the unknown side lengths can be calculated using proportions.

This is a relatively straightforward application of algebra. Typically, we could solve this triangle by setting up proportions and solving the ensuing equation algebraically.

In asserting that the triangles here are similar, and providing you with at least one side length, you are empowered to find the other lengths — despite knowing nothing about the angles! In producing valid answers, though, you have demonstrated *belief* that each of these three triangles are unique, and that the missing values could be found.

We propose the idea of a reference triangle, then. For any combination of angles, we can set the length of the longest side to a common value — say, 120 — and then somehow measure the lengths of the remaining sides. Repeating this process for every practical combination of angles, we could produce a book of such reference triangles. Then, when presented with a triangle having certain characteristics (sides in a particular ratio, for instance), it should be possible to flip through the book, find the triangle possessing those same traits, and solve the original triangle by scaling appropriately.

There are two problems with this proposal. If we were to use an interval of 1 degree, there would be 2700 different possible combinations of angles (178-1-1, 177-2-1, 176-3-1, 176-2-2, ...), which would be a prohibitively arduous undertaking. Further, there would be no easy way to measure the sides of each triangle, short of constructing each one. We will have to find a different way to characterize enough triangles to be useful, in a way that can be managed.

There are three types of angles, as shown in Figure 4.51. Given a random triangle ABC, angle B is either acute, obtuse, or right. If you drop an altitude from vertex A to the line extended from BC, then the altitude AD meets BC at one of three places. If angle B is acute, AD meets BC inside the triangle. If B is obtuse, AD meets BC outside the triangle. And if B forms a right angle, then D and B coincide.

In this way, we can say any triangle either is a right triangle, or can be broken up into the sum or difference of right triangles. So, if we were to find a way to pre-compute the characteristics of a set of right triangles, that reference would enable us to solve any triangle whatsoever.

Further, as you can see in Figure 4.52, an arbitrary polygon can be decomposed into triangles. For, in any polygon you can find three points such that the middle point is outside the line drawn between

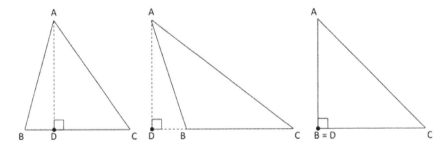

Figure 4.51. Given an arbitrary triangle and vertex, you can draw an altitude to the opposing side. In this way, every triangle is either a right triangle (right), or composed of the sum (left) or difference (center) of two right triangles.

Figure 4.52. A (non-convex) five-sided figure (left) can be broken into triangles, each of which can be solved independently.

the other two. Then, you can remove that point and consider the triangle so formed separately from the rest of the figure. If you keep repeating the same procedure, the main figure will lose one vertex after each step, until eventually you are left with nothing but triangles. Each of these can be solved as before — assuming, of course, that it's possible to solve arbitrary right triangles!

Ptolemy saw this, and understood that possessing a reference table of right triangles would greatly simplify his grand undertaking. Thus, he constructed a machine for building trigonometry.

Building a trigonometry machine

What would a trigonometry machine look like? There are many ways to construct right triangles. One would involve straight rods, marked at fixed intervals, that could be arranged in such a way that for any given angle the lengths of the other two sides could be measured. Ptolemy did, in fact, use such an instrument — illustrated in Figure 4.53 — but he did not use it to generate his tables. For, trying to use such an instrument to construct right triangles presupposes the ability to measure angles with the utmost precision, which is not a safe assumption to make.

To reduce the chance of errors in such a foundational reference, the best approach would be to use geometric proofs to construct ideal triangles whose sides could be computed exactly, and methods that would enable you to add, subtract, multiply, and divide ideal angles without ever having to deal with fickle and imprecise physical machinery.

Our goal, then, is to demonstrate how to measure the sides of specific triangles from basic principles, using the smallest number of arithmetic steps. To do so demands that we first produce several tools.

Trigonometry machine

Parts required:

- The Inscribed Angle Theorem.
- Thales' Theorem.

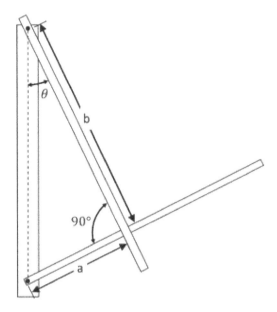

Figure 4.53.   One way to produce a trigonometry table. Two rods fixed at opposite ends of a common base can be rotated freely. If they are positioned at right angles to one another, then (for each angle θ), a and b can be measured and recorded.

- The Pythagorean Theorem. (Proof not included.)
- Four (4) precisely measured angles.
- Ptolemy's Theorem.

Beware! This is the longest demonstration of this book so far, and it will be quite easy to lose sight of our goal. As before, the purpose of putting this all together is to be able to generate right triangles with almost any angles, with which we can use proportionality to solve any triangle whatsoever.

Note, too, that we are not beginning here with first principles. Every mathematician who lived prior to the 19th century assumed his audience was familiar with Euclid, and Ptolemy was no exception. Here, we presume familiarity with some common principles, without proof, and otherwise hope that vague recollections of school geometry will allow you to believe in the remaining steps.

Prelude: Some reminders

The sum of the measures of the three angles of a triangle is equal to 180 degrees, as is the sum of the measures of angles that lie alongside one another with a common vertex along a line. These angles are called supplementary angles.

Triangles with two equal legs are called *isosceles*, and the two angles opposite these equal sides have equal measures.

By definition, an angle drawn from the center of a circle, called a *central angle*, has the same measure as the determined arc of the circle. For instance, a right angle drawn at the center of a circle measures 90 degrees of arc on the outside (Figure 4.54).

A line segment drawn through a circle, called a *chord*, touches the circle at exactly its two endpoints, and divides the circle into two arcs (Figure 4.55, left). We can make some observations about chords. First, radii can be drawn from the center of the circle to the points of intersection with the endpoints of the chord, thus producing a triangle. This triangle, as noted in footnote 21, will always be isosceles, as by definition all radii of a particular circle are equal in length.

Moreover, all equal chords of a given circle cut the circle into equal arcs. For instance, no matter how you choose to draw a chord of

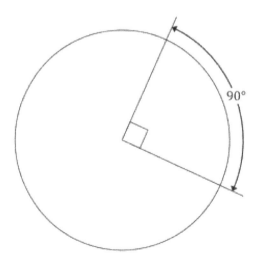

Figure 4.54.   The size of a central angle and the amount of arc cut off a circle are, by definition, equal.

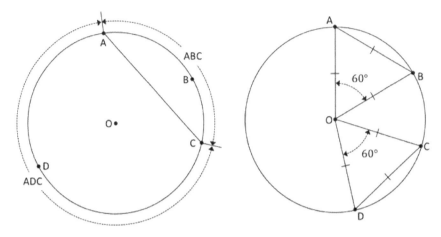

Figure 4.55. (Left) Chord AC cutting through the circle centered at O cuts two arcs: the short arc ABC and the longer arc ADC. (Right) A chord of a given length will always cut equal arcs off a circle of a given size, no matter its location.

length 1 in a circle with radius 1, it will always cut an arc measuring 60 degrees from the circle (Figure 4.55, right).

An *inscribed angle* is an angle drawn on an edge of a circle through its interior to the edges.

Tool 1: The Inscribed Angle Theorem

Here, the goal is to find a way of measuring inscribed angles.

Start with a circle containing the points A, B, and C, with center O. There are three cases, as illustrated in Figure 4.56: the angle CAB spans the center; one line goes through the center; or the entire angle sits to one side of the center. We address each of these three cases separately.

First, consider the case where one line of angle ABC coincides with a diameter of the circle. (Follow along by referring to Figure 4.57.) Without loss of generality,[29] suppose O lies on AB. Draw the radius OC,

---

[29] This is a canned phrase commonly seen in proofs. What it means to say is that we can choose one option or another without making the proof any less valid. Here, whether O lies on AB or BC doesn't change the validity of the proof, nor does our choice of the points' names.

Like Juliet said to Romeo, "A rose by any other name would smell as sweet."

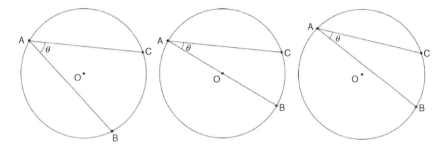

Figure 4.56. The Inscribed Angle Theorem, in three cases. We wish to describe the measure of angle θ, whether angle BAC spans the center O, intersects it, or lies completely to one side of it.

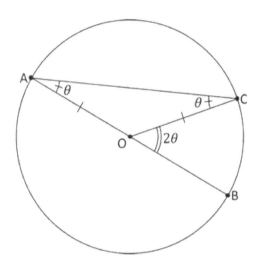

Figure 4.57. The case where AB intersects the center O.

producing isosceles triangle OAC. Note that angle BAC has the same measure as angle OAC, since they overlap each other.

The two angles OAC and OCA are equal to one another, and together with angle AOC they add up to 180 degrees. Angles AOC and COB also add up to 180 degrees, since they are supplementary. Since these two sums are equal and they share a term, the sum of angle OAC and angle OCA has the same measure as angle COB.

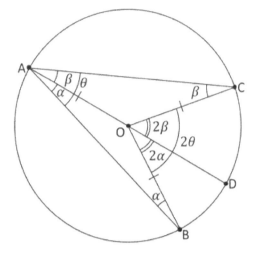

Figure 4.58. The case where angle BAC spans the center O can be solved by using the previous result.

Thus, the inscribed angle BAC is equal to one half of the central angle COB.

Next, consider the case where the center of the circle lies inside the inscribed angle (Figure 4.58). Then we can draw point D, the reflection of point A across the center O. AD is a diameter of the circle.

DAC and DAB are both inscribed angles with one line running through the center, which means we can apply the result we just proved. That is, the measure of angle DAC, marked β, is one half the measure of angle DOC, and angle DAB (marked α) is one half the measure of angle DOB.

But angle BAC is equal to the sum of angles DAC and DAB, and angle COB is equal to the sum of angles DOC and DOB, thus the inscribed angle BAC is equal to one half of the central angle COB.

Finally, consider the case where the center lies to one side of the inscribed angle. We can use the same method as in the second case, this time subtracting instead of adding the angles. Proceeding as before, you will find that once again angle BAC is equal to one-half angle COB (Figure 4.59).

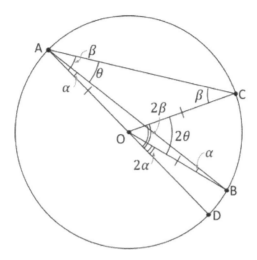

Figure 4.59. The case where angle BAC lies completely to the side of the center O requires subtracting angles instead of adding them.

These three cases are the only possibilities, and thus an inscribed angle has half the measure of the central angle that measures the same arc.

This is a widely useful theorem, but it is quite possible that you will miss some of its relevance. So, imagine that a circle has been drawn on a piece of wood, and two nails have been hammered into it in random positions, indicated as A and B in Figure 4.60. If you were to attach a rubber band to these two nails, no matter where you stretched the rubber band on the larger arc, the interior angle, and the length of the chord, would be the same. Further, if you were to add additional nails anywhere along that same arc and stretch multiple rubber bands around these new nails and the original two, all of these angles would have the same measure.

Note, here, that while we looked at points all along the longer arc, we didn't address the smaller arc between A and B. Can you guess how the angles there are related to the central angle?

The relationship there is the same as before, but here you must consider the *external* central angle — measuring greater than 180 degrees!

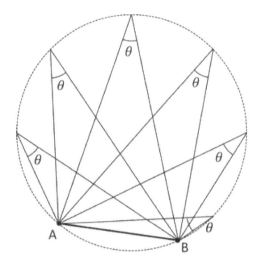

Figure 4.60. Given a fixed chord AB of a circle (bold), no matter where you stretch a rubber band on the larger arc the inscribed angle will have the same measure.

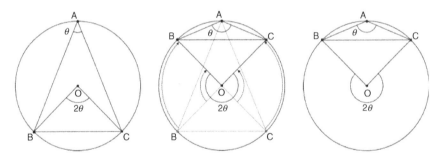

Figure 4.61. Inverting an inscribed angle. The Inscribed Angle Theorem still applied when considering an external angle (right), but you must alter your perspective to view it relative to the measure of the *external* central angle.

In Figure 4.61, imagine a chord BC and inscribed angle BAC such that angles BAO and CAO are equal. Now, imagine the radii OB and OC moving at equal speeds around the circle. Angle BAC grows wider and wider as points B and C move around the circle, until they get very close to point A, when angle BAC is almost equal to 180 degrees (that is, it is almost a straight line), and the central angle is almost 360 degrees. If you ever forget the Inscribed Angle Theorem, this visual might help you recall the principle: an inscribed angle can take any value between *almost* 0 degrees and *almost* 180 degrees, while the central angle can vary between 0 and 360 degrees.

Tool 2: Thales' Theorem

We have made a big deal about Thales' discovery of the Intercept Theorem, but the theorem most commonly associated with him is the one that bears his name.

Thales' Theorem states that an inscribed angle that spans a diameter is a right angle. We have the tools here to provide two separate proofs of this (Figure 4.62).

Proof 1: Note that angle BOC is a straight line, and thus measures 180 degrees. By the Inscribed Angle Theorem (Figure 4.63), angle BAC measures half this arc, or 90 degrees. ■

That's not very satisfying though, is it? The second proof is a bit longer, but ultimately richer.

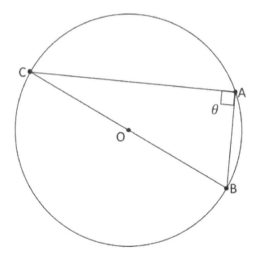

Figure 4.62.   Thales' Theorem. If an inscribed angle (here angle BAC) spans a diameter of the circle, then it is a right angle.

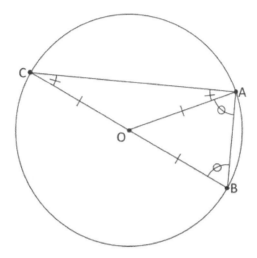

Figure 4.63. The classic proof of Thales' Theorem. OAC and OAB are isosceles triangles, and thus the measure of the angle at A must be equal to the sum of the angles B and C. The only way this is possible is if angle A is a right angle.

Proof 2: BC is a diameter of the circumcircle of triangle ABC, centered at point O, and A lies on some other point on the circle. We can draw OA, which, as a radius of the circle, is equal in length to OB and OC.

Triangles OAB and OAC are both isosceles. Therefore, angle OAB is equal to angle OBA, and angle OAC is equal to angle OCA.

The sum of angles OAB and OAC is equal to the measure of angle BAC, and is also equal to the sum of angles OBA and OCA. Since the sum of the angles of triangle ABC equals 180 degrees, angle BAC must be half that, or 90 degrees. ∎

Both of these proofs are equally valid, but Proof 2 is the more satisfying of the two. According to legend, the first time Thales proved this theorem, he sacrificed an ox to the gods in thanks! A good proof should be elegant and memorable, and this easily qualifies as both.

Tool 3: The Pythagorean Theorem

The Pythagorean Theorem is probably the most remembered relationship from school mathematics. We won't spend time proving it here, since it is no doubt familiar to you. Instead, we will use this space to connect it to the previous theorem.

By drawing a diameter, then connecting its endpoints to any point whatsoever on the circle's circumference, we have a means to generate all the right triangles we need. If there were a way to find two of that triangle's sides, finding the third would become a matter of arithmetic.

But we have a method of finding two sides, at least for some triangles! First, since you already know that the goal here is to use proportions to solve arbitrary triangles, you can set the length of the diameter to any convenient value! As we hinted before, Ptolemy chose to use the value 120 for the diameter of his circle, for two main reasons. First, 120 highly divisible, since it is a multiple of 60. Second, as you may recall, the Babylonians probably came up with the number 360 degrees by filling up the circle with equilateral triangles — each with a side length of 60! In Figure 4.64 you can see what this looks like, and you can spot three different diameters of the circle, each of length 120.

Choose one of those diameters, and one of the adjacent triangles. Using two of these points, you can draw a third point at the opposite end of the diameter, and thereby generate a triangle with two sides of

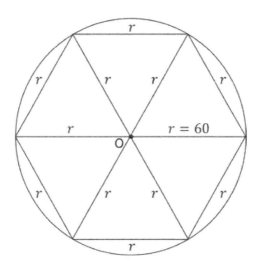

Figure 4.64.   A regular hexagon inscribed in a circle, and split into six equilateral triangles.

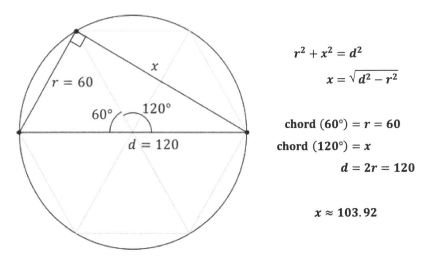

$$r^2 + x^2 = d^2$$

$$x = \sqrt{d^2 - r^2}$$

$$\text{chord}\ (60°) = r = 60$$

$$\text{chord}\ (120°) = x$$

$$d = 2r = 120$$

$$x \approx 103.92$$

Figure 4.65. Using the Pythagorean Theorem to find the values of supplementary chords.

known length (see Figure 4.65). The third side can be computed using — you guessed it — the Pythagorean Theorem!

This is the first and easiest-to-understand component of the trigonometry machine. No matter which chord length you compute, you can find the length of its supplementary chord, the chord connecting its end to the other end of the diameter, by applying the Pythagorean Theorem.

To get this machine running, we will need a few more precise angles, and some way to combine them together to produce more chord-angle pairs.

Tool 4: Four Precise Angles

The specific arrangement of six triangles used by the Babylonians anticipated a construction described by Euclid in the *Elements*. In their case, it was a convenient way to measure a circle, but to Greek geometry this construction was thought of as a method of inscribing a *regular*[30] *hexagon* in a circle.[31]

---

[30] In this case, *regular* means *equiangular* and *equilateral*; that is to say, a figure where all the angles are the same size and all the sides are the same length.

[31] Proposition 4.15.

We can inscribe a regular hexagon in any circle by first finding its diameter. Then, using our compass to measure the distance from one edge of the diameter to the center, we trace an arc that intersects the edges of the circle. Each of these points of intersection defines one of the vertices of the regular hexagon.

The method for inscribing a regular hexagon is perhaps the most well-known of the regular inscribed figures from Euclid's *Elements*, but it is far from the only one. Ptolemy uses three others — the square,[32] the regular pentagon,[33] and the regular decagon[34] — as bases for the rest of the trigonometry machine.

He was able to compute the length of a side of an inscribed square using nothing more than the Pythagorean Theorem, and the side lengths of the remaining two regular Euclidean shapes using only a little more effort.

The angles corresponding to each of these measures are, of course, the number of degrees in a circle divided by the number of sides of a polygon. This is because each side of an inscribed regular polygon is a chord of the circle that divides it into equal portions. As you've already seen, one side of the regular hexagon corresponds to the chord of 60 degrees. Further, one side of the square corresponds to a chord of 90 degrees; a pentagon, 72 degrees; and a decagon, 36 degrees.

Again using the Pythagorean Theorem, Ptolemy computed the lengths of the chords for the supplements of the pentagon and decagon, producing entries for angles measuring 108 and 144 degrees, respectively.

At this point, Ptolemy had seven measures in hand, but he needed one more tool to make get his trigonometry machine up and running.

Tool 5: Ptolemy's Theorem

Ptolemy used geometry extensively, but he didn't break a lot of new ground in the field, with one exception. His main workhorse for producing trigonometric tables was this, his eponymous theorem.

---

[32] Proposition 4.6.

[33] Proposition 4.11.

[34] Proposition 13.10. A decagon is a 10-sided polygon.

He first drew a random quadrilateral with all four vertices lying on a circle, which is known as a *cyclic quadrilateral*, along with its two diagonals. He then proposed to prove that the product of the diagonals is equal to the sum of the products of the two pairs of opposite sides.

On the left in Figure 4.66, you can see the general form of his theorem, which seems somewhat arbitrary and almost unrelated to the matter at hand. On the right, though, we re-draw the figure in a specific way, foreshadowing what's to come. First, though, the proof.

Using a construction from Euclid, we duplicate angle ABD on the line BC, thereby producing point E in such a way that angle ABD is equal to angle EBC.

By the Inscribed Angle Theorem, angles BCA and BDA span the same chord (segment AB) and thus are equal.

Since the triangles ABD and EBG have two angles equal, they must be similar, and similar triangles have sides in common ratios (Figure 4.67, left side). Thus $\frac{BC}{CE} = \frac{BD}{DA}$, and cross multiplying, you can see that BC = DA = BD · CE.

If angles ABD and EBC are equal, you can subtract the internal angle EBD from both and thus show that angles ABE and DBC must be

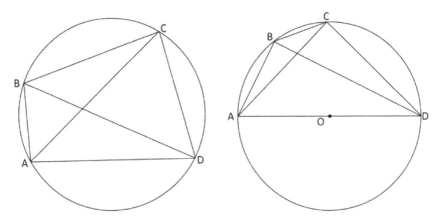

Figure 4.66.   (Left) The basis of Ptolemy's Theorem: a method of relating the lengths of the sides and diagonals in an arbitrary cyclic quadrilateral. (Right) How Ptolemy would end up using his theorem. This should remind you of previous diagrams.

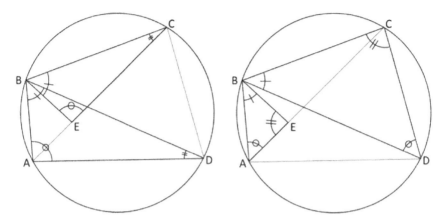

Figure 4.67.   Finding similar triangles by constructing point E such that angles ABD and EBC are equal.

equal. Again using the Inscribed Angle Theorem, angles EAB and CDB are equal; therefore, once again, triangles ABE and DBC must be simi-lar (Figure 4.67, right).

Thus $\frac{AB}{AE} = \frac{DB}{DC}$, and AB · DC = DB · AE.

The final connecting step takes some effort to see, so take a moment to name the segments in the two portions above. First, we demonstrated that the product of one pair of opposing sides, BC and AD, is equal to the product of the diagonal BD and some short seg-ment, CE. Next, we demonstrated that the product of the other pair of opposing sides, AB and DC, is equal to the product of the same diago-nal BD, and the remainder of the diagonal AC.

You can perform algebraic manipulations to show that the sum of these two products is equal to the product of the diagonals, that is: BD · CE + DB · AE = BD(CE + AE) = BD(AC). The Greeks preferred to think geometrically, though!

So instead, imagine a rectangle with sides BD and CE, and another with sides BD and EA (Figure 4.68). If you combine these together, do you believe the sum of their areas is equal to the product of the diago-nals? (And, do you see why the Greeks' perspective, viewing arithme-tic geometrically, had such staying power?)

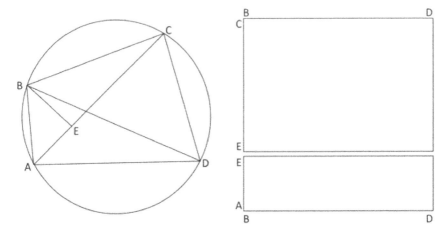

Figure 4.68. Multiplying and adding like a Greek. The rectangle formed by BD and CE can be added to the rectangle formed by BD and EA, yielding a rectangle equal to the one formed by BD and AC.

Then, we can add the results from the two previous equations, demonstrating that

$$AC \cdot BD = AB \cdot CD + BC \cdot DA$$

That is to say, the product of the diagonals of a cyclic quadrilateral is equal to the sum of the products of opposing sides.

If you were to stumble across Ptolemy's Theorem on your own, it might at first glance appear to be one of those rarefied theorems that mathematicians prove for their own ego, only to leave a mark on the field. If anything, it is the opposite.

Again, look at Ptolemy's circle. Draw any two known chords and their supplements, whose lengths we must know already (or can find via the Pythagorean Theorem). We know the length of the diameter of the circle because we have fixed its length at 120. Looking at Ptolemy's Theorem, the only unknown remaining is the length of the chord between the two original chords, which corresponds to the difference of their angles, and which can be calculated directly.

Choosing two chords at random, the chord of 60 degrees equals 60 and the chord of 72 degrees equals 70.5342 (which Ptolemy calculated by applying Euclid).

The supplement of 60, 120 degrees, corresponds to a chord of 103.9230, and the supplement of 72, 108 degrees, corresponds to a chord of 97.0820, both of which can be found by application of the Pythagorean Theorem:

$$a^2 + b^2 = c^2$$
$$60^2 + b^2 = 120^2$$
$$b^2 = 120^2 - 60^2$$
$$b \approx 103.9230$$

$$a^2 + b^2 = c^2$$
$$70.5342^2 + b^2 = 120^2$$
$$b^2 = 120^2 - 70.5342^2$$
$$b \approx 97.0820$$

Filling in numbers accordingly, we find

$$AC \cdot BD = AB \cdot CD + BC \cdot DA$$
$$\text{chord}(72°) \cdot \text{chord}(120°) = \text{chord}(60°) \cdot \text{chord}(108°) + x \cdot d$$
$$70.5342 \cdot 103.9230 = 60 \cdot 97.0820 + 120 \cdot x$$
$$x = 12.5434$$

In this case, x corresponds to the chord of 12 degrees. By the Pythagorean Theorem, we can find the chord of 168 degrees, and from there, repeat the process for other combinations of angles.

Using his theorem, Ptolemy identified two other minor results (Figure 4.69). The first allowed him to find the chord of half a given

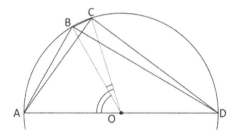

Figure 4.69.   Using Ptolemy's Theorem to find the measure of BC, corresponding to the chord of 12 degrees (angle BOC). Angles AOB and AOC are based on constructions of the hexagon and pentagon, respectively.

angle, and the second provided a way to find the chord of the sum of any two angles. The half-angle theorem allowed him to reduce 12 degrees to 6, then 3, then 1.5; the usefulness of the sum formula should be obvious.

By grinding away in this manner — adding, doubling, subtracting, and halving — Ptolemy's trigonometry machine was able to spit out a table of values measuring every chord between 0 and 180 degrees at half-degree intervals.

Fine measurements

Table 4.20 illustrates a portion of his final table, for a circle with diameter 120 and angles between ½ and 12 degrees.

The first two columns, marked "arc" and "chord," should be self-explanatory by now. The third column, marked "sixtieths," requires a bit of explanation.

Choose two chords of similar size at random, as in Figure 4.70. Here, instead of drawing chords rooted at the same point, as we did when applying Ptolemy's Theorem, we illustrate both centered on the same diameter. By drawing a line from point A to C, we can create a tiny triangle which approximates the amount of change between the two chords.[35] This is enlarged in the right-side of Figure 4.70.

For a large difference between AB and CD, this triangle is a very rough approximation of the difference between the two chords, but as they get closer to one another, the shaded triangle helps approximate chords of intermediate lengths more and more accurately.

This is the general principle Ptolemy applied to measure intermediate angles. The sixtieths column indicates how quickly a chord grows as you change the angle near an indicated value.

For instance, in Table 4.20 you can see that the length of the chord for an arc of 60 degrees is 60, and in the sixtieths column the value is 0;54,21. This value indicates the amount the chord length increases for each increase in the span of the arc by one sixtieth of one degree.

---

[35] Actually, half the change, since there would be another triangle on the other side of the circle between points B and D.

Table 4.20. A sample of Ptolemy's Chord Table, in both sexagesimal (base-60) and decimal.

Arc (degrees)	Sexagesimal		Decimal	
	Chord	Sixtieths	Chord	Sixtieths
½	0;31,25	1;2,50	0.5236	1.0472
1	1;2,50	1;2,50	1.0472	1.0472
1 ½	1;34,15	1;2,50	1.5708	1.0472
2	2;5,40	1;2,50	2.0944	1.0472
2 ½	2;37,4	1;2,48	2.6178	1.0467
3	3;8,28	1;2,48	3.1411	1.0467
3 ½	3;39,52	1;2,48	3.6644	1.0467
4	4;11,16	1;2,47	4.1878	1.0464
4 ½	4;42,40	1;2,47	4.7111	1.0464
5	5;14,4	1;2,46	5.2344	1.0461
5 ½	5;45,27	1;2,45	5.7575	1.0458
6	6;16,49	1;2,44	6.2803	1.0456
6 ½	6;48,11	1;2,43	6.8031	1.0453
7	7;19,33	1;2,42	7.3258	1.0450
7 ½	7;50,54	1;2,41	7.8483	1.0447
8	8;22,15	1;2,40	8.3708	1.0444
8 ½	8;53,35	1;2,39	8.8931	1.0442
9	9;24,54	1;2,38	9.4150	1.0439
9 ½	9;56,13	1;2,37	9.9369	1.0436
10	10;27,32	1;2,35	10.4589	1.0431
10 ½	10;58,49	1;2,33	10.9803	1.0425
11	11;30,5	1;2,32	11.5014	1.0422
11 ½	12;1,21	1;2,30	12.0225	1.0417
12	12;32,36	1;2,28	12.5433	1.0411
...				
60	60;0,0	0;54,21	60.0000	0.9058

For an arc one minute greater in size than 60 — 60;1,0 — the length of the chord would grow by 0;54,21 *minutes*, to 60;0,54,21. Using decimal values instead, we can say that for each *tenth of a degree*, the chord grows by 0.9058/10 = 0.09058. Thus, the chord for

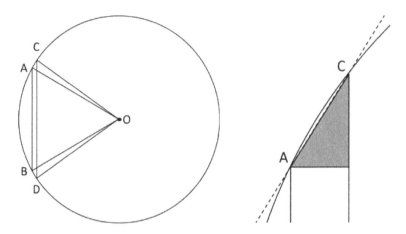

Figure 4.70. Starting with two chords of angles of similar measure (left), we can approximate values between them by joining them along a small triangle (right). For small changes in angle, the measure of the corresponding chord approximates this triangle.

Table 4.21. Interpolating from 60 to 60.5 degrees by adding 0.09058 repeatedly.

Angle	Chord
60.0°	60.0
60.1°	60.09058
60.2°	60.18116
60.3°	60.27174
60.4°	60.36232
60.5°	60.4529

an arc of 60.1 degrees is 60.09058. We can approximate the other intermediate values between 60.0 and 60.5 degrees by adding constant multiples of 0.09058 as seen in Table 4.21.

Similarly, this process can be applied in reverse. Again using decimals, if Ptolemy measured a chord of length 60.2, he could approximate the measure of the unknown angle by performing a little arithmetic.

If we start with a value measured from the sky and subtract from it the chord that is closest to it in the reference table, we find a

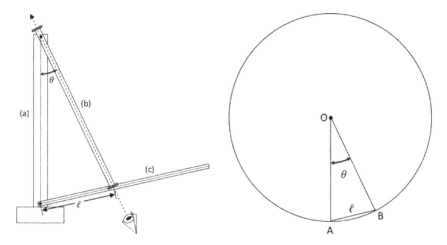

Figure 4.71.   Ptolemy's method of measuring angles in the sky.

difference. Then, dividing that difference by the number of "sixtieths" from the table, we discover a *change* in angle from the reference angle. We can add that change to the reference angle to approximate the value of the original angle.

$$\text{chord}\left(\angle X\right) = 60.2$$
$$60.2 - 60 = 0.2$$
$$\frac{0.2}{0.9058} \approx 0.2208$$
$$m\angle X \approx 60° + 0.2208° = 60.2208°$$

This technique is called *linear interpolation*. The "linear" part of the name refers of course to the manner of approximating curves using straight lines and triangles, which are far more easy to work with while producing results that are good enough for most purposes. While it is not well known to most people today, until the invention of the calculator it was an essential part of every student's toolkit.[36]

Figure 4.71 illustrates one device Ptolemy used for making angle measurements of celestial bodies. (This should look familiar, as we presented a similar version of this in Figure 4.53.) Part (a) is a fixed

---

[36] The idea of approximating curves using straight lines and triangles is also a central concept in calculus.

vertical rod, carefully positioned so as to be at exact right angles to the earth. At either end of this rod, similar rods (b) and (c) are fixed at pivots in such a way that an observer can sight an object along (b) and set the center of (c) at the exact center of the line of sight along rod (b), thus producing the geometrical configuration on the right in Figure 4.71.

Rod (c) is carefully marked from 0 to 60, allowing the observer to measure precisely the length ℓ of the chord along this rod. By referring back to a table of chords, and using the method of linear interpolation, the angle θ can be computed with a great deal of accuracy.

How accurately, though? In the discussion accompanying Figure 4.53, we mentioned that constructing a trigonometry table with such a device could be error prone. To reduce the chance of errors, Ptolemy recommended that this device be large — at least 4 cubits in height (that is, around 6 feet, or nearly 2 meters)!

Assuming rod (c) is six feet in length and marked from 0 to 60, each foot would be divided into 10 parts. Each of these should, in turn, be divided as finely as possible. But, to divide each part into 60 minutes would require producing 600 marks per foot! In comparison, rulers are typically marked at $\frac{1}{16}$ inch intervals, which works out to 196 marks per foot. To measure angles at one minute accuracy (that is, one sixtieth of a degree, denoted in sexagesimal as 0;01,00) with the same facility as a ruler, Ptolemy's device would have to be over 18 feet tall! Moreover, keep in mind he was performing most of his measurements at night while watching the stars, exacerbating his difficulties.

Nevertheless, he was able to make accurate measurements with this device, by combining his trigonometry table with the method of linear interpolation.

Practical trigonometry

If there is a single theme to our discussion of Ptolemy's achievements, it concerns the application of practical observations and measurements to drive the development of mathematics. This is the difference between theory and practice; and the difference between mathematics as a goal in itself versus useful mathematics. If you sit in an otherwise empty room and think about philosophy and mathematics, any system you dream up might seem reasonable. Once you look outside and begin to observe how mathematics fits (or does not fit) the real world, all bets are off.

Certain facts about the cosmos were widely known in Ptolemy's time. First, it was well known that the distance to the moon is not fixed, nor is its speed across the sky. By measuring the size of the moon (with, for instance, a diopter) one could approximate the difference in distance between the moon's nearest point (call the *perigee*) and its farthest point (the *apogee*).

Second, by observing the motions of the planets relative to the fixed stars across several nights, it appeared that they would slow down and reverse direction for a period of days before resuming their normal courses, a phenomenon known as *retrograde motion.*

How could these facts be reconciled with a belief in the uniform circular movement of the planets? We can consider the process step-by-step, illustrating each one in Figure 4.72.

The top left diagram illustrates what the moon's orbit (M) would look like if it were uniform, circular, and centered on the earth (E). We already know that the moon's orbit is not centered on the earth, and that is it possible to measure its size at different points in time. By comparing the size of the moon at its closest and farthest points, it would be possible to specify an offset between the true center of the moon's orbit (O) and earth's position, as illustrated at the top right.

This change is not sufficient to explain the observed changes in the moon's speed over time, but if you imagine that the moon orbits a point C that in turn orbits the center O (middle left), then when the moon is farthest away from the earth it would appear to move slower, speeding up as it approaches perigee. Remember, the Greeks thought that circular motions were aesthetically pleasing, and thus this model of circles on circles (called *epicycles*) seemed to be the best possible compromise between a belief in a well-ordered cosmos and the astronomical observations made by people like Ptolemy.

The beauty of this system is that it could potentially explain the apparent retrograde motions of the planets. Think about the planet

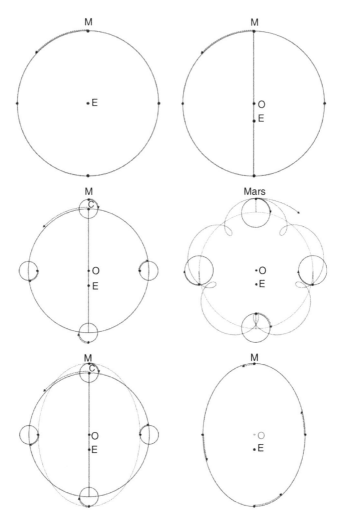

Figure 4.72.   Different models of orbits according to the ancient Greeks. (Top left) The moon, or Mars, orbits the Earth in a circle. (Top right) Over time, it became clear that heavenly bodies were not all at a uniform distance from the Earth, which could be accounted for if the Earth were slightly off center. (Middle left) Sometimes, planets seem to move backward in the sky. The Greeks accounted for this by saying that the planets circled around a center C, which circled the earth. (Middle right) This model described very effectively the apparently motion of the planets, as seen from the Earth. (Bottom left) If we use a similar model for the Earth and the Moon, and then trace the Moon's path in the epicycle model, it traces out a path that should be familiar to every amateur astronomer: the elliptical orbit of the Copernican model (Bottom right).

Mars: if, instead of making one turn about its epicycle per orbit, it made multiple turns, its retrograde motion would be explained quite simply! You can trace the motion of that planet in the middle right diagram of Figure 4.72; these combined motions produce a remarkably lovely graph.

There is a punchline to all of this, of course. Ptolemy's system wouldn't have persisted as long as it did if it weren't somewhat accurate. At the bottom left in Figure 4.72 we have overlaid the path the moon would take according to Ptolemy's model, while at the bottom right we have removed everything but this path and the position of the earth itself. If you have ever studied astronomy or physics, this diagram should appear remarkably familiar.

After all this work, it is only fair that we demonstrate an application of trigonometry on a practical problem. In Figure 4.73 we have reproduced a section of the orbital arrangement of the earth and moon according to Ptolemy, naming the respective points as we did in Figure 4.72. His goal was to calculate the relative positions of the earth and moon in order to predict heavenly events, such as solar eclipses. To achieve this goal, he would need to produce another table of values, and thus set out to determine the distance between the earth and the moon when the latter was at any point along its epicycle.

Using better techniques and instruments than his predecessors, Ptolemy was able to measure the mean distance to the moon (that is, the length of line EC) more accurately than before. Further, he had

Figure 4.73.   A close up view of one section of the moon's orbit, using the theory of epicycles.

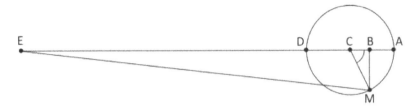

Figure 4.74. By building several right triangles, the moon's precise distance and motion at any point in time can be predicted.

already determined that the ratio between the radius EC and the radius CM was 60: 5.25 — again, working in his preferred base, 60.

Here, his goal was to find the measure of EM, the distance between the earth and the moon, in terms of values he already knew. You can see the right triangle EBM in Figure 4.74; here we describe how he went about constructing this figure. First, he extended line EC through to the far edge of the moon's epicycle and marked points A and D, then drew the radius between the moon's epicycle and the moon itself. Finally, needing right triangles to perform his calculations, he dropped line BM perpendicular to line AE.

To produce a table of distances to the moon, he would need to go through the same kind of laborious work used to produce his table of chords, performing hundreds of calculations. So, this section merely demonstrates a method of computation, and is by no means exhaustive. However, knowing that, we are free to choose any convenient value for the measure of angle ACM, and as we have seen, 60 degrees is quite convenient!

Pretend for a moment that CM measures 120. Since BM is a right angle, a circle drawn around triangle BCM would have CM as a diameter (by Thales' Theorem). We already computed the respective chords for a triangle of these proportions (cf. Figure 4.65), and thus we can say that if CM = 120, then CB = 60 and BM = 103.92.

CM and CA are radii of the same circle, and earlier we noted that EC: CM = 60: 5.25. We can use proportions to find the ratio of the values in the previous paragraph to the value of radius EC. That is, by scaling BCM down until CM is equal to 5.25, we can find the lengths of CB and BM in terms of EC.

Thus, we have the following proportions:

$$\frac{5.25}{120} = \frac{CB}{60}$$

$$\frac{5.25}{120} = \frac{BM}{103.92}$$

Therefore, if EC = 60 and CM = 5.25, then BM = 4.055 and CB = 2.63.

Finally, adding CB to EC, we know that the longer leg of triangle EBM, and we have calculated the value of BM directly. We can now use the Pythagorean Theorem to solve for the value of EM; you can verify that this value is approximately 62.80.

But, 62.80 what? Once again, Ptolemy wasn't primarily concerned with absolute measures, so this dimensionless value was more than adequate to indicate the distance from the earth to the moon *relative to* the median value. By setting the median value to his favorite base, 60, he could easily see that at this angle, in this configuration, the moon would be more than its median value away.

Perhaps an easier way to grasp this idea is to think in terms of percentages. In these terms, the distance EC from the earth to the center of the moon's epicycle is 100%. The radius CM measures $\frac{5.25}{60} = 8.75\%$; this implies that the distance from the earth to the moon varies between 91.25% and 108.75% of its median value. Finally, at the chosen demonstration angle, the moon must be at a distance of 104.67%.

We hope you can see how practical needs drove the development of mathematics of all kinds. Ptolemy's astronomy was by no means the final word in astronomy, but it was *good enough* that it stood as the high water mark for astronomy for over 1000 years.

His work found its way to Islamic mathematicians, who refined his methods. They also expanded on it to help solve problems that were of particular importance to faithful Muslims, such as how to find Mecca from any point on earth, at any time of the year.

As Europe began to emerge from the long night of the Middle Ages, it re-learned classical mathematics via the Arab world, and learned for the first time new fields of math, including decimal arithmetic and algebra. European scientists and mathematicians would eventually apply these learnings to more accurately model the heavens, enabling travel and trade around the world. Today, many things we take for granted — GPS, global trade, and our economic system as a whole — find their roots in these *important* and *useful* problems.

In these pages you have witnessed the development of three fields: geometry, trigonometry, and the beginnings of algebra. Each of these branches of mathematics arose in response to definite needs — even when that need was mere curiosity! — and in turn made solving difficult problems much easier.

Further, you have seen how to prove mathematical theorems using three different methods. We proved Thales' Theorem via *direct proof*, the Inscribed Angle Theorem using *proof by exhaustion* (that is, enumerating all possible cases and proving each one individually), and the irrationality of the square root of 2 using *proof by contradiction*.

This is yet another example of how applying names to things can help you think about them. You have seen how thinking about the names of numbers differently helps to create better notations, and how defining terms empowers geometry to prove more and more complex results. Extend this idea, and you can begin to name methods themselves, as above, and begin to ask questions about what else exists. What other kinds of proof might there be? Are there any primitive objects in geometry that no one has yet bothered to name? Are there other types of geometry? How might they be useful? Might there be a better number system, beyond the ones you're familiar with, that would shed light on new and useful properties of numbers heretofore unknown?

It might seem that we've taken an easy way out by filling a chapter with the broadest categories of mathematics to demonstrate how important and useful mathematical problem solving can be. Regardless, we hope you have begun to develop a deeper appreciation for these topics. And, if you understood the underlying theme of the preceding pages, and have begun to ask questions like those in the previous paragraph on your own, you are already halfway to understanding how to solve any mathematical problem.

Chapter 5

Visual Mathematical Problems

Picture This: The Purpose of Visual Problem Solving

There is no limit to the variety of problems that can be expressed in mathematical language. When solving a problem, the first challenge we face is translating it into an appropriate mathematical representation. This can be a significant hurdle, especially when dealing with unfamiliar problems. Mathematics is not just one language; an effective solution can be algebraic, geometrical, or something else entirely, and the best solution to a particular problem might rely on a completely *ad hoc* approach. In the most general sense, mathematics deals with objects and rules, but the objects we consider, and the rules we apply, can be anything at all.

Mathematical problem solving has a long and rich history, so oftentimes the rules and objects we learn about have been already studied in great detail. Problem solving would be an easy feat if it simply demanded the memorization of these discoveries! However, the spirit of problem solving is best understood not through studying famous historical accomplishments, but rather by taking a step back and considering what they all have in common. When exploring a problem that would not yield to existing approaches, the great mathematical minds of the past made connections to new ideas. Indeed, we have already followed in their footsteps: in previous chapters, we invented rules and definitions that were useful to us *in that particular context*. Euclid, Descartes, Gauss, and others are famous today because many of their discoveries happened to have broad applications, but creating new mathematics is not solely the domain of the brilliant. This is something that each of us must do, moment by moment, when solving problems.

Problem solving requires a minimum of three steps: translating the problem into a mathematical form, applying appropriate rules, and translating the result back into something that has meaning in the original context. During each of these phases, you must be able to justify (to yourself, at least) why your chosen model represents the problem, and why each action you perform is allowable.

While the previous paragraph is a gross simplification, it highlights two important steps with which learners often struggle. A mathematical model should mirror the structure of the underlying problem so that as we perform each step, it tells us something about the thing we actually care about. By contrast, when we don't understand either the problem or the mathematics, such as when we apply rules by rote, we are like the blind men discussing the elephant.

According to the parable of the blind men and the elephant, there was a town in which six blind beggars lived. One day, they heard that a tame elephant had come to town. The beggars, having no idea what an elephant was, decided to go experience it for themselves. Each one took a turn touching the wondrous creature. Later that day, they all shared their impressions.

The first beggar claimed that the elephant is like a snake: muscular and sinuous. The second disagreed; he said that the elephant is like a tree trunk, thick and straight. The third knew they were both wrong, for an elephant is sharp and pointed, like a spear. Each one, having felt a different part of the beast, knew the others had it all wrong: an elephant is flat and wide, like a wall; an elephant is bristled, like a broom; an elephant is flat and thin, like a leaf.

Of course, we know that each of the beggars knew the truth, but only part of it. The trunk of an elephant is like a snake, the legs are like tree trunks, and an elephant's tusks are sharp like spears. An elephant's body is flat and wide, like a wall; its tail is like a broom; and its ears are like large leaves.

It is unlike any other animal, and defies description by analogy to anything other than itself. If you look at a picture of an elephant, it's instantly clear what each beggar was describing.

Only when you have a clear mental image of an elephant do the six different descriptions fit together into a logical whole!

In mathematics, we often "feel" different parts of a problem using words, formulas, tables, and so forth. Each view can shed light on a particular aspect of the problem, but these views are complementary: no one perspective can tell us everything we might need to know to understand a problem deeply. In these cases, a picture can often help make all the other pieces we know come together, producing a deep, holistic understanding of the problem!

This chapter is about translations to and from mathematical models, and the techniques we can use to make sense of complicated ideas. Mathematics is the stuff of thought, so it's often difficult to understand when a translation *has* taken place if our only guide is a change in notation. When we convert an idea to or from a picture, though, the change is undeniable!

The problems we explore will help make you *cognizant* of the mathematical translations that you can perform, whether they be between similar models (for instance, from an algebraic formula to a geometrical graph), or between models at differing levels of complexity and abstraction.

Maps and Territories

Maps are a common type of visual representation. They range from highly representative (for instance, aerial photos of terrain), to schematics that emphasize political boundaries, roads, waterways, or any other kind of data. Mapping is a natural way to present data because it leverages human senses — in this case, visual and spatial cues — to communicate information.

Unless you're an astronaut, you have never seen your entire home country all at once. You have certainly seen maps of it, though! When thinking about where you live, you probably mentally picture one of these maps. The scientist and philosopher Alfred Korzybski famously remarked:

> *A map is not the territory it represents, but, if correct, it has a similar structure to the territory, which accounts for its usefulness.*

Mathematical models are like maps in that they are abstract representations of the things (or places) they depict. But if maps are not the places they represent, then what does that say about mathematics? To put it bluntly, "all mathematical models are wrong." A model never represents every facet of a problem — but this is by design!

Figure 5.1 presents a topographical map of the Contiguous United States. Is this what you think of when you think of the US? What does it show? What is missing? How is it useful?

Figure 5.1. A topographical map of the lower 48 United States. Is this how you visualize America? What information is present? What is missing?

This picture does not show internal political boundaries, nor does it show roads, or population information. And yet, to someone unfamiliar with the United States who nevertheless knows how to interpret maps, it says a lot! You can plainly see mountains, plains, and waterways across the country, and if you are skilled at making inferences, you can begin to guess where farms and population centers are located. This map is not the country itself, but it tells you certain information about it.

In Figure 5.2, we've removed all details from the previous map, leaving only the silhouette of the lower 48 states. What can this map tell you? Certainly, it contains strictly less information content than the previous figure, and yet this map is useful for one specific task!

If you wanted to know the geographic center of the country, you could print a map and cut its shape out of a piece of cardboard. Then, by hanging the shape from its edge, as in Figure 5.3, you can draw a line straight down from where the string meets the cardboard. After repeating this process several times, the lines you draw will intersect at the center of the country!

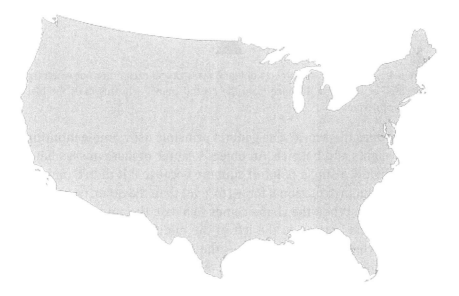

Figure 5.2. The silhouette of the Contiguous United States. Is this the simplest possible version of the previous map?

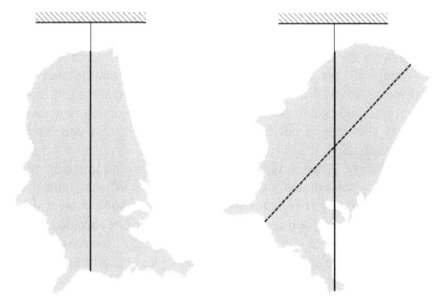

Figure 5.3.   Hanging a cutout of the United States to find its geographical center.

Figure 5.4.   Intuitively, two objects of equal mass cancel each other out when they are located at the same distance from the central point — in this case, the black triangular fulcrum.

Why does this work? The general principle uses simple intuitions about weights and balance. An object's center of mass always hangs directly below a single point of support because if it didn't, one side would exert more rotational force (*torque*) than the other, causing the shape to turn. When the shape comes to a rest, the amount of torque exerted by the portion to the left of the middle line must equal the torque exerted by the right half — the definition of the center! And since it is the center, once you locate the intersection point, you could place it on the tip of a nail, and it should balance completely!

Figure 5.4 illustrates this principle using the idea of a balance. There are two weights of equal mass resting on a long bar

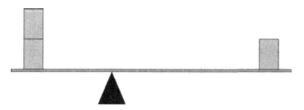

Figure 5.5.   If the mass on one side is doubled, then the fulcrum must move closer to that side to re-establish equilibrium, demonstrating that the system's center of mass has moved left.

(of insignificant mass). Your intuition should tell you that the center of mass of this system is at the point where the fulcrum sits, exactly halfway between the two weights.

If we add weight to one side, then the center of mass shifts toward that side, as shown in Figure 5.5. The single weight to the right exerts enough force to counteract the two weights on the left, because it lies twice as far away from the fulcrum.

Similarly, one square mile near Bangor, Maine (1600 miles from the center of the U.S.) "weighs" as much as two square miles near the Grand Canyon (800 miles from the center, but in the opposite direction).

You can prove this using calculus, and find the same center using complex integral calculations. If your data and computations are accurate enough, you can find a result that is far more precise than the intrinsic inaccuracies of scissors and pen will allow. Or, you can do what we have demonstrated here, and get a result that is good enough to satisfy your curiosity!

Why would you choose to do this? Beyond mere curiosity, there are several good motivations. First, geographical centers often become tourist destinations; small towns with a defensible claim of being at the center of the country can attract welcome visitors.

Second, consider the meaning of a geographical center: it is the point that has the lowest average distance to every other point. That is, if people selected from random locations throughout the United States were to fly to the center of the country, the average distance they would have to travel would be minimized. Placing regional capitals or other common resources near the center of a territory can defuse claims of bias. Famously, State College, Pennsylvania, the home of Penn State University, is located near the geographical center of that state. Since the school is funded by the state for the benefit of all residents of Pennsylvania, it's only fair that it be located at the center!

Third and perhaps most relevant to our discussion of mathematical models, this technique can be applied to many different kinds of data, not just physical maps!

In Figure 5.6, we have produced an arbitrary statistical distribution. (Perhaps it represents the average age of visitors to a museum that's popular with school groups: the peak to the left would then represent children between the ages of 13 and 17, and the smaller peak to the right would represent their teachers, who are typically in their 30s and 40s.) As you can see, the distribution is resting on a fulcrum like the one in Figure 5.5 showing that the left and right sides of the graph are perfectly balanced. The point where the fulcrum meets the graph is its center of mass, just as you would expect. Curiously, you already know the name of this point — this center of mass is the *expected value* you learned about in Chapter 4!

We can also combine statistics with geography. In Figure 5.7, you can see a map of the United States that only displays population. Each dot represents 7,500 people. If you could figure out a way to print this map such that each dot had the same weight, but the spaces between

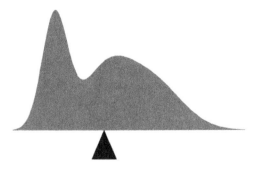

Figure 5.6. A hypothetical statistical distribution balanced on a fulcrum. If a distribution's center of mass — its expected value — is too complicated to calculate directly, you can find it by balancing the distribution on a point!

Figure 5.7. A "map" of the lower 48 United States showing only population. (Each dot represents 7,500 people.) The gray star (to the right of center) represents this map's center of population. Is this still a map even though the borders are missing? (Modified from US Census Bureau data.)

the dots had none, you could repeat the earlier experiment to locate the geographical center of population. This center (the star located near southern Missouri) is arguably fairer than the geographical center of the country, because it is the location with the shortest average distance to each *person*, rather than each piece of land.

In other words, if we wanted to move the United States Congress to a location that is fair to more people, Figure 5.3 shows where we might want to place the Senate (since each state has the same number of senators) while Figure 5.7 shows where we would want to place the House of Representatives!

Did you notice anything peculiar about the previous figure? We removed even the silhouette of the country, but nevertheless were able to talk about the map as if it were still there! The density of dots (as well as your memory and ability to match patterns) probably allowed you to "fill in" the white space where the map should be, and yet it's clear that there is no "map" *per se*!

These three examples all exist at a similar level of abstraction. In each case, we have removed all but two pieces of information: the mass and relative location of many points.[1] Once that was done, we were able to describe the figures using the same principle: that of center of mass! The last step in each case was to translate the mathematical solution into a meaning that makes sense in the original context, and it is at this point that the problems diverged.

This example demonstrates the spirit of this chapter, and indeed of all mathematical abstraction. Much too often, students learn how to manipulate concepts like this without internalizing that any model, if it is useful, has a similar structure to the thing it represents. Mathematics gives us common rules to discuss problems that appear superficially different, but it is always our duty to discover the meaning of the answers we discover.

Bending the Rules

If you have spent any time studying maps, you know that not all are alike. We stated that all maps are wrong, but your vision of the world is no doubt shaped by the maps you have seen. It is impossible to map the sphere of the earth to a flat piece of paper without distorting some information, and thus depending on what type of map you consider "best," your perspective of the world will differ from that of others.

Each type of map, called a projection, presents the same information flattened in a different manner. Perhaps the most well-known, and most maligned, projection is called the Mercator Projection (Figure 5.8). Under this transformation, each line of latitude is stretched to the same width, resulting in a great deal of distortion far from the equator.

One of the desirable properties of the Mercator Projection is that it preserves angles, meaning that any angle measured between two points corresponds to the same bearing read from a compass. On a

---

[1] This is an abstraction in itself, for in two of the previous examples — the first map and the statistical distribution — it might not be obvious that a solid shape can be thought of as a collection of points!

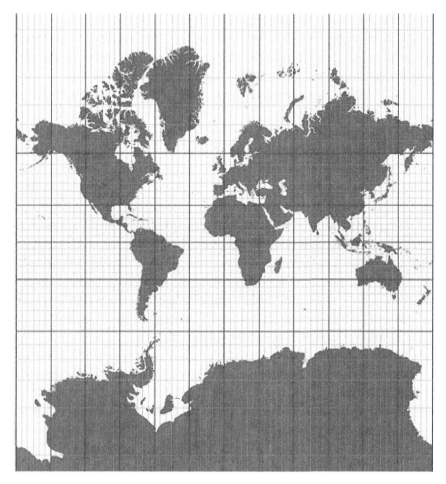

Figure 5.8.   A Mercator Projection map of the Earth. This projection is useful for navigation, since compass directions correspond to straight lines drawn across the map.

Mercator Projection map, north is always up, south is always down, left is always east, and right is always west. You can trace a straight line between any two points and measure the compass bearing between them using a protractor.

This property causes an unfortunate side-effect, though: it results in vertical stretching at the poles. The end result of this distortion is that Greenland is displayed as having the same area as the entire continent of Africa, despite being only 1/15 of its size. Humans tend to perceive larger things as more important, so this distortion skews peoples' perspectives about the size and importance of less equatorial regions, such as the United States, Europe, and Australia.

Nevertheless, the Mercator Projection has a long history, and continues to be used to this day. Why? There are two main reasons. First, it was the dominant projection used during the age of exploration due to the aforementioned property of preserving angles. Sailors could navigate from point to point at sea by reading compass directions directly off their maps, avoiding some cumbersome trigonometric conversions.

Second, the Mercator Projection maps the earth to a rectangle. If you truncate it at well-chosen latitudes, it can be reduced to a square that covers most of the inhabited regions of the globe. With the rise of computer mapping, it is useful to have all map data fit inside a square for the purpose of display and storage: each tile of the world map can be retrieved and displayed on a user's screen without having to perform computationally expensive transformations.

There are countless projections and each has its own downsides. Some projections preserve particular distances, such as those along meridians, making it easier to compare those distances at-a-glance. Others (such as the Gall–Peters, shown in Figure 5.9) preserve areas, guaranteeing that the relative area of every country is the same, even as it distorts local shapes and distances. Most projections strike a balance between these and other factors, including aesthetics, to meet the demands of the task at hand.

In each case, the skin of the globe is stretched and twisted to play a role, and to make answering other questions easier. No one map can answer any question, but some maps can answer specific questions very well.

The lesson for us here is that maps are pliable. Given a particular map, we can do more than add dots for population or remove political outlines, as we did in Figure 5.7 — we can change anything we need to, especially the shape of the map itself!

In mathematical representations, there is no "one true" visualization. Different "maps" tell us different things about the problem, and

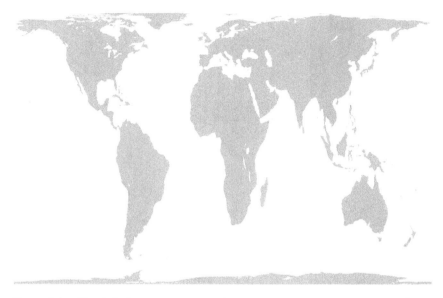

Figure 5.9.   The Gall–Peters projection. This projection preserves measures of area around the world, so that Africa is correctly shown to be fourteen times larger than Greenland. However, shapes on this map are distorted.

each representation has its own particular advantages and disadvantages. If a map has useful properties, we can use those to our advantage to solve problems!

To illustrate this principle, we will start with a few "straightforward" problems.

A tale of two cities

There are two cities on opposite sides of a river, shown in Figure 5.10. You want to connect them using the shortest road possible, with one twist: for structural reasons, the bridge must cross the river perpendicular to its banks. What route should the road take?

Normally, the shortest path between two points is a straight line. Given the perpendicular restriction on the bridge, it might seem that this is no longer true. Imagine, though, that this map has been printed on a loose sheet of paper. By folding the paper to remove the river, like

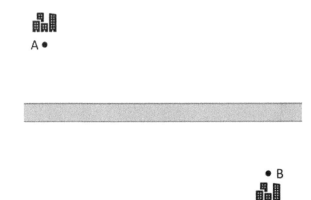

Figure 5.10.   Two cities separated by a river. What is the shortest road connecting these two cities, given that the bridge must cross the river perpendicularly?

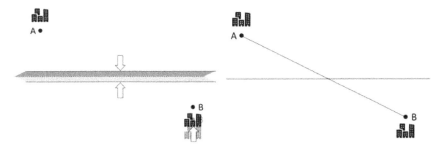

Figure 5.11.   (Left) We can ignore the river at first by "folding" it away. This has the effect of moving B closer to A, although the distance between B and the riverbank remains the same. (Right) Once the river is out of sight, the answer leaps out: The shortest road between A and B is a straight line!

in Figure 5.11, we can ignore it momentarily. (This also has the effect of moving city B closer to city A.) We no longer need to consider the bridge, and we can connect the two cities with a straight road!

Once we unfold the sheet, we can connect the two lines with a bridge (Figure 5.12), producing a road that follows the shortest possible path!

The "trick" we used here might seem unnatural, but compared to "normal" maps, this alteration is actually quite small! Critically, we restored the map to its original shape before completing the solution.

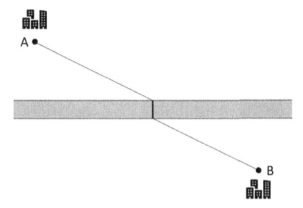

Figure 5.12. After unfolding the map, we can connect the two half-roads — one between A and the river, and the other between the river and B — via a bridge that satisfies the original constraint.

There are countless ways to draw maps, but most world maps use common visual idioms. Find any map of the world: north is probably up, and south is down; and since it is impossible to map a sphere to a square without breaking it, there are discontinuities along the edges. Most commonly, these breaks occur in the middle of the Pacific Ocean and at the poles.

Sticking with these standards saves cartographers time, and makes it easier to everyday readers to understand the information the map is trying to communicate. Occasionally, these standards create confusion. For instance, the shortest route between New York and Beijing flies almost directly over the North Pole, but there is no easy way to draw this path on an average map. If you ask a child to draw a path between the cities, he very well might draw a straight line connecting them, regardless of the map's distortions!

There is more than one way to cut the globe. Instead of cutting through the Pacific Ocean and along the poles, what if we chose a starting city — New York, in this case — and poked a hole in the globe at the point opposite. Then, we could stretch the hole out until the entire map could be laid flat, with the chosen city at the center, as in Figure 5.13. In this case, we actually *can* connect the starting city to any destination using a straight line!

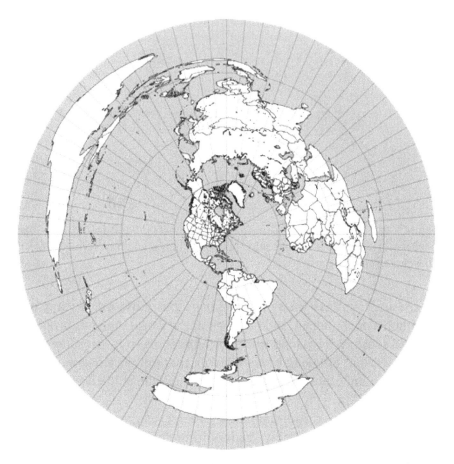

Figure 5.13. A map in azimuthal equidistant projection, centered on New York City. This kind of map is useful for finding angles and distances from a fixed location, such as an airport or a radio antenna.

This map can help us answer another question. Which US state is closest to Africa?

If you look at a map in a typical schoolbook, you might be inclined to say it's Florida. But look again at Figure 5.13 it appears that Maine is actually closer to Africa than Florida! Clearly, different visual representations can make answering some problems easy.

When faced with problems, it is often helpful to look at things from a different angle — either metaphorically or literally. The next two problems explore this idea.

The spider and the fly

A standard shipping container is 10 feet tall, 8 feet wide, and 40 feet long. In an otherwise empty container, there are two creatures: a spider, which is centered horizontally 1 foot off the floor on one wall, and a fly, centered 1 foot from the ceiling on the opposite wall (Figure 5.14; the near wall has been rendered transparent for clarity).

The spider can crawl on any surface and in any direction, but cannot jump or fly. What is the shortest path it can take to catch the fly?

Your first guess might be a path something like the one shown in Figure 5.15. You can add up the lengths and find that this path is 50 feet long: 9 feet up one wall, 40 feet across the ceiling, and 1 foot down to the fly.

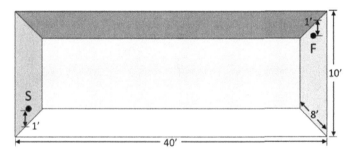

Figure 5.14. A standard shipping container, containing a fly and a spider. The spider is at point S, centered 1 foot up from the floor; the fly is at point F, centered and 1 foot below the ceiling. (Figure not to scale.)

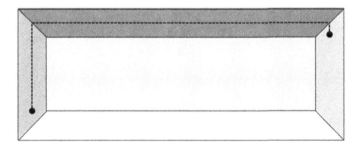

Figure 5.15. One path the spider could take to catch the fly. This path is easy to see, and its length is easy to compute, which should make you suspect that maybe there is a shorter path.

But you know that this is probably not the shortest route, especially if you think back to the lessons of Chapter 3! The fact that we're asking this question at all suggests that there *must* be a shorter path; the question is, how can we find it?

We will start by labeling the five visible surfaces with appropriate letters in Figure 5.16. The wall currently occupied by the spider is labeled S; the one occupied by the fly is of course labeled F. The three remaining surfaces are labeled A, B, and C.

Now, we can cut along the edges of the container and flatten it out, as illustrated in Figure 5.17. Notice how we have duplicated wall F in two locations, but curiously, they are oriented in two different directions. In the lower position (illustrated using dashed lines), the fly is

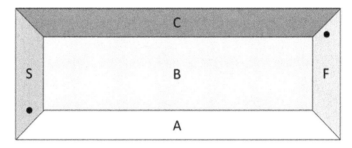

Figure 5.16.   Labeling the five visible walls to keep track of their relative positions as we disassemble the container.

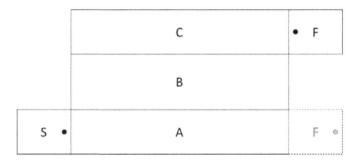

Figure 5.17.   The same space, unfolded and drawn to scale. When unfolding the container, you should pay attention to the location of the spider and the fly. The spider is 1 foot away from the junction between side S and side A, while the fly is far from the floor, 1 foot away from the corner where side F and side C meet.

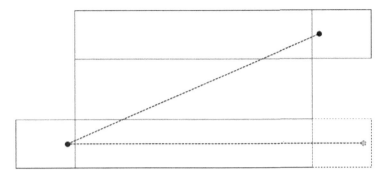

Figure 5.18.   Drawing both paths under consideration between the spider and the fly. You should be able to see that the diagonal path is shorter, by nearly 10 percent!

far from the floor, labeled A. In the upper position, though, the fly is close to the ceiling C. You can confirm that this makes sense by examining the previous figure!

Already, it appears that there is a shorter path than the one we identified first, but we should confirm that it is indeed shorter.

In Figure 5.18 you can compare these two possible paths. If you had any lingering doubts that the diagonal route is shorter, this should resolve them!

Nevertheless, finding the length of this route will make the answer unambiguous. According to Figure 5.19, the spider must climb 1 foot down, 40 feet across the container, and then 1 foot *down* from the ceiling to catch the fly, for a total of 42 feet. At the same time, it covers half the width of the container, two times; and it travels the container's entire height, yielding a distance of 18 feet. We can use the Pythagorean Theorem to demonstrate that this path is about 45.7 feet long, which is more than 4 feet shorter than our first guess![2]

In Figure 5.20 we trace the route taken by the spider inside the container, once again folded into its original shape. Do you think you would have ever guessed that this is the shortest path between those

---

[2] These are not the only two possible routes. For instance, a third type of route takes the spider across four surfaces, for a total of 46.84 feet. Can you find it?

What other ways can you find to unfold the container and connect the spider and the fly?

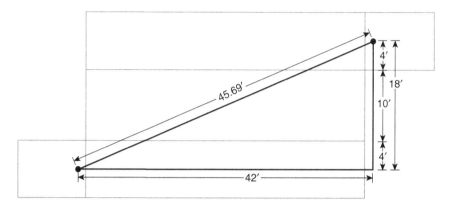

Figure 5.19. Using the Pythagorean Theorem to calculate the distance traveled by the spider.

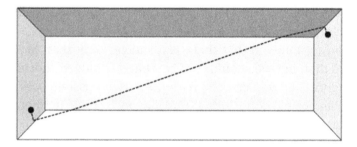

Figure 5.20. The shortest path between the spider and the fly loops around several walls. This is also the path that would be followed by a rope strung outside the container and pulled taut.

two points? Do you think you would be able to calculate this distance without unfolding the box? By unfolding the box in the right way, the problem becomes very "straightforward"!

Finally, if you have trouble imagining a situation where this knowledge might be useful, plug in a corded vacuum cleaner and move it as far away from the outlet as possible. As the cord snags on objects and the corners of walls, it will follow a path that looks remarkably like this. Ropes, cables, and cords under load try to follow the shortest possible path, just like our hypothetical spider; and while the shortest path might be mysterious in three dimensions, it's quite clear in two!

A trick shot

At Ralph's Mini-Golf, there is a special course set up as shown in Figure 5.21. Customers can win a free round by hitting a hole-in-one that bounces off all four walls before it goes in. You are a big fan of mini golf and love a challenge. Using your knowledge of mathematics, can you figure out the angle at which you need to hit the ball to make the shot?

It might not be obvious yet, but this problem is very similar to the one we just solved. We will make one change to the previous figure to help you think about the answer; in Figure 5.22 we have placed the logo for Ralph's Mini-Golf in the center of the course.

How does this help? To make the shot, you will have to bounce the ball off the walls. Each time the ball meets a wall, it will reflect off at

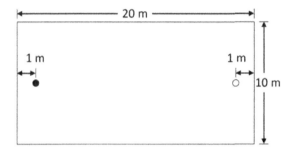

Figure 5.21. The dimensions of the miniature golf course used for our trick shot. The black dot on the left is the tee and the white circle on the right is the hole.

Figure 5.22. The same miniature golf course, with the letter R printed in the center. What's special about this letter?

the same angle it had when it made contact. If the ball meets the wall at an angle of 22º, it will reflect back at an angle of 22º.

There are two ways to think about a bounce. You can view it either as the ball reflecting off the wall (the solid line in Figure 5.22), or as the ball continuing straight through the wall into *a reflection* of the course, as shown with a half-dashed line in Figure 5.23.

To make the trick shot, you will need to hit the ball in such a way that it is reflected twice horizontally and twice vertically, corresponding to the four bounces. We use the letter R — which is neither horizontally nor vertically symmetrical — to help us identify that we've returned to the "real" course, which will happen after four flips.

If we flip the course repeatedly in every direction, covering the plane like floor tiles, then we can see by examination where our candidate targets are, as shown in Figure 5.24.

Here, the golf ball starts out in the gray course in the center — the "real" course — and our target is any one of the holes on the "virtual"

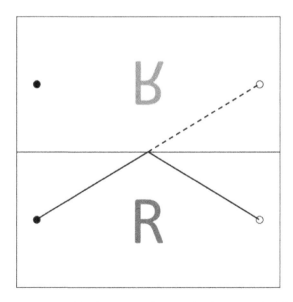

Figure 5.23. Two ways of looking at a shot. In the first way (solid line) the ball bounces off the wall and enters the hole. But this is equivalent to the ball passing through the wall into a course that has been reflected across the wall, like a mirror!

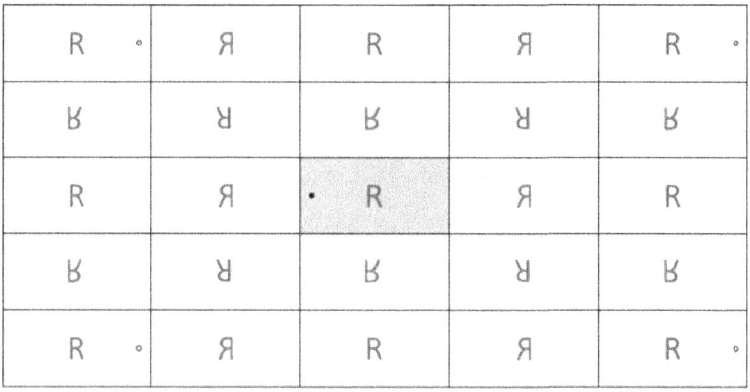

R ○	Я	R	Я	R ○
Я	Я	Я	Я	Я
R	Я	• R	Я	R
Я	Я	Я	Я	Я
R ○	Я	R	Я	R ○

Figure 5.24.   Starting from the tee in the center (shaded), a successful trick shot must cross two horizontal and two vertical lines. Each time the ball passes through a wall (bounces), we trace the reflection by reversing the orientation of the course, as indicated by the Rs. The only valid targets are the holes in the courses in the corners, where the Rs have returned to their correct orientation.

courses that are displaced twice horizontally and vertically from the starting point. Once again, this is because a reflection *within* the real course is mathematically equivalent to traveling *through* a reflected version of the course.

This figure is vertically symmetrical, so the holes on the bottom are at a similar angle to the holes at the top. We need only consider two cases, then — one to the left and one to the right — to find all possible solutions.

In Figure 5.25, we trace both of these shots both across our grid of reflected courses to find the shortest path. Next to each diagram, we show what path the two golf balls would take in the real world.

We can use the same method we used earlier to measure the horizontal and vertical distances traveled. Here, one ball moves 20 m vertically and 58 m horizontally; the other moves 20 m vertically but only 22 m horizontally.

In each case we can calculate the angle of the shot using the inverse tangent function. If we say that the line joining the tee and

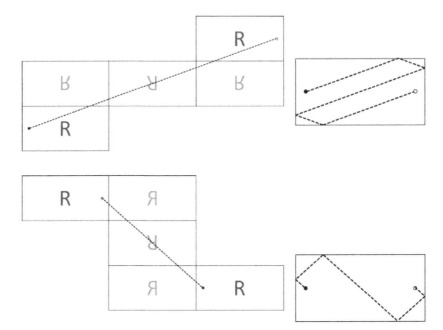

Figure 5.25.   Two trick shots across reflected courses and their corresponding paths in the real world.

hole measures 0 degrees, then the two shots are hit at the angles shown in Figure 5.26.

Note, here, that we ignored the signs in each of these two equations. It should be clear from looking at Figure 5.26 that the first shot is made *toward* the hole, while the second shot needs to be made *away from* the hole. That is, if the first shot must be made at an angle of 19.02°, the second should be hit at $(180° - 42.27°) = 137.73°$!

If you simply plugged the numbers into your calculator without considering the mappings above, you might be confused when your shot missed the hole by a mile — demonstrating why we must take care to think about what our answers are telling us!

So, the answer to this problem is that there are two angles at which we can make a trick shot that bounces off all four walls once and only once: 19.02° and 137.73°.

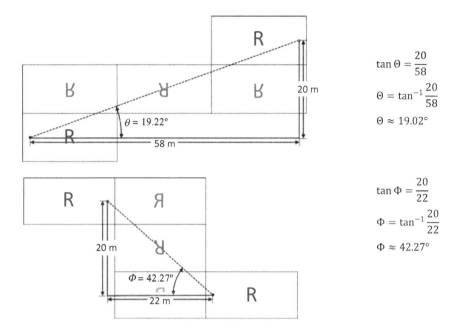

Figure 5.26. Calculating the angles of the two trick shots using trigonometry.

Making Connections

Mapping information geographically makes intuitive sense, but not all maps need to bear a close correspondence to geography to be useful. We are free to manipulate and transform data until nothing inessential remains. Of course, what is "essential" may not always be clear at first!

Each of the problems we have considered so far in this chapter have involved lengths and distances in some way. But when we don't care how far away things are, we can even toss out that information!

In Figure 5.27 we once again present the lower 48 United States, this time also displaying the individual states' borders.

Let's begin by making this map a bit more abstract.

Figure 5.28 presents the same map, this time overlaid with a *graph* showing how the states are connected. If the only kind of graph

Figure 5.27.   A map of the lower 48 United States, with each state's outline shown.

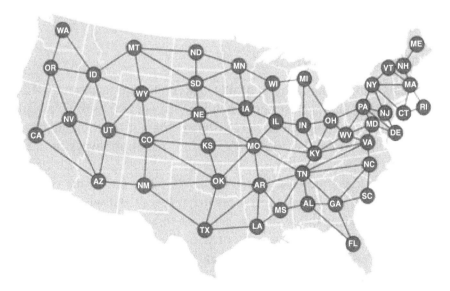

Figure 5.28.   Each state can be represented by a circular *node*, containing that state's postal code for clarity; and multiple *edges*, representing the shared borders, or connections, between states.

you have seen before are x–y plots, the use of the word graph here might seem confusing. This is a different type of graph that happens to use the same term (much to the chagrin of students), but in time you should be able to easily distinguish between the two from context alone.

The distinguishing features of this type of graph are *nodes* (or *vertices*)[3] and *edges*. If these terms are unfamiliar, think of "nodes" as dots, and "edges" as lines that connect two dots.

The nodes here represent states, and are labeled with the states' respective two-letter postal code; the edges are the lines representing shared land borders.

We are going to investigate properties of the connections between states, so there is no need to include the background map. Now liberated of geographical constraints, we are free to layout the nodes to improve readability, moving the eastern states farther apart and the western ones closer together (Figure 5.29).

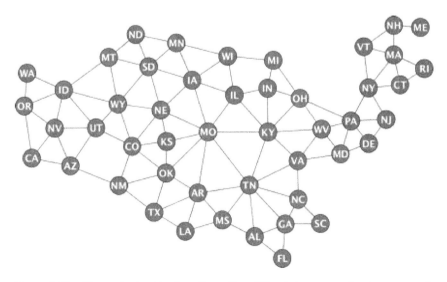

Figure 5.29.   Unconstrained by the silhouettes of the states, we are free to move each state's node to improve readability.

---

[3] In general, we will use "node" to refer to vertices that have some other information, such as postal codes, associated with them.

Already, we can see new information emerging. There are two highly connected states that have eight neighbors apiece: Missouri (MO) and Tennessee (TN). Meanwhile, the least connected state, Maine (ME), has only one neighbor.

Let's say we wanted to answer a simple yes or no question: is it possible to travel through all 48 of these states and return to our starting point without traveling through any state twice? You should be able to quickly see that this just isn't possible. Maine *only* connects to New Hampshire, so if you started there, there would be no way to return to Maine without passing back through New Hampshire! Furthermore, New York creates a bottleneck: there is no way to pass from New England to the rest of the US without passing through New York. So, even ignoring Maine, the problem is impossible from the start.

Consider a slightly different question. Is it possible to travel through each of the lower 48 states once in a single trip? Surely it is; there are countless routes you can take, as you can demonstrate by tracing your finger along the lines of Figure 5.29. This would be much more difficult to see, just by looking at the map from Figure 5.27!

In Chapter 1 we mentioned a problem called the *Four-Color Problem*. As a reminder, that problem asks if it is possible to color a map with only four colors, such that no two neighboring regions are colored the same. The kind of maps considered by this problem are simplified compared to the real world, where boundaries can be quite complex — just look at Michigan's Upper Peninsula for an example of this! The Upper Peninsula is separate from the Lower Peninsula, but the two parts together make up the state of Michigan.

The Four-Color Problem considers idealized maps, where countries and states don't have disconnected regions.[4] Happily, even though we include Michigan, the political boundaries illustrated in Figure 5.29 are simple enough to demonstrate this problem.

---

[4] Examples of such disconnected regions include the Kaliningrad Oblast, which touches Poland and Lithuania but not Russia, to which it belongs; French Guiana, an overseas department of France located in South America; and, of course, Alaska. Political maps can be as complicated as history itself, while the Four-Color Problem is concerned with idealized maps in which all countries are self-contained.

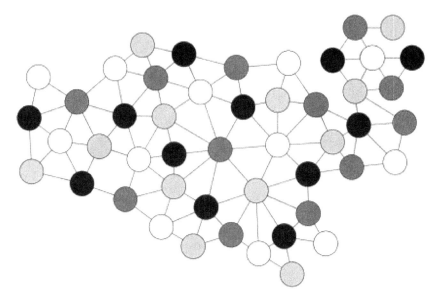

Figure 5.30.   A demonstration that the lower 48 states can be colored using no more than four colors. Each shade is used to color exactly 12 states.

In Figure 5.30 we have colored each of the "states" with one of four colors. This is not a particularly difficult task (there are many solutions!), but it illustrates a further transformation of earlier maps to help us answer a question. It is, in fact, possible to color a map of the United States with only four colors. Here, we have even constrained ourselves in another way: there are exactly 12 states of each color!

If you have never seen this problem before, you might wonder if it's possible to color the same map using only three colors. It turns out this is not possible. To prove this, you need not look at the entire map — just one well-chosen sub-section will suffice!

The left side of Figure 5.31 displays the smallest possible counter-example to this three-color hypothesis. If the node in the center is colored black, we can color the node above it white, and the node to the bottom left gray... leaving no other color for the bottom left node!

There's no part of our map of the United States which is in this exact configuration, but on the right of Figure 5.31 we have

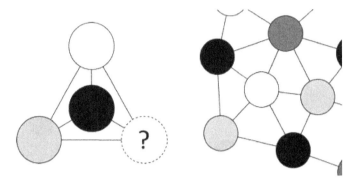

Figure 5.31   (Left) The simplest possible example of why you cannot color many graphs using three colors. (Right) The state of Nevada, in white, is surrounded by an odd number of states. There is no way to color this portion of the graph using only three colors showing that the map as a whole requires *at least* four colors.

highlighted a section of the map that has an odd number of states in a ring around another state. The same principle applies to both of these graphs: once we color the center node, we need at least three more colors to color the nodes in the ring. Therefore, there is no way to color either of these graphs with only three colors.

In short, if a node (or state) is connected to an *odd* number of nodes which all connect to each other in a loop, that area of the map requires *at least* four colors, and thus the map as a whole requires at least four colors.[5]

We will make two final transformations to demonstrate the limits of this approach. First, we will move the states to points on a uniform grid, then we will remove the states completely (Figure 5.32).

You might wonder if this final graph is still the same as the original, or if it has any use — and rightfully so! First, the answer is — yes! The last graph presents the same connectivity information as Figure 5.27, despite the fact that these two are completely different at

---

[5] This is not the only possible criterion that can force a graph to require four colors. It is intended to demonstrate that there are planar graphs that are *not* three-colorable.

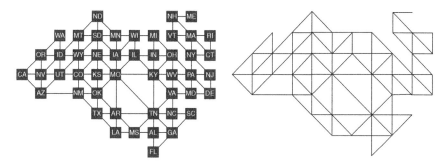

Figure 5.32: Further abstracting the previous graphs of the lower 48 states produces equivalent graphs that almost look like abstract art.

first glance. The nodes are still present as before, in the form of the vertices at the intersections between edges. They're not as easy to recognize, but they're still there!

Also, none of the earlier maps or graphs could be easily represented using integral coordinates, but these two could. This can be handy in certain circumstances, like when storing data in a spreadsheet. The graph on the right in Figure 5.32 also surfaces a characteristic of graphs like these which may have been difficult to perceive in earlier iterations: the triangular (or in some cases, quadrilateral) regions between states, called *faces* in graph theory. Depending on what questions you want answered, regions like these can be very useful.[6]

Here, we are using successive transformations to demonstrate the principle of abstracting one type of information to a minimal form. The advantage of performing these types of transformations to and from different kinds of problems is that when you find a solution to one type of problem, it can be used to solve another problem, even when at first they appear superficially different.

Let's see how in the following problem.

---

[6] Here, the faces indicate regions where multiple states share a boundary. It's a bit of a stretch, but one circumstance where you might want to know about these sorts of three- or four-state borders would be when negotiating water rights along a river.

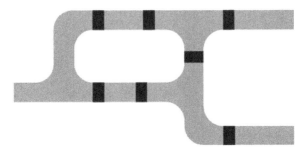

Figure 5.33. The city of Königsberg in the early 18th century. The Pregel River contained two islands, and there were seven bridges connecting the four bodies of land.

The bridges of Königsberg

The origins of graph theory are relatively modern. During the early 18th century, in the city of Königsberg, Prussia,[7] there were seven bridges that crossed the Pregel River (Figure 5.33).

Residents of the city would stroll around the city on Sunday afternoons, crossing the bridges, and idly wonder if there was any way to cross each bridge once, at the end returning to where they started. This problem was explored and puzzled over for several decades, until it caught the attention of Leonhard Euler.

Euler was the first person to point out that the shape of the city, and the precise route taken within it, doesn't matter. He boiled down the problem to its essence: in the city, there are four landmasses, separated by water. The landmasses can be replaced with nodes and the bridges can be represented using edges connecting nodes. But the nodes themselves don't particularly matter, only their degree of connectivity, so we can remove the labels to simplify the diagram.

A complete loop that crosses every bridge will look the same no matter where you start. That is, were you to begin at one vertex and exit by traveling some edge, then whenever you completed a full loop, you would enter the initial vertex via another edge. Similarly, during the tour, every arrival at a vertex must be paired with a departure. No matter where you begin your tour, there must be an even number

---

[7] Modern-day Kaliningrad, Russia.

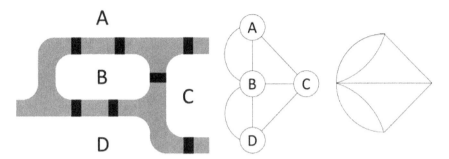

Figure 5.34.   Three ways of viewing the same information. (Left) The same map as before, with labels on each body of land. (Center) A graph that preserves the labels. (Right) The bare minimum graph; we only need to know how many edges enter into each vertex — the *degree* of the vertex — to determine if a solution is possible.

of edges at each vertex to make a complete tour of all the edges possible.

In graph theory, we refer to the number of edges at a vertex as that vertex's *degree*. You can find the degree of each vertex in Figure 5.34 by counting. In the graph under consideration, there are three vertices with degree 3, and one with degree 5 — so a tour is clearly not possible!

Perhaps we can relax the restrictions a bit by asking if it is possible to start at one vertex and travel every edge, regardless of where we end up at the end. In this case, the starting vertex would have one unpaired exit and the ending vertex would have one unpaired entrance; every other vertex must still have even degree. So, if *exactly two* vertices were of odd degree, then such a path would be possible. Once again, however, we can see that we have not met the criteria, so even a one-way trip is not possible!

These two types of route were named after their discoverer. The first route, which returns to the starting point, is called an *Eulerian circuit* or *tour*, and the other is called an *Eulerian path*, or *Euler walk*.

With minimal changes, we can make an Euler walk possible. By adding a bridge that connects any two landmasses, we would increase the degree of two of the vertices, giving them even degree. Then, by starting at one of the other two vertices, a walk would be possible.

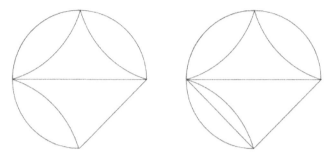

Figure 5.35. (Left) Adding an additional edge from top to right makes an Euler walk possible; the walk would start from the bottom or left vertex. (Right) Adding an edge between the bottom and left vertices makes an Eulerian tour possible. Here, all vertices have even degree.

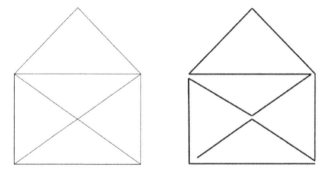

Figure 5.36. A puzzle commonly reproduced as a children's activity. The challenge: draw the "house" without lifting your pencil or retracing any lines.

Finally, we can create an Eulerian tour by connecting the last two vertices of odd degree (Figure 5.35).

Curiously, there are some equivalent expressions of these solutions that make this method leap out. Consider the Euler walk: what criteria make such a thing possible? An Euler walk is equivalent to drawing a figure without lifting your pencil from the paper, which is a puzzle you may recognize from children's books (Figure 5.36). A continuous line only has two ends; from every other point, you can go either of two directions. You can easily see how this relates to the calculation for determining if an Euler walk is possible!

By extension, an Eulerian tour is a path that can be traced by a tangled loop of string. In that case, at *every* point you can go in either of two directions.

The lesson to learn here is that the same problem can either be incredibly difficult, or easy enough that a child can solve it, with the main difference being how the problem is presented visually!

Graph theory originated in response to questions of walking paths, but this is far from the only type of question that can be addressed using the ideas Euler developed.

Consider the following problems.

Building a wire sculpture

An artist wants to construct a cube out of wire. Her first plan is to cut twelve pieces of wire of equal length, and solder three of them together at each corner (Figure 5.37). Then she has an idea: what if she cut longer wires, bending them at the corners. Will this save her any time or solder?

Each of the eight corners — that is, each vertex — has a degree of 3. But as we have just seen, there is no way to traverse every edge of a graph in a continuous path if more than two vertices have odd degree!

In particular, a single wire can be bent to pass through all eight vertices, but it will pass through six of them only twice. This means that three more wires will be necessary to complete the cube — and she will still need eight drops of solder!

There are many ways to construct such a cube, but none of these requires fewer than four wires. In Figure 5.38, we present two ways to construct this sculpture. On the left, we use one piece of wire of the

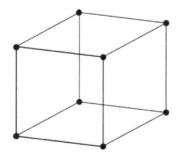

Figure 5.37. A wire cube consisting of twelve pieces of wire of equal length connected by eight drops of solder.

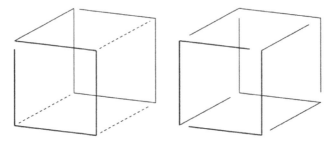

Figure 5.38. Two different ways to make a wire sculpture of a cube. There is no way to make a cube using fewer than four pieces of wire. (Left) An example of a cube made of the longest possible length of wire. (Right) A cube made of four pieces of wire of equal length, all bent in the same way.

greatest possible length; on the right, we demonstrate a method of constructing the cube using four pieces of wire of the same length, all bent in the same manner.

This provides a nice segue into our next topic. A cube is one of the five *Platonic solids*[8] — a three-dimensional shape composed of congruent, regular polygonal faces. These were first studied by the Greeks, and have been used and misused[9] endlessly since then. The Platonic solids determine how many objects with regular shapes form naturally, most notably in chemistry. The structures of certain crystals can be expressed in terms of cubes, and certain complex molecules form naturally into Platonic shapes, since sides of uniform length have the lowest possible chemical energy.

Euler used the insights gained from the Bridges of Königsberg problem to investigate more complex graphs, including those related to three-dimensional shapes. Before we follow his footsteps, let's learn a bit more about these shapes.

---

[8] The five Platonic regular solids are the tetrahedron, the cube, the octahedron, the dodecahedron, and the icosahedron.

[9] The Platonic solids have inspired all sorts of mystical investigations. The Pythagoreans and other Greek philosophers thought they had mystical properties, and Copernicus attempted to construct a model of the solar system that consisted of planetary orbits defined by Platonic solids.

The shape of space

The definition of a Platonic solid (beyond having been named by Plato) can be broken into four parts. First, every face must have the same shape. Second, these shapes must be polygons — no curved edges are allowed. Third, each of the shapes' side must have the same length. Finally, the solids must be *convex* — any straight line connecting two points of the solid must pass completely through the interior of the solid.

These may not seem like particularly difficult criteria to meet, and since there are an infinite number of regular polygons, it seems possible that there would be many of these solids. In truth, there are only five! Why is this?

Constructing the Platonic solids

Can you show that there are only five Platonic solids? How?

We will begin by considering the example of the cube. It's one of the easiest Platonic solids to visualize; you have probably idly drawn wireframe cubes while sitting in class. What defines a cube?

In a cube, three squares meet at each corner. If only two squares met at a corner, they would fold flat, and the resulting shape would not be a solid. If four squares met at each corner, there would be no way to fold them into three dimensions, since their angles already sum to 360 degrees.

What about other shapes? Three triangles can be folded together, as can four or five, but six triangles add up to 360 degrees and cannot be folded. Similarly, three pentagons can be folded together, but three hexagons cannot. And if three hexagons cannot be folded into three dimensions, then there is no point in investigating polygons with more sides!

In Table 5.1, we illustrate all candidate arrangements, as well as terminal arrangements — the configurations which show that it is not possible to add more polygons.

By folding each set of shapes, adding more of each shape as necessary, we find that there are indeed five, and only five, Platonic solids (Figure 5.39)!

Table 5.1. To make a vertex in three-dimensional space, three or more polygons must meet at an angle of less than 360 degrees. If they meet at 360 degrees, they can tile the plane.

	3 polygons	4 polygons	5 polygons	6 polygons
Triangle				
Square				
Pentagon				
Hexagon				

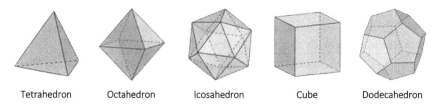

Tetrahedron   Octahedron   Icosahedron   Cube   Dodecahedron

Figure 5.39.   The five platonic solids, ordered by face shape and number of shapes per vertex.

Characterizing shapes

Each Pythagorean solid has a certain number of faces, vertices, and edges. For instance, the cube has six faces (think of the six sides of standard dice), eight vertices, and twelve edges.

Can you determine a relationship between these three numbers, for each of the five Pythagorean solids?

Let's start by counting each of these values for the five solids. It's easy to lose track when counting edges and nodes, so we will demonstrate one "trick" for keeping track of these counts. The octahedron is composed of eight faces, each of which has three sides. These triangles have twenty four edges. But each edge shows up as part of two different triangles, so you must divide the previous total by 2, which results in a count of twelve total edges.

Similarly, each triangle has three vertices, and there are eight triangles, for a total of twenty four vertices. Now, since four triangles meet at each point of the polyhedron, each vertex is part of four different triangles, so there are 24/4 = 6 unique vertices.

You can confirm for yourself the remaining results in Table 5.2.

You might notice some interesting properties of this table. First, every vertex of the octahedron has even degree, so there *must be* an Eulerian tour around this polyhedron! Second, there are some interesting symmetries between these solids. For instance, the cube and the octahedron have the same number of edges, but their numbers of faces and vertices are the reverse of one another. As it turns out, if you draw a vertex at the center of each face of a cube and connect them together, they form an octahedron — and vice versa!

Table 5.2.   Characteristics of the Platonic solids. Here, we list the shapes used to make each polyhedron, as well as the degree at each vertex, and the numbers of faces, vertices, and edges in the completed polyhedral.

	Tetrahedron	Octahedron	Icosahedron	Cube	Dodecahedron
Base shape	Triangle	Triangle	Triangle	Square	Pentagon
Degree	3	4	5	3	3
Faces	4	8	20	6	12
Vertices	4	6	12	8	20
Edges	6	12	30	12	30

This transformation, from faces to vertices and vertices to faces, is called finding the *dual* of a graph. The icosahedron and dodecahedron are duals of one another, and the tetrahedron is its own dual. Duals show up in many corners of graph theory, so while we won't spend any more time discussing them here, it can be handy to have this concept in your toolkit so that you can recognize them when you see them.[10]

Less obvious here is a key observation that Euler made. Let the numbers of faces, vertices, and edges be represented by the letters F, V, and E, respectively. Then, for any Platonic solid:

$$F + V - E = 2$$

Can you prove this relation?

A sketch of the proof of this follows. Imagine that we have drawn the lines of a cube on a balloon, and then poked a hole in one of the faces and flattened the cube out, similar to the transformation performed on the globe in Figure 5.13. The resulting shape, shown in Figure 5.40, either has one less face than the original cube, or that face has moved to the entire outside of the figure, filling the entire plane. Depending on your point of view, these perspectives are equally valid; in the former case, we would consider the formula $F + V - E = 1$ instead. There are useful reasons to use the previously stated value, so try to imagine that there are six faces here: five inside the graph, and one outside, encompassing the entire graph to infinity.

---

[10] Note, however, that the graph in Figure 5.28 forms the dual of the map of the 48 states!

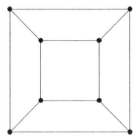

Figure 5.40. The graph of a cube stretched to lay flat on a plane. After stretching, the sixth "face" surrounds the graph.

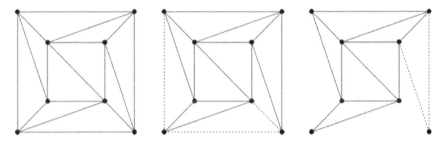

Figure 5.41. Removing edges, faces, and vertices while preserving the Euler characteristic. (Left) The cube, after triangulation. (Center) When we remove one edge, one face is also removed. (Right) When we remove a vertex, we remove two edges and one face. The characteristic remains unchanged after each of these operations.

We continue by dividing any face with more than three sides into triangles, as shown in the first part of Figure 5.41. This is possible because for each new edge we add, one new face is created, but no new vertices are added. This preserves the values on both sides of the equation.

$$F + V - E = 2$$
$$\left(F + 1\right) + V - \left(E + 1\right) = 2$$
$$F + 1 + V - E - 1 = 2$$
$$F + V - E = 2$$

We now remove edges from the outside of the figure, making sure to never break the graph into two pieces. We remove edges one-by-one;

each time an edge is removed, the number of edges and the number of faces are reduced by one. Since the number of edges and the number of faces both shrink at the same rate, the equation remains balanced at all times.

$$F + V - E = 2$$
$$(F - 1) + V - (E - 1) = 2$$
$$F - 1 + V - E + 1 = 2$$
$$F + V - E = 2$$

Eventually we will find ourselves in a configuration like the configuration at the right of Figure 5.41. Here, we remove two edges, one face, and one vertex. Again, the two sides of the equation remain balanced.

$$F + V - E = 2$$
$$(F - 1) + (V - 1) - (E - 2) = 2$$
$$F - 1 + V - 1 - E + 2 = 2$$
$$F + V - E - (1 - 1 + 2) = 2$$
$$F + V - E = 2$$

Eventually, no matter which polyhedron we began with, we will find ourselves with a single triangle composed of three vertices, three edges, and two faces. You can plug those numbers into the original formula and find that the equation remains true — as it did every step of the way!

This number, called the *Euler characteristic*, seems to be a property of these solids. But is it a property of *only* these?

If you draw any graph on a sheet of paper where no vertices cross, or any shape in space where no face crossed through another, you will quickly discover that all these graphs demonstrate the same property. What's going on?

It turns out that the Euler characteristic is a property of a *space*. That is, if a graph can be drawn on a sphere *or a plane* without any edges crossing (not always an easy task!), then the relationship between the number of faces, vertices, and edges will *always* hold.

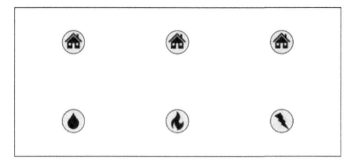

Figure 5.42.   Three houses and three utility companies. Each utility needs to connect to each house without crossing lines. Is this possible?

As is often the case in mathematics, we have a term for planar graphs, because they have many useful properties. For one, the Four-Color Map Theorem only applies to planar graphs. Every graph we have explored so far has been planar.

What does a graph that is not planar look like? How can we tell when a graph is not planar? One classic problem will help us illustrate exactly how.

The three utilities

In a particular town, there are three houses and three utility companies: water, gas, and electricity (Figure 5.42). The town's laws require that all utility lines be buried underground, but after an ugly incident involving one unlucky backhoe operator, two fire departments, and several dozen lawyers, the utility companies refuse to lay their lines across one another.

Nevertheless, each homeowner wishes to be connected to all three utilities. Further, a utility line can't run through a home — each line must begin at a source, and end at a single destination.

Is it possible to connect each of the utilities to every one of the houses while meeting all of these restrictions?

Feel free to spend a few minutes playing around to getting a feel for this problem.

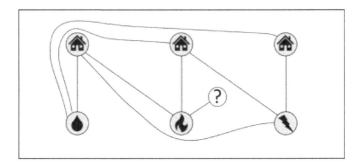

Figure 5.43.   Eight edges drawn between utilities and houses. The ninth and final edge, starting at the gas company, cannot be connected to the third house without crossing another edge.

In Figure 5.43 we illustrate a typical attempt at a solution. We start out by drawing straight lines, but soon find ourselves tracing longer and longer paths around the rest of the graph, until finally we are stuck with one last connection to make, and no way to complete it.[11]

Even having read the chapter up to this point, you may think there's a solution to this problem. It's tempting to keep breaking out new sheets of paper and testing one idea after another!

There are some common observations you may have noticed while drawing graphs. For one, the sizes and shapes of the nodes don't affect the shape of the graph. Even if you reduce the houses and utilities to mere points, it's still not possible to make all the connections.

Also, the positions of the nodes don't affect the outcome. If you spend long enough drawing the same graphs over and over, eventually you should strike upon the idea of moving the houses around. What you should notice is that while certain configurations are "neater," the end result is always the same — deadlock!

To demonstrate that no solution is possible, we can use properties of the Euler characteristic that we just finished proving for 3-D solids,

---

[11] While constructing this diagram, the artist found himself trying to make the last connection, despite knowing that there is no solution. This puzzle can be very seductive!

and apply it to this 2-D graph. We can identify the pertinent numbers, and solve the equation algebraically to see what comes out.

First, note that there three utilities and three houses, for a total of six vertices. Each utility must connect to each house, which tells us that a complete solution would involve nine edges. The number of faces required here is still a mystery, so that will be the value we solve for.

If this graph has a planar solution, the following equation holds:

$$F + V - E = 2$$
$$F + 6 - 9 = 2$$
$$F - 3 = 2$$
$$F = 5$$

Now, trace edges from one utility and back to form a loop. One such loop might be electric-house 1-water-house 2-electric; the four edges connecting these vertices define a face. You can trace just such a loop in Figure 5.43, and recognize the face as the closed region defined by the oddly-shaped edges.

Each face must have at least four edges, because two utilities can't connect, nor can two houses. Imagine a face with three edges: the edges would have to go utility-house-house-utility, or house-utility-utility-house, both of which violate our rules.

But if there are five faces with four edges each, and (recalling the "trick" from earlier) each side is shared by a pair of faces, then there must be at least ten edges in the graph — which contradicts our assumptions! Therefore, this graph cannot be planar.

So, it is not possible to connect three house with three utilities according to the rules set up in this problem.

If this graph isn't planar, than what is it? Can we draw this graph according to our original rules on some other shape?

The Euler characteristic pointed the way towards spaces with different shapes. If the plane and the sphere both have an Euler characteristic of 2, that suggests they are in some way identical. And, as is typical in mathematics, people took note of this weird constant and started looking for spaces with other shapes, or other characteristics. What might they look like?

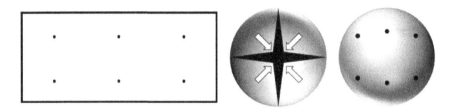

Figure 5.44. Folding the corners of a plane to a single point changes the plane into a sphere. The solid borders all meet up as invisible seams along the back of the sphere.

We can consider the transformation between the plane and a sphere, and examine alternate transformations by varying our approach. First, consider the case where we stretch the entire border of the sheet that the graph resides on to join it at a single point (Figure 5.44). As you well know from the chapter introduction, this turns the flat sheet into a sphere.

Starting on the left of Figure 5.44, with a flat sheet, we stretch the corners toward each other "behind" the graph. As these corners get closer, as shown in the middle image, we can sew together the sheet to make one continuous surface. When we re-examine the front of this shape, as shown in the rightmost image, we can see that the six points are now sitting on a sphere.

If we mark only one pair of edges and join them together, the sheet instead becomes a cylinder, as demonstrated in Figure 5.45. A cylinder isn't much different from a plane, since anything we might accomplish by wrapping an edge around the seam could be accomplished just as well by looping around the edge of the sheet. That is, of course, as long as we disallow going over the top edge of the paper and around the back side.

What happens if we do allow edges to pass over the edge of the sheet? Or, more accurately, what happens if we connect the top and bottom of the sheet together, and the left and right edges together?

Here, we form a new space, called a *torus* (Figure 5.46). The torus is often used in example problems in topology, because it is a space that is easy to visualize, but has a shape that is markedly different from our own.

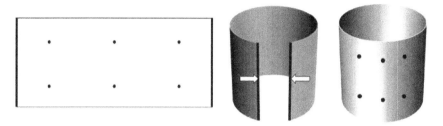

Figure 5.45.   Folding just one edge of the plane together (bold) changes the plane into a cylinder. If we have to remain on the outside of the cylinder, the shape of the space is no different than the plane.

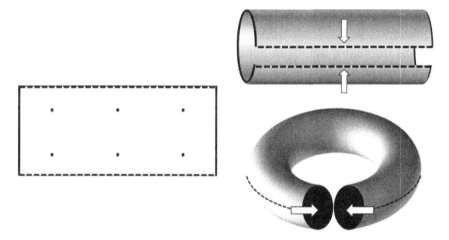

Figure 5.46.   Folding dashed edges together makes a cylinder; joining solid edges turns the cylinder into a torus. This surface has a different shape than any of the other spaces we have seen so far.

We still need to demonstrate that this shape is different than the sphere. To do so, we will provide you with a new rule: you can draw lines off the edge of the paper; when you do, they re-appear on opposite side. Do you agree that this is a valid rule for the torus? And, with this, can you connect the three utilities to the three houses?

In Figure 5.47, we have done just this. You can trace each line to find its counterpart on the opposite side of the sheet. The edge providing water to house number 3 passes off the left side of the figure and re-appears on the right, while the edge connecting the gas

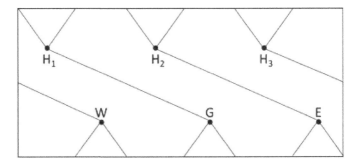

Figure 5.47.   The three utilities problem solved on a plane that follows the rules of a torus. Lines that extend off one edge reappear on the opposite edge.

company to house 2 starts off going down and to the left to the bottom side of the figure before re-appearing at the top. You can easily confirm that all connections have been made here, and that no edges cross — providing us with a solution! On a torus, it is in fact possible to solve the three utilities problem!

We have stated that this space is different than the normal spherical (or planar) spaces we are most familiar with. Can you demonstrate that it differs, though? We can start by determining the Euler characteristic of this space, and see if it is not equal to 2.

We can find the Euler characteristic of this space by the same method we used before, but this time we won't assume its value in advance. The faces are a bit hard to pick out, so we have highlighted them in Figure 5.48. You should be able to confirm that there are, in fact, three of them!

We can plug all three of these values into the formula for the Euler characteristic to find its value:

$$F + V - E = \chi$$
$$3 + 6 - 9 = \chi$$
$$0 = \chi$$

On the sphere, the Euler characteristic is 2, but on the torus it has a value of 0!

To satisfy your curiosity, in Figure 5.49 we have folded the space into a torus to show you what a solution to the three utilities problem looks like in three dimensions!

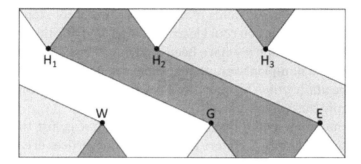

Figure 5.48.   To calculate the Euler characteristic, we have to have a count of the number of faces. Coloring faces is a good way to count them; here, there are three faces total.

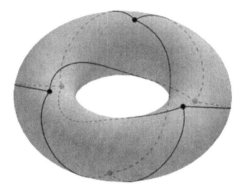

Figure 5.49.   Folding Figure 5.47 into a torus shows how each edge connects without crossing other edges.

If you have been following along so far, you should have spotted a thread running through this problem. Naming ideas, and defining them clearly, makes new mathematical insights stand out. We began this problem by defining a property then exploring what happens when that property no longer holds, and many of the subsequent insights have had a similar character.

Graph theory is a good introduction to this type of mathematical thinking because everything you are likely to encounter is discrete, finite, visual, and has a name. (What's more, graphs are *very* forgiving — a graph is still the same whether it's drawn cleanly or sloppily!) We have shown you plenty of new terms in this chapter, each of which could lead you to new realms of thought. We need not stop here — there are more terms you could learn just from studying this solution! In this case, the graph we have been studying is one of the smallest examples of a non-planar graph: the *complete, bipartite* graph called $K_{3,3}$, or the *utility graph* for short. And yes, the nickname for this graph comes from this very problem!

The notation should be easy to understand: K is just the name for this type of graph; 3,3 refers to the number of vertices of each type (3 houses and 3 utilities).

Bipartite means that the graph can be split into two groups, such that there are no edges between vertices that both belong to the same group. Here, bipartite means that the gas company hasn't run a pipe to the water company, and no two houses share pipes.

Another way of thinking about bipartite graphs is that they can be painted with only two colors. This follows from the definition of bipartite, but it might be helpful to think through exactly why this is true. If there are two "classes" of vertices that have no edges between them, then we can be assured that white vertices only connect to black vertices (for instance, utilities only connect to houses), and vice-versa.

Finally, *complete* means that every possible connection between nodes has been made. Figure 5.43 is *not* complete because we weren't able to make the final connection, but it *was* bipartite. Other examples of complete graphs include the K_n graphs, a few of which we illustrated in Chapter 3 when discussing handshakes. K_5, for instance, is the graph of all handshakes that can take place in a group of five people.

The utility graph is notable because it is (alongside the complete graph K_5) one of the most basic ways a graph can "break" the conditions for planarity. If you want to know if a given graph is planar, and you find that a subset of its vertices is identical to the utility graph, you can be assured that the original graph is non-planar.

You can visualize the process we've been using as a graph of concepts. Think of an empty space called "graph theory." Inside that space, there are regions of unified characteristics, such as "planar," "complete," and "bipartite." Sometimes, these regions overlap (some graphs are complete *and* bipartite). There are nodes inside this space linked with edges that represent relationships. A bipartite graph *is* 2-colorable. A planar graph *is* 4-colorable. Vertices *have* degree. Some graphs *have* a Euler characteristic or an Euler walk. And there are edges that link nodes in this space to other spaces: a planar graph *is equivalent to* a map on a sphere, which *represents* a political map.

When we study math, we walk the edges of these graphs, learning the answers to questions, such as: "How are spaces related to the Euler characteristic?" or "What is the degree of each vertex in a complete graph?" Mathematical problem solving is the process of finding a region of this "mathematical concept space" that adequately parallels the structure we're interested in, and then following connections — or creating our own — to find the answers we seek.

If there is one thing that theoretical mathematics excels at, it is fleshing out the connections between nodes in these conceptual graphs. But there is no limit to the number of edges you could draw between concepts, and thus, no limit to the amount of time you could spend studying mathematics! If, in twelve years of schooling, you learned 1,000 distinct concepts, then the next time you learned something new there would potentially be 1,000 new connections — 1,000 new edges — you could make within your existing knowledge. How could anyone possibly say they "know mathematics" in these impossible circumstances?

Luckily, we don't have to create the world of mathematics from scratch. We learn to recognize patterns, and experiment with new problems, and (thanks to the internet) search for prior work that mirrors the problems we want to solve. Becoming good at mathematical problem solving is not about learning everything there is to know. Rather, it is about learning the locations of enough conceptual landmarks that you can navigate between points, even when you find yourself somewhere you have never been before.

If theoretical mathematics is about learning to navigate conceptual space, then applied math — problem solving — is about mastering the process of translation from the real world to conceptual space, and back.

Fluency in mathematics is closely related to fluency in these types of translations. Sometimes, you are fortunate enough to just "get" a problem, and the process doesn't even feel mathematical. Other times, you might be provided with a representation that makes the solution "jump out" at you. The other 99% of the time, you will need to learn to bend, stretch, and twist problems (and your own point of view!) to bring the most productive path forward into sharp focus.

In the second half of this chapter, we will practice different types of translations. Each of the following sections will have a theme, demonstrating a variety of ways a single type of problem can be translated into new forms. Remember, this chapter is not *comprehensive* — no math book could ever be! — but by the end, we will have explored many new landmarks to help guide our problem solving.

Ways of Seeing: Chessboard Problems

In this section, we will examine a variety of puzzles and problems inspired by the game of chess.

Chess attracts many of the same kinds of people who are pulled to mathematical puzzle solving. As such, chess and chessboards are popular themes for puzzles!

Fundamentally, a chessboard is a simple grid of squares, eight rows and eight columns in size. If you think of maps as visual representations of real countries, then a chessboard can also be thought of as a map for representing any problem that relates to a grid.

We often choose to look at the squares on a board as a list of coordinate pairs, like an x–y plot — and, if you look at a typical chessboard, they are commonly labeled with what is (appropriately) called "algebraic notation," where each square is assigned a pair of coordinates. Typically, they use the letters A through H for files, or columns; and the numbers 1 through 8 for ranks, or rows. This is far from the only way to see a board though.

Continuing with the theme of the previous section, you can see a chessboard as a graph that is connected under a specific move. So while you might look at a board as a grid of squares, you can also

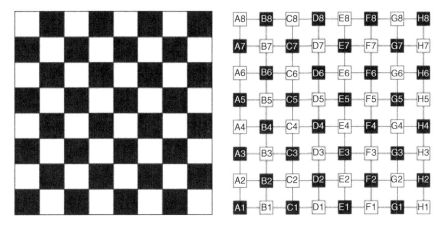

Figure 5.50. (Left) A standard chessboard consists of eight ranks and eight files. (Right) Another perspective on the same board. We link neighboring squares — represented by their position, written in algebraic notation — with edges representing adjacency.

transform it into planar graph of connectivity (Figure 5.50), just like the one we constructed for the United States in Figure 5.29.

When would you do this? Not when playing the game of chess, certainly — no chess piece moves in the way we have just shown! The legal moves available to a rook (which can move any number of squares horizontally or vertically as long as it's not blocked by another piece) resemble the previous figure, but if we tried to draw the entire graph of moves available to a rook, that graph would have 448 edges. For instance, a rook in position A1 could move to seven different spaces vertically, and seven horizontally, leading to 14 edges linking A1 to other vertices. Expand that to the other 63 squares on the board, and you arrive at $\frac{64 \times 14}{2} = 448$. That's far too many edges to make out any detail!

On the other hand, if we consider the King's Pawn, the graph remains legible, as you can see in Figure 5.51. If you remember the rules of chess, a pawn can only move diagonally when capturing another piece, so it's unlikely that this pawn would ever make it to the far corners of the board. However, this is a graph of all possible moves, not only likely moves.

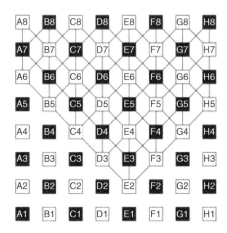

Figure 5.51. The King's Pawn can move to any square forward or diagonal from its starting position. Not all vertices in a graph have to be connected.

A graph's connectivity is determined by the relation under consideration. Most of the time, only one type of graph "naturally" falls out of a problem, but it is often handy to consider other ways of connecting the same graph, as we can see when examining a chessboard.[12]

One piece lends itself particularly well to graphical solutions: the knight. Recall that a knight's move is shaped like the letter "L": a knight can move two squares out, and one square over, in any direction. At any time, a knight has at most eight different moves available (Figure 5.52).

On a normal sized chessboard, the graph of all possible knight's moves is cluttered, but still surprisingly readable (Figure 5.53).

A curiosity that has been explored by countless problem solvers over the years is the Knight's Tour. This puzzle asks if it is possible for a knight to visit every square on a chessboard exactly once? With a

---

[12] Recall the solution to the problem of Filling Buckets, presented in Chapter 2. In that solution, we considered only the exterior points of a rectangle of size 5×3, which were joined with edges representing the state transitions as we filled and emptied buckets.

That answer was presented in the language of Cartesian coordinates, but it was actually a graph theory problem in disguise!

Figure 5.52.   From a given spot on the chessboard, a knight has at most eight available moves.

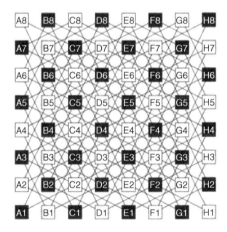

Figure 5.53.   A standard chessboard connected by knight's moves.

little patience you can no doubt find such a tour, as there are literally *quadrillions* of them.

A slightly different version of this puzzles asks if you can find a *closed* tour. A closed tour visits every square on the board, but it starts and ends on the same square. The advantage to exploring closed tours is that there is no "end," so you can start drawing the tour anywhere, in practically any direction. One such tour is shown in Figure 5.54.

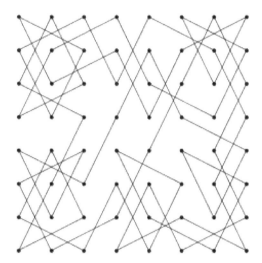

Figure 5.54.   An example of just one closed knight's tour on a standard chessboard.

There are fewer closed tours than open ones, but closed solutions still number in the trillions. We couldn't possible tackle all of these trillions of solutions in a book, but some simpler forms of this problem can teach us fascinating insights into visual problem solving.

The minimal knight's tour

It is possible to find a closed knight's tour on a standard chessboard. What is the smallest square board that permits a closed tour? What is the longest tour possible on each of these smaller boards?

First, consider the smallest possible board: one square. Simply placing a knight on such a board "visits" every square — but that hardly seems to be in the spirit of the problem!

So, the first board we will seriously consider is the 2×2 board.[13] We quickly discover that the knight cannot move at all in such a small space.

The next size to consider, 3×3, seems promising at first, but the center square is unable to connect to any of the outer squares. By examination, we find that there is really only one path available to the knight, of length 8, as illustrated in Figure 5.55.

---

[13] That is, 2×2; the one consisting of two rows and two columns.

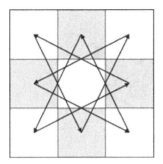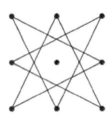

Figure 5.55. The longest possible knight's tour on a 3 × 3 board cannot visit the center square.

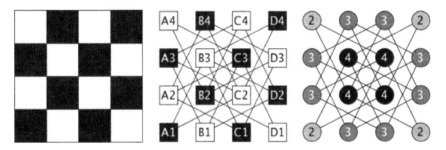

Figure 5.56. Three different ways of looking at a 4 × 4 chessboard. (Left) A simple grid. (Center) Connected by knight's moves. (Right) The number of moves available — i.e., the degree — of each vertex is displayed.

The 4 × 4 board seems promising, and is worth examining in closer detail. We provide three perspectives on this board in Figure 5.56. The first is an illustration of the board itself; the second shows the same kind of connectivity information as in Figure 5.53; and the third shows the degree at each vertex. Try to decide if a closed tour is possible by considering these diagrams.

After a few trials, you might begin to notice that the corners don't behave nicely. Often, you trace a path into a corner only to realize that there is no available route back out. Each of them has degree of 2 — or, in other words, only one way in, and only one way out! If you have already visited both of their neighbors earlier in an attempt, there's no way to continue on from that corner.

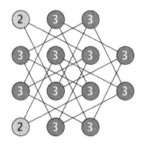

Figure 5.57.   The degree of each vertex once two corners have been removed.

Look at A1 and D4 in Figure 5.56. Both of these corners *only* connect to the same two squares, B3 and C2. There can be only one closed
knight's tour that includes these two corners — namely the one that
goes A1, B3, D4, and C2.[14] The longest possible tour must exclude
at least one of these corners. Therefore, there is no closed tour that
visits every square on this board.

By the same logic, there is no way to visit both of the white corners, so we can remove one of them from our board. By symmetry, it
doesn't matter which ones we remove: all boards of this size which
are missing one black and one white corner have identical graphs.
Under these constraints, it seems that the longest possible tour we
might find on this board has a length of 14 — but we still have to
prove such a path exists.

When illustrating this new board, we are free to choose the most
useful representation. The third form — showing each vertex's
degree — is particularly useful here. The new version of this graph is
displayed in Figure 5.57.

There are a couple of observations about this graph that are
worth highlighting. First, there are 20 total edges. You can confirm
this by adding up the number in each circle and dividing by two, since
each edge connects two vertices.

A closed tour of length 14 must encompass all of the vertices and
14 of these edges. You might recall from Chapter 4 that there is a

---

[14] For the sake of simplicity, we consider identical any two tours that visit the same
squares in a given order, or in the reverse order.

formula for counting the number of ways to take objects from a collection of size n. To count all the possible ways to select k edges from a group of n, we can use that formula:

$$\frac{n!}{k! \cdot (n-k)!} = \frac{n \cdot (n-1) \cdot \ldots \cdot 1}{k \cdot (k-1) \cdot \ldots \cdot 1 \cdot \left((n-k) \cdot (n-k-1) \cdot \ldots \cdot 1\right)}$$

Here, a naïve approach would require at most 38,760 tests to *disprove* the existence of a tour of our chosen length:

$$\frac{20!}{14! \cdot (20-14)!} = \frac{20!}{14! \cdot 6!} = \frac{20 \cdot 19 \cdot 18 \cdot 17 \cdot 16 \cdot 15}{6 \cdot 5 \cdot 4 \cdot 3 \cdot 2 \cdot 1} = 38,760$$

However, any closed tour *requires* the inclusion of the four edges connecting the corners to the rest of the graph, so we actually only have to choose 10 edges from a possible 16. This cuts the worst number of cases to test down to 8,008:

$$\frac{16!}{10! \cdot (16-10)!} = \frac{16!}{10! \cdot 6!} = \frac{16 \cdot 15 \cdot 14 \cdot 13 \cdot 12 \cdot 11}{6 \cdot 5 \cdot 4 \cdot 3 \cdot 2 \cdot 1} = 8,008$$

When solving problems, occasionally a solution seems likely only to later prove impossible, and it can take a significant number of trials just to discover this. While it isn't likely to happen here, this method of analysis is worth keeping in your back pocket when you want to check how long the direct approach might take.

We can check to see if there is a valid tour using trial-and-error. You may struggle to keep track of which vertices have been visited already. One way around this is to number the vertices in pencil and, if you get stuck, to backtrack to the last number that had two paths available and take the other path.

Figure 5.58 demonstrates this process. The bottom left corner is marked with the number 1, and its two neighbors are numbered 2 and 14, since any closed tour must visit all 14 vertices before returning to the start. We begin selecting vertices at random, numbering as we go, until we get stuck with no way to proceed at the ninth

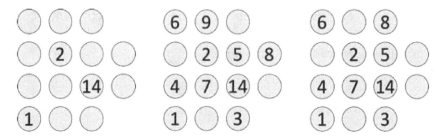

Figure 5.58. (Left) Beginning to order the nodes in a closed tour. (Center) Choosing one-by-one at random, we eventually get stuck at the number 9. (Right) We can backtrack, erasing numbers, until we find another way to move forward. Neither 9 nor 8 offered any alternative paths, but when we reach 7, we can continue by writing 8 in a different location.

vertex. We backtrack, erasing numbers as we go, until we reach the number 7, which allows us to test a different route.

If a tour exists, this method will reveal it eventually. But there is still a possibility that there's an easier method to see, and demonstrate, that the tour we seek exists.

Figure 5.57 is suggestive of one method. Think back to our exploration of Platonic solids. What did most of those solids have in common? (You may want to flip back to Table 5.2 and compare it to Figure 5.57.)

Many of the vertices in that diagram had degree of 3, as does our graph! This suggests that there might be a way to re-arrange this graph to make it look like a three-dimensional shape. After some experimentation, you should end up with something like Figure 5.59, and now the length 14 tour[15] should practically leap out at you!

Reversing this transformation, we can transfer the final tour back onto the chessboard, as seen in Figure 5.60. So there is, indeed, a way to form an Euler tour on 14 squares of a 4×4 chessboard.

The next size board, at 5×5, seems like a likely candidate for the smallest board that has at least one closed tour (Figure 5.61). But there is one problem. Have you noticed anything about the types of

---

[15] We say "the tour" and not "a tour" because, after considering symmetries, all other possible tours are functionally equivalent to this one.

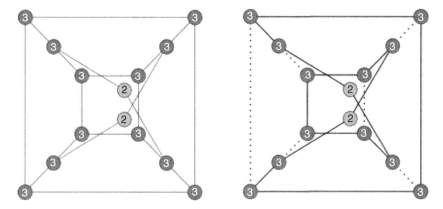

Figure 5.59. (Left) Transforming the previous graph into an arrangement resembling the graph of the cube. (Right) Here, the closed tour stands out. We remove 6 edges (dotted lines), leaving behind a tour of length 14.

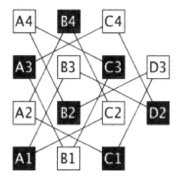

Figure 5.60. The longest closed tour possible on a 4 × 4 board passes through 14 vertices.

moves available to a knight? If you want a hint, try flipping back to Figure 5.52 (the diagram of a knight's possible moves) and see if you spot any patterns.

A knight can only ever move to squares of the opposite color. Therefore, a successful tour will always alternate white and black squares.[16] If we started on a white square, the second square visited,

---

[16] In graph theoretic terms, the graph created by connecting chessboard with knight's moves is *bipartite*.

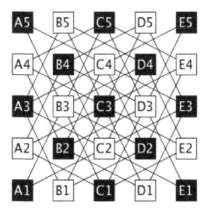

Figure 5.61. A 5 × 5 board connected by knight's moves.

and the second-to-last square, must both be black. To produce the tour we're looking for, it is necessary to have an even number of squares — but this board has 25 squares! In fact, this observation is enough to prove that a board with an odd number of squares *never* has a closed tour, and that all closed tours must have an even length!

What square should we remove? It must be a black square, but not just any black square. We can extend the logic from the previous example to see that the only tour which incorporates all four corners visits eight squares in total: A1, B3, A5, C4, E5, D3, E1, and C2. So we *must* remove one of the corners.

The connectivity of this graph is much higher than any of the previous ones, so there aren't any obvious tricks we can use to make the process of finding a closed tour deterministic. What we can do is use a trick discovered by Euler to turn an open tour into a closed one.

Begin by linking up the seven required vertices: three corners and the four squares to which they connect. Extend the path you've just created using any knight's moves whatsoever. If you get stuck, extend the other end of the tour instead. It might require a bit of backtracking, but eventually you should be able to draw an open tour that visits all 24 vertices (Figure 5.62).

Next, choose one of the endpoints of the tour, and connect it to another point somewhere in the middle of the tour. Making this

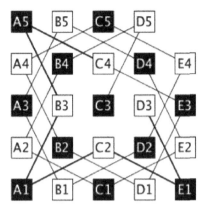

Figure 5.62. An open knight's tour on a 5 × 5 board. The seven vertices linked with bold lines have no degrees of freedom and cannot be altered; the remainder of the tour is arbitrary. Note the open ends terminating at C3 and D3.

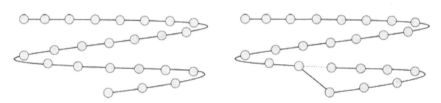

Figure 5.63. If you take a chain of links and break it in the middle, then connect the broken part to the end of the chain, the result remains a single chain. We can use this method to try to turn an open tour into a closed one.

connection will necessitate removing the edge that leads toward the end of the graph, otherwise we will produce two *separate* tours, one closed and the other open. As long as we do this, though, we are no worse off than before, as this operation will always preserve open tours (Figure 5.63).

If we choose judiciously, or are simply lucky, the new end of our open tour will be one move away from the opposite end (Figure 5.64), and we'll be able to close the tour!

The 6 × 6 board has none of the restrictions of the previous boards. All of its squares have at least one available move, it contains an even number of squares, and none of the corners connect to the same squares as each other. We can modify the result from the 4 × 4 board

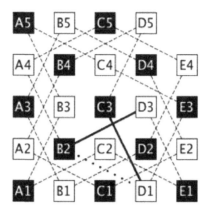

Figure 5.64.   We can connect D3 to B2, removing the edge between B2 and D1. D1 is a knight's move away from C3, so we can link them and close the tour.

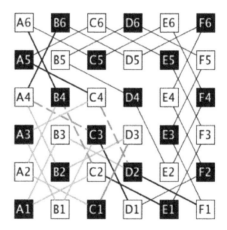

Figure 5.65.   Starting with an earlier solution (gray lines), we can connect the outer two layers of vertices together using three paths, then "splice" these into the inner graph (dark lines) by deleting three edges (dashed).

to find one valid closed tour that visits every square, thereby answering the original question: the smallest square board that permits a closed tour that visits every square has side length 6 (Figure 5.65).

The variety of approaches required to solve this problem offered us an excellent opportunity to develop mental flexibility. If that's all this problem had to offer, that might be enough justification to include it! But it has a more important role to play. We can apply many of the techniques we just demonstrated in other contexts; after practicing these, you will be able to say you have mastered them!

A covered chessboard

Using 32 dominoes laid either vertically or horizontally, you can cover an entire chessboard (Figure 5.66).

If we remove two opposing corners, as shown in Figure 5.67, can you find a way to cover the chessboard with 31 dominoes?

You may see the solution to this problem right away. If so, congratulations! Most people can only find the answer to this problem with a lot of experimentation, or by having seen the correct "trick" applied before. If you don't see the solution — that's exactly why we're tackling this problem now!

Play around a bit. What do you notice? Before we go into the solution, what other approaches or other perspectives do you suppose you could take?

In the previous problem, we explicitly asked about smaller boards. Even though this problem isn't phrased in a way that invites us to look at smaller examples, nothing prevents us from trying to learn from them!

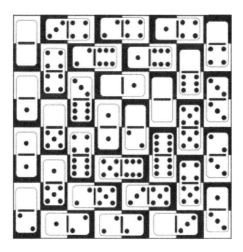

Figure 5.66.   Covering a standard chessboard with 32 dominoes.

Figure 5.67. A degenerate chessboard; that is, one that is missing two opposing corners.

Figure 5.68. Degenerate chessboards of sizes 2×2, 3×3, and 4×4. What do these have in common that definitely makes a tiling with dominoes impossible?

In Figure 5.68, we have produced three smaller examples of chessboards marred in the same manner as the one above. The first of these degenerate chessboards, made from a 2×2 board, could hardly be called a "board" at all! It should immediately be obvious that there's no way to place a single domino in a way that covers both these squares. What might not be obvious is whether there is some characteristic these boards have in common that makes this problem unsolvable.

Think about the nature of any two squares covered by a single domino. What do all pairs of squares have in common — or, more aptly, how do they differ?

Every domino must cover both a black and a white square. By removing opposing corners, we have reduced the number of squares of a single color by 2 while leaving the number of squares of the opposite color untouched. In Figure 5.67 there are 32 black squares, but only 30 white squares, which makes full coverage impossible.

This problem isn't exactly the same as the knight's tour, but we made use of a similar observation there. It's rare for a mathematical trick to manifest itself in **exactly** the same way in two different problems, but we can often benefit from our prior experience, despite that.

A visit to the museum

You're in a foreign city on a whirlwind tour of the cultural attractions. The city is known for its world-famous art museum, and you are intent on visiting as many galleries as you can in the shortest possible time.

The museum is shaped like a plus sign, and each section of the building is broken into a grid of smaller rooms (Figure 5.69). Every two adjoining rooms has a door running between them, and you can

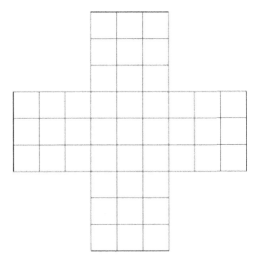

Figure 5.69. The floorplan of the famous museum. Each room (squares) is linked to its neighbors, or the surrounding property, by a doorway.

Figure 5.70.   Coloring the rooms of the museum like squares of a chessboard.

enter and exit through any exterior wall. Is it possible to see every room in the museum in one pass?

This problem doesn't involve a chessboard, but we can definitely borrow some ideas from earlier to find a solution. Begin by coloring the museum like a chessboard, starting with any room and using the opposite color to fill in each neighbor. Eventually, you should have a drawing like the one in Figure 5.70.

Count the number of rooms of each color. Here, we have 24 black rooms and 21 white rooms. Once again, every time we pass through a doorway (i.e. travel through an edge of the graph), the color of our starting room and ending room must differ. The best open tour we could find in this museum *might* pass through all 21 white rooms, and at most 22 black rooms. It is impossible to visit every room in one tour!

This also suggests a strategy. Each wing has 5 black rooms and 4 white rooms. If you intend to visit every room in a given wing, you must enter and exit that wing through black rooms. In contrast, if you want to visit every room in the center section, you will have to skip one black room in each of two different wings. Finally, the route that passes through the greatest number of rooms must start and end in a

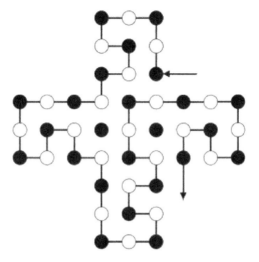

Figure 5.71.   A path of the greatest possible length through the museum. There are three more black nodes than white, so by starting and ending on black nodes we can visit all but two rooms.

black room. One possible route, passing through 43 rooms, is shown in Figure 5.71.

Graph colorings aren't only useful for finding paths, but you could be forgiven for coming to that conclusion. The next time we use this approach might come as a surprise! We're not done with chessboards quite yet, but the next two problems will head in a slightly different direction.

The haunted labyrinth

You have been put in charge of building this year's Halloween Haunted House. You decide to build a labyrinth, full of twisting turns and dead ends. To make it as scary as possible, you place secret doors throughout the maze, allowing actors to disappear into corridors running between the arms of the maze, only to reappear elsewhere.

After several days of planning, you have finally finished drawing the labyrinth's layout, and now it's time to order materials. The design calls for a vertical post every six feet in both directions, and each post

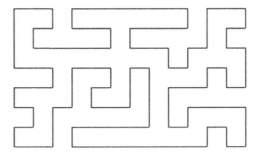

Figure 5.72. A labyrinth, drawn between the points of a uniform grid. In this problem, the hallways are six feet wide.

is either part of a straight wall or a corner (Figure 5.72). What's the total length of wall material you will need to order in order to construct the maze shown?

The first order of business is to determine the size of the maze. Since you're thinking about regular grids already, it should be no great feat to split this drawing into boxes. You will find (Figure 5.72) that there are 7 squares counting from bottom to top, and 12 running from left to right.

In theory, it should be possible to count each length of wall and produce an answer. That process would be error-prone, and you wouldn't want to end up with too much (or too little!) material for building the maze. Is there another way to look at this problem to ensure the count is as accurate as possible?

Instead of thinking about the number of walls, think of the posts at each corner. There is one more post than there are walls in each direction, so there are 8 posts in each vertical line, and 13 horizontally, for a total of 104 posts throughout the labyrinth.

The shape of the labyrinth forms a closed tour, connecting every post with two others. As you recall from our study of the knight's tour, in a closed tour, for every vertex there is exactly one edge. This labyrinth forms a closed tour of the posts, meaning that for every post, you will need one length of wall. But we already know the number of posts by multiplication! Therefore, you will need 104 *units* of wall, or a total of $104 \cdot 6 = 624$ feet of wall.

Our final chessboard problem should be an easy one; all you have to do is count. Of course, these problems are never *easy*…

Squares and rectangles

How many different squares and rectangles can be constructed from the squares on a chessboard?

Looking at a chessboard, it's tempting to begin counting right away. There are 64 small squares, and 1 large square (the whole board), and 1×2 rectangles, and 2×1 rectangles… quickly, your head might start spinning from all the numbers you have to track!

If you continue this way, staying organized is essential. Figure 5.73 displays one rectangle of each size from 1×1 to 3×3. Of course, there are more rectangles than just these; we have left out all the rectangles from 1×4 up to 8×8.

There are many ways to organize your data. Beginning with the squares only, each 2×2 square can have one of 49 different positions, each 3×3 square can have one of 36 different positions… then, you might begin to look at rectangles of width 1, then rectangles of *height* 1, then 2, … If you cut out a rectangle of each size, you could then write

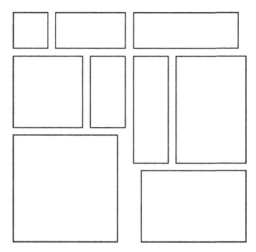

Figure 5.73. Some of the various kinds of rectangle that can be found on an 8×8 chessboard.

the number of different rectangles of that size on the piece of paper that represents it (Figure 5.74). This might simplify keeping track of all this information.

This looks like a table of data, which suggests we can record the subtotals as a table and skip the intermediate step of cutting out cards. Table 5.3 collects the results of this process.

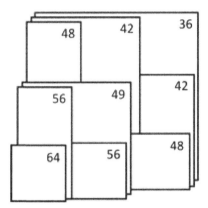

Figure 5.74. We can cut out each kind of rectangle and write the total quantity of that type in the corner. Stacking these cards suggests that we might benefit from recording this data in a table.

Table 5.3. The number of rectangles of each height and width that can be found on an 8×8 chessboard. Note that this produces a multiplication table, mirrored across a diagonal.

	1	2	3	4	5	6	7	8
1	64	56	48	40	32	24	16	8
2	56	49	42	35	28	21	14	7
3	48	42	36	30	24	18	12	6
4	40	35	30	25	20	15	10	5
5	32	28	24	20	16	12	8	4
6	24	21	18	15	14	9	6	3
7	16	14	12	10	8	6	4	2
8	8	7	6	5	4	3	2	1

There are definite patterns to the quantity of each type of rectangle! The first one you should notice is that the number of different squares is equal to — you guessed it — the square numbers. The larger pattern that emerges when you consider the entire table is that it is a multiplication table, but mirrored across a diagonal!

You could add up the number in each cell to simply produce a total, but understanding *why* these numbers appear will shed light on the general solution, and help us develop a deeper understanding of the problem.

Next, then, we can look at how many rectangles can be found in smaller boards. From the table, you can begin to calculate sums by reading the values in each cell. Starting from the bottom right corner, a 1×1 board contains 1 rectangle; a 2×2 board contains $2 + 4 + 2 = 8$ more, for a total of 9; a 3×3 board contains $3 + 6 + 9 + 6 + 3 = 27$ more, for a total of 36.

These numbers are all cubes and squares (respectively); a bit of experimentation will allow you to figure out the patterns they follow. Each increase of size (by one row and one column) seems to add n^3 more rectangles. So, the grid of side length 1 has $1^3 = 1$ square, the grid of side length 2 adds $2^3 = 8$ more squares and rectangles, and the grid of side length 3 adds $3^3 = 27$ more.

The totals form the series $1^2, 3^2, 6^2, 10^2$.... Each successive base is n greater than the previous one. That is, $0 + 1 = 1$, $1 + 2 = 3$, $3 + 3 = 6$, and $6 + 4 = 10$.

Following this series, the eighth value is 36. This suggests that the total number of rectangles contained on an 8×8 board is $36^2 = 1,296$.

You might recognize this as the definition of a triangular number, which has appeared in each of the previous chapters. It's not yet clear why it should appear here!

We will place that question aside for now. There's no better way to check our work than to work out another, independent solution for this problem, and see if it produces the same results. Perhaps a different solution will also resolve this mystery!

We can apply the general idea from the previous problem. Instead of looking at squares, we can look at *corners*. A chessboard is formed of squares, but if you trace the lines that run between each rank and

 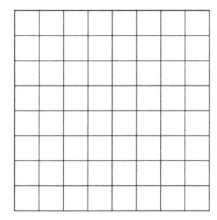

Figure 5.75. We can transform the "map" of a chessboard into a grid of squares. There are nine lines running vertically and nine lines running horizontally, and these meet at 81 different corners.

file, you can see that it also resembles a grid of lines (Figure 5.75). The line that defines one side of a square meets the line that defines the top of that square at a corner; this is also a corner for up to three other squares on the board. Let's see if we can find a relationship between the numbers of squares and the numbers of corners.

A standard chessboard has 9 corners in each row and 9 in each column (that is, one more than the number of squares). Choose a corner at random. In theory, we can pair it with any one of the 80 other corners and make a rectangle. There are 81 different ways to choose the first point, and 80 points we can pair it with, producing $81 \cdot 80 = 6,480$ different pairs of points.

In actuality, we have to make certain of two things when counting.

- First, each rectangle has to actually *be* a rectangle.
- Second, we need to ensure that we don't count any shapes multiple times.

We can address these corrections individually. In Figure 5.76, we have chosen a representative point at random. If we choose any point in the

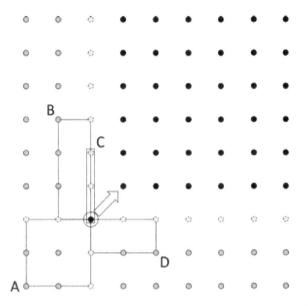

Figure 5.76. When pairing points, we must take caution so as to neither count impossible rectangles (straight lines, like the one to point C), nor to count the same rectangles multiple times. If we start pairing points from the bottom left, moving up and to the right, then each of A, B, and D produce rectangles which duplicate ones that have already been counted.

same row or column as this one (for example, point C), we will produce a line instead of a rectangle. To eliminate these lines, we need to subtract all these points from our count. Then, there are at most $(80 - 8 - 8) = 64$ points that can be paired with this one to produce a rectangle.

We can generalize this result. If there are m rows and n columns of points, $(m \cdot n)$ represents the total number of points. We won't pair a point with itself, so we subtract 1: $(m \cdot n - 1)$. Each point can't be paired with any of the other points in the same row, so we can remove $(m - 1)$ more points from the total. We also don't want to pair a chosen point with another point in the same column, so we remove the $(n - 1)$ points that lie in the same column as our chosen point.

Then from each point, there are $(m \cdot n - 1) - (m - 1) - (n - 1)$ possible ways to pair that point with another to form a rectangle. This can be simplified to: $m \cdot n - m - n + 1$ pairings.

Next, look at the rectangles formed in conjunction with points A, B, and D. If we make the pairings in a singular direction (starting at the bottom left corner and moving up and right), then the rectangle formed by A and our chosen point would have been accounted for when we looked at point A. The other two rectangles here would have been accounted for, not when drawing rectangles at B and D, but rather when looking at the lowest, leftmost point in each rectangle. By the current method, each rectangle will be counted exactly four times: once starting from each corner.

If we divide the resulting number of rectangles by 4, we correct for this overcounting. So, there are 81 (that is, $m \cdot n$) different starting points, and we have accounted for the necessary subtractions and divisions. Putting these three pieces together, the final formula produced for an *m*-by-*n* grid of points is:

$$\frac{(m \cdot n)(m \cdot n - m - n + 1)}{4}$$

We can insert the current values for and into this formula to confirm our result.

$$\frac{(9 \cdot 9)(9 \cdot 9 - 9 - 9 + 1)}{4} = \frac{(81)(81 - 9 - 9 + 1)}{4} = \frac{81 \cdot 64}{4} = 1,296$$

We have confirmed our result, but we still have to solve the mystery of the triangular numbers. Think back to the meaning of triangular numbers; you have seen them applied twice already. They are the numbers formed by adding consecutive integers in order, or the number of (say) bowling pins contained in triangles of increasing size.

In particular, if you say that the first triangular number is 0, then the formula for generating the *n*th one is

$$\frac{n(n-1)}{2}$$

There is... *something* like this in the formula we generated, but it will take a change in perspective to see it. First, look at group of terms in the right set of parentheses. If your algebra is rusty, you might not see the identity that's hiding in there; we will draw it out for you, and you can check our work.

$$m \cdot n - m - n + 1 = (m-1)(n-1)$$

Now, we can use the rules of multiplication to split apart the terms in the left parentheses, and use the same trick in the denominator to transform this formula.

$$\frac{(m)(n)(m-1)(n-1)}{4} = \frac{m(m-1) \cdot n(n-1)}{4} = \frac{m(m-1)}{2} \cdot \frac{n(n-1)}{2}$$

How remarkable! The number of rectangles contained in a grid is *definitely* related to triangular numbers! But why?

Consider the case of a 1×4 chessboard. (That is, 2 columns and 5 rows of points.) You can imagine a rectangle of each size *sliding* along the board. The largest rectangle, 1×4, cannot move at all, so it has one valid position. The next smaller rectangle, 1×3, can be either low or high. The 1×2 rectangle can be in a low, middle, or high position. And, of course, the 1×1 square can take four distinct positions. In total, there are $1 + 2 + 3 + 4 = 10$ possible rectangles on this board (Figure 5.77), which is the definition of the fourth triangular number.

If we think of these rectangles not as having a width of 1, but rather being of the board's full width, then this accounts for the entire

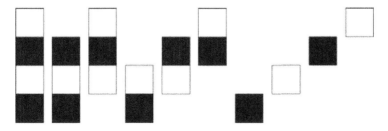

Figure 5.77.   On a 1×4 chessboard, the number of positions each rectangle can occupy relates to the triangular numbers. Here, there are 1, 2, 3, and 4 different positions a rectangle of each size can have, yielding 10 total rectangles.

number of full-width rectangles on *any* chessboard that is 4 rows high. A 2×4 chessboard must have 10 full-width rectangles; and there are again 10 rectangles of width 1, each of which can take 2 different *horizontal* positions, for a total of $(\mathbf{1} + \mathbf{2})(10) = 30$ rectangles. By the same rationale, a 3×4 chessboard must contain $(\mathbf{1} + \mathbf{2} + \mathbf{3})$ $(10) = 60$ rectangles, and so on. You can again see the triangular numbers emerging from this problem.

The appearance of triangular numbers here is no coincidence!

Now that you know *why* this problem relates to triangular numbers, you might have an easier time remembering how to solve this problem again. If, like many learners, you have trouble remembering formulas, this has only moved the difficulty from one place to another!

There is no substitute for having a necessary piece of information ready right when you need it. Even if you continue studying mathematics for years to come, particular facts might not be readily available in your memory when you need them next, ten or twenty years down the road.

The challenge now becomes, how can you improve your retention and recall of all the bits and pieces of information you learn about math, when the two facts you need to put together to solve a problem might have been learned at vastly different times?

Once again, visual problem solving can help. Memory is fickle, and the more connections you make to a particular memory, the easier it can be to recall the details later on, or to reconstruct it from pieces. Just as a specific smell can bring back rich memories of times long past, a specific mental picture can open the floodgates of mathematical memories.

Our most recent example offers a good entry point into the study of this principle.

A Gateway to Memory

Some learners believe that since everything can be looked up on the internet, there is no benefit to memorization. Not true! Just in the previous problem, we were able to benefit directly from the ability to recognize squares and cubes of numbers to identify a pattern, and used that pattern to craft a formula.

Their arguments do have some merit, though. A fact memorized separate from the context that gives it meaning is hard to use, easy to misuse, and easy to forget.

Sometimes we remember things on our own, due to repeated exposure. If you see the powers of two, or the quadratic formula, or the statement of the Pythagorean Theorem frequently enough, they are bound to stick in your head.

Sometimes the motivation to memorize is caused by repeated frustration. This is akin to trying to learn a new language, and finding yourself looking up the same word in a paper dictionary for the fifth time on a single day. Eventually, it's easier to just force yourself to memorize the word than it is to look it up again!

Here, we're not going to spend time on these cases. Instead, we are going to examine some ways of seeing problems that result in deeper understanding, more connections, and a single strong visual that acts as a focus for our work.

Triangular Numbers

We begin where we left off in the previous problem: Here our challenge is to derive a formula for the triangular numbers and thereby get a better feel for these structures.

The word "triangular" should instantly cause you to imagine an image like Figure 5.78. The first "triangle" is formed by the top row, and consists of one object (e.g. coin, orange, can of soda). The second triangle consists of the first and second rows, and *is* actually a triangle — there are three points, and we could draw an equilateral triangle around them. The entire figure, consisting of 21 objects, has six rows, illustrating the sixth triangular number.

The key to deriving the formula is to "push" the circles to one side so they lie on the points of a grid (Figure 5.79). If it wasn't clear before, you can see now that there are the same number of rows and columns. (We are using a specific value here, but this diagram works for all triangular numbers.)

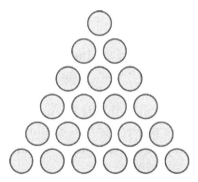

Figure 5.78. The sixth triangular number can be represented as an equilateral triangle with sides of length six.

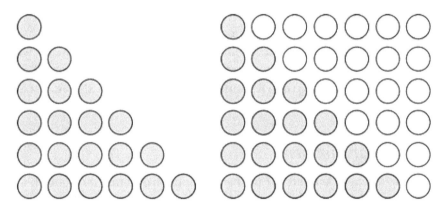

Figure 5.79. (Left) Shift the previous triangle over to one side. (Right) Fit this triangle with a copy of itself; the area of the rectangle is equal to $n(n + 1)$, and the triangle's area is half that.

Mirror the triangle over a line to create a rectangle. This rectangle's width is one greater than its height, so it contains $(n) \cdot (n + 1)$ objects. But the rectangle has twice the area of the triangle, so we must divide this total by 2, yielding the formula for the nth triangular number: $\frac{n(n+1)}{2}$.

Hexagonal numbers

Here are problem is to determine how we can do packing in three dimensions. The most efficient way to pack large numbers of round

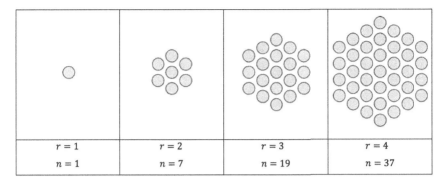

$r = 1$	$r = 2$	$r = 3$	$r = 4$
$n = 1$	$n = 7$	$n = 19$	$n = 37$

Figure 5.80.   Hexagonal packings of successive "radii." A radius 4 hexagonal packing can fit 37 circles.

(or spherical) objects into a space is to place them in layers, with each object touching six others in its layer. Because of this characteristic, this is called *hexagonal* packing.

The first four iterations of a hexagonal packing are illustrated in Figure 5.80. If we say the first of these has a radius of 1, what is the formula for a hexagonal packing of radius r?

(Here, by radius, we mean the maximal number of circles when counting from the center to the edge.)

If you squint your eyes a bit, later hexagons begin to resemble a three-dimensional shape. We can display these circles instead as spheres to make the relationship more obvious (Figure 5.81).

The hexagonal packing of radius 4 is related to the shell (a *gnomon*) of a cube of side length 4! This suggests we can relate the packing to the value of r^3.

We can frame this as a visual equation. A large cube, minus the next smaller cube, produces the gnomon of a cube (Figure 5.82).

We've found the solution, although it doesn't hurt to clean it up with a little bit of algebra:

$$n = r^3 - (r-1)^3$$
$$= r^3 - (r^3 - 3r^2 + 3r - 1)$$
$$= 3r^2 - 3r + 1$$

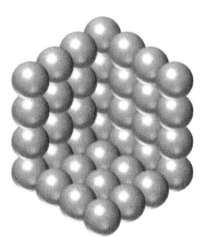

Figure 5.81.   A hexagonal packing can be viewed as the gnomon of a cube.

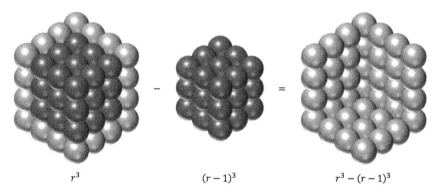

$$r^3 \qquad\qquad (r-1)^3 \qquad\qquad r^3 - (r-1)^3$$

Figure 5.82.   A visual representation of how to produce the formula for the volume of a gnomon: from a cube of a given size, subtract the next smaller cube.

In Chapter 4, we emphasized the connections between *squares of numbers* and literal squares; you can use this final formula to produce a useful mental image that again adds up to the same result (Figure 5.83).

On the subject of cubes, there is another problem we can investigate in this same vein. Back in chapter 2, we mentioned that, with the card trick we used, it was possible to identify one of 27 cards in three iterations, but we didn't explain why. We now have the skills needed to solve that problem.

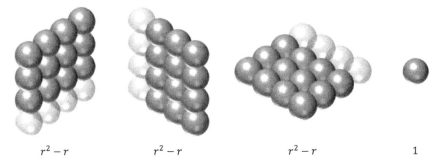

$$r^2 - r \qquad\qquad r^2 - r \qquad\qquad r^2 - r \qquad\qquad 1$$

Figure 5.83.   Breaking an algebraic formula down to shapes. Each side of the gnomon can be seen as a square minus a line; they fit together perfectly when you add one more sphere to the corner.

A 27-card trick, redux

The card trick from Chapter 2 involved laying out 27 cards in three rows and nine columns. The volunteer then indicated which of three rows contains their card, three times. After following the procedure spelled out there, our "magician" was able to identify which card the volunteer had selected!

Our problem here is to determine why the greatest number of cards that can be reliably used with this trick is 27.

Think of the volunteer's selections as three independent coordinates. Each time the magician deals out the deck, the volunteer provides one new piece of information from a set of three options. If the magic card was selected at random, it can appear anywhere in the set, so our procedure has to account for all possible locations.

Three dimensions, each with a size of three, corresponds to a $3 \times 3 \times 3$ grid of points. These points can be represented as a cube. Each time the magician deals the cards, he scoops up the cards by rows and deals them one column at a time, "rotating" the cube. In the first deal, all the cards in a single row have the same z-coordinate, then all the cards in a single row have the same x-coordinate, and finally they have the same y-coordinate.

We highlight the volunteer's selected card in black (Figure 5.84). You can see how each of her selections (gray) correspond to a slice of the cube along a single axis. There can only be one grid point (card) at

Figure 5.84. Three perpendicular slices through a cube. There is only one object that belongs to any three chosen slices.

each ordered triple (x, y, z), so the greatest number of cards that can be used with this particular trick is 27.

Figure 5.85 makes these relationships explicit. You can trace the movements of the cards between the card trick and the cube. The first deal lays the cards out flat in the order: x, y, z, and the volunteer indicates their card has a z-coordinate of 1. When the magician picks up the cards by rows and deals them in columns, the cards trace the axes of the cube in the order y, z, x. This time, the volunteer points to the row corresponding to an x-coordinate of 1. The final deal is performed in the order z, x, y, and the magic card has y-coordinate 0. (This is also exactly the card's value in ternary, or base-3, notation!)

There is no way to add more cards that does not also increase the size or the number of dimensions of the corresponding cube!

If you have a good visual memory, this transformation can make all the possible variations on this card trick easy to remember, and to understand.

We have spent a fair portion of this chapter looking at discrete mathematics problems, including graphs, paths, and grids. Discrete mathematics is not the only field where you can gain the advantages of visual problem solving — the variety of figures in previous chapters should demonstrate that! — but, again, it is a field that is particularly suited to visual solutions.

Another field that is particularly suited to visual problem solving is, of course, geometry. In Chapter 4, we demonstrated this in great detail. If we were successful, you should still remember the principle of parallax, and what "completing the square," among others, entail.

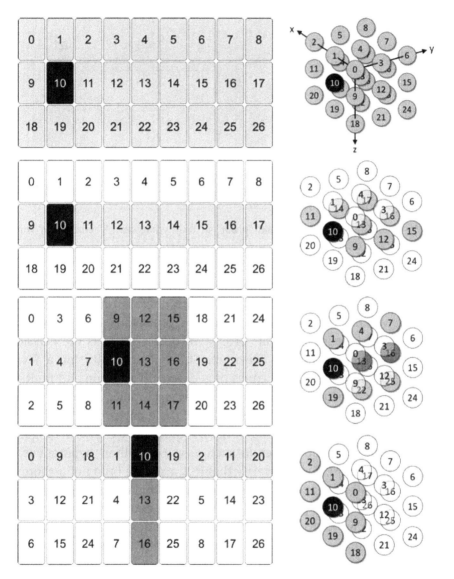

Figure 5.85. Successive iterations of the 27-card version of this trick can be viewed as slices of a 3 × 3 × 3 cube. Each time the volunteer selects a row (light gray), the magician learns one coordinate; he then picks up the cards row-by-row and deals them column-by-column, which effects a rotation.

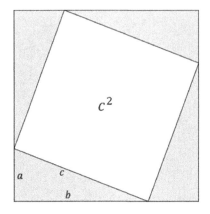 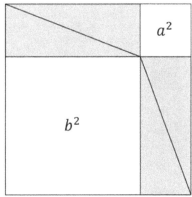

Figure 5.86. A well-known visual "proof" of the Pythagorean Theorem. Do you agree that this is a proof?

Nevertheless, in this chapter we have made the point that visual problems can improve memory and comprehension, and it's only fair that we validate this claim with problems from outside discrete math.

One of the easiest visual "proofs" to demonstrate is the classic proof of the Pythagorean Theorem. Study Figure 5.86 for a few moments, recalling the principles of geometrical addition and multiplication, and see if you are convinced this *is* a proof.

Our problem here is to explore one of the many proofs that justifies the Pythagorean theorem. In the left diagram, we have arranged four identical right triangles in such a way that their hypotenuses describe a square. You can verify that it is a square (and not a rhombus) by recalling that the angles of a triangle add up to 180 degrees, so the smallest angle plus the next larger angle equal a right angle, the three angles on each straight line sum to 180 degrees, and thus, the corners of the region marked c^2 must all be right angles.

Both of the outer squares have sides of length $a + b$, so they have equal areas. There are the same number of shaded triangles in each diagram, so the sums of those regions have the same areas. And the two regions marked a^2 and b^2 are clearly squares, with side lengths are a and b, respectively. Subtracting the shaded region from the entire square shows that the sum of the areas of the right squares equals the area of the left square. Taken altogether, we have shown that $a^2 + b^2 = c^2$, proving the theorem.

It takes a bit of mental effort to convince yourself that the diagram proves what it claims to — but this is no different than reading, and understanding, a written proof! Mathematicians generally believe that elegance is a virtue, especially when it comes to writing new proofs. So it is *always* the reader's responsibility to do enough work to convince themselves that a demonstration shows what it claims to.

There are many examples from Euclidean proofs contained in the previous chapter. You may wish to look at some of those, in light of the lessons of this chapter, and see if you can discover any unexpected insights therein.

Descartes showed once and for all that any geometric problem can be converted into algebra, and any algebraic equation can be represented graphically. The most effective demonstrations of this fit thematically into the next chapter, where we will examine them in detail. However, there are problems that straddle the line between algebra and geometry, where it might not be initially clear if the best approach is analytic or visual. Students typically spend more time in school learning algebra than they do geometry, so frequently their first impulse is to apply the tools they know best, rather than take a step back and try to see the problem from another perspective.

The final section of this chapter presents situations where the numerical approach seems best initially — but a change of perspective shows us an easier way.

Visual Solutions to Analytical Problems

The first of our analytical problems straddles the line between analytic and Euclidean geometry. Its presentation suggests taking a numeric approach, but the synthetic solution is both easier to understand, and more powerful.

Kissing quadrilaterals

A rectangle of sides of length 5 and 8 shares a vertex with a parallelogram with one side of length 5. The two quadrilaterals each have a vertex touching a side of the other, as shown in Figure 5.87. What is the parallelogram's area?

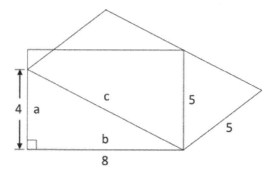

Figure 5.87. A rectangle and parallelogram share one vertex, and each one has a vertex that "kisses" the other. The point where the rectangle and parallelogram meet is 4 units from one corner of the rectangle.

The analytic approach here might start by applying the Pythagorean Theorem to the tempting right triangle on the bottom left.

$$a^2 + b^2 = c^2$$
$$4^2 + 8^2 = c^2$$
$$80 = c^2$$
$$c = \sqrt{80} = 4\sqrt{5} \approx 8.94$$

This is a bit distracting, as the lengths of a parallelogram's sides don't, by themselves, tell us anything about the area.[17]

If you're comfortable applying trigonometry, there is a way to find the area by applying certain trigonometric functions. Follow along with Figure 5.88 as we demonstrate.

The angle theta can be calculated as the inverse tangent of two known sides.

$$\tan\theta = \frac{4}{8} = \frac{1}{2}$$
$$\theta = \arctan\frac{1}{2} \approx 26.57°$$

---

[17] You can demonstrate this to yourself by imagining a parallelogram with fixed side lengths, and then "squishing" it down. The largest it can get is when all four angles approach 90 degrees and it becomes a rectangle, while the smallest area it can have is practically 0, whenever one pair of angles approaches 0 degrees!

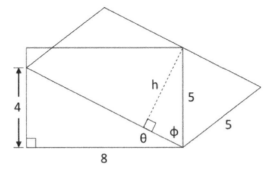

Figure 5.88. Using trigonometry to find the parallelogram's altitude. Angle theta can be found by taking the inverse tangent of the known measures 4 and 8; theta and phi equal a right triangle; the sine of phi can be used to find the altitude.

Its complement, ϕ, is equal to a right angle minus θ.

$$\phi = 90^\circ - \theta \approx 90^\circ - 26.57\phi^\circ = 63.43^\circ$$

The altitude of the parallelogram h is related to the sine of ϕ.

$$\sin\phi = \frac{h}{5}$$
$$5 \cdot \sin\phi = h$$
$$h = 5 \cdot \sin(63.43^\circ) \approx 4.47$$

Finally, we can multiply the parallelogram's base by its altitude to find its area.

$$\text{Area} = \text{base} \cdot \text{altitude}$$
$$A = 8.94 \cdot 4.47 \approx 39.98$$

That wasn't very pretty! But the final answer, after repeated rounding, seems *awfully* close to the area of the rectangle. Is that a coincidence, or is there something deeper going on here?

We can answer this question with the aid of Figure 5.89. Connect the two "kissing points" to create the shaded triangle. The area of a triangle is equal to half of its base times its altitude. We have drawn two specific altitudes here, each corresponding to a different base.

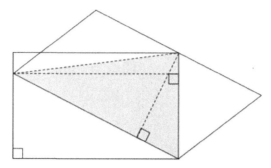

Figure 5.89. The triangle (shaded region) shares a base and altitude with both the rectangle and the parallelogram. The triangle's area must equal half the area of both these figures; therefore, they have equal areas.

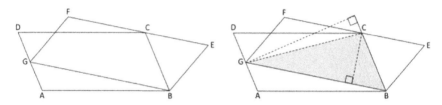

Figure 5.90. The general case of the previous solution. When parallelograms share a vertex and kiss at two other vertices, they have equal areas.

First, the triangle shares one side with the rectangle. The altitude perpendicular to this side is parallel to the rectangle's width, and runs between parallel sides, so they must have the same measure. Thus, the triangle's area is half of that of the rectangle.

The triangle also shares one side with the parallelogram. The triangle's altitude is the *same* as that of the parallelogram, so the triangle's area must be half of that of the parallelogram.

The triangle's area is half of the rectangle's, and half of that of the parallelogram, so the two quadrilaterals must have equal area!

Curiously, this problem is actually a special case of a more general theorem. In particular, and for the same reasons we have just shown, any two parallelograms (including rectangles, rhombuses, and squares, see Figure 5.91) that "kiss" in the way we just saw have equal areas (Figure 5.90).

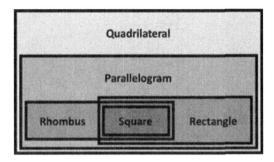

Figure 5.91.   Squares are both rectangles and rhombuses, and all three of these are parallelograms. A proof demonstrated using parallelograms applies equally to rhombuses, rectangles, and squares.

This is a yet another demonstration of the power of naming things, and having detailed definitions associated with the names we use. Whatever we can say about the general class, we can say about the specific case — if we are certain that specific case is *always* a member of the general class. A quadrilateral always has four sides; a parallelogram always is a quadrilateral where pairs of opposite sides are parallel. A rectangle is a special kind of parallelogram that has four right angles, so whatever we can say about a parallelogram, we can say about a rectangle — but the reverse isn't generally true. Figure 5.91 doesn't account for all types of quadrilaterals, but summarizes the relationships that are relevant here.

Euclidean geometry is a powerful tool to have available, especially when exploring problems in two dimensions. In higher dimensional geometries, it becomes more difficult both to visualize problems, and to find elegant solutions. Most of the general techniques for solving problems in three dimensions require calculus. Occasionally, though, a change in perspective makes problem solving a breeze.

An angle in a cube

Given the cube in Figure 5.92, what is the measure of angle *ABC*?

Method 1. There is a technique for calculating angles given points in space. We will illustrate it here to show you that there is an analytic

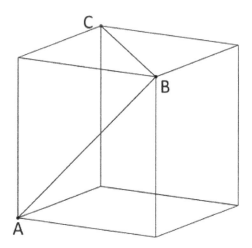

Figure 5.92.   What is the angle between three vertices of a cube?

solution, but this approach will be unfamiliar to most people who haven't taken physics or calculus.

We can express the two legs of angle *ABC* as vectors originating at point *B*. In three-dimensional space, vectors are usually expressed as an ordered triple (x, y, z). If we treat point *B* as the origin, and arbitrarily assign the x, y and z axes in directions parallel to components of the cube, we can say that *BA* is equivalent to the vector $v = (1, 0, -1)$. That is, it points out 1 unit in the x-direction, does not stick out in the y-direction, and is 1 unit *down* on the z-axis. Similarly, *BC* is equivalent to the vector $\mathbf{u} = (1, 1, 0)$.

The length, or magnitude, of a vector is its Euclidean distance from the origin. Euclidean distance is a generalization of the Pythagorean Theorem, applied to spaces with any number of dimensions. In this example, the magnitude of *BA*, $\|v\|$, is computed as follows:

$$\|v\| = \sqrt{x^2 + y^2 + z^2} = \sqrt{1^2 + 0^2 + (-1)^2} = \sqrt{2}$$

Two vectors and the angle between them can be related using the definition of the dot product. The dot product is calculated by taking the sum of the products of each pair of coordinates of two vectors.

Therefore,

$$v \cdot u = \left(x_v \times x_u\right) + \left(y_v \times y_u\right) + \left(z_v \times z_u\right)$$
$$= \left(1 \times 1\right) + \left(0 \times 1\right) + \left(-1 \times 0\right)$$
$$= 1$$

The dot product is also equal to the magnitude of the two vectors, multiplied by the cosine of the angle between them.[18]

$$v \cdot u = \|v\| \|u\| \cos \theta$$
$$1 = \sqrt{2} \times \sqrt{2} \times \cos \theta$$
$$\frac{1}{2} = \cos \theta$$
$$\theta = \arccos \frac{1}{2} = 60°$$

Method 2. Observe that both AB and BC are diagonals of faces of the cube. Observe that CA is also a diagonal (Figure 5.93). AB, BC,

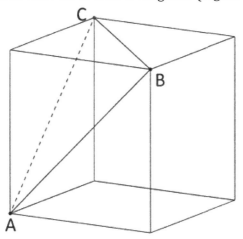

Figure 5.93.   By drawing a third diagonal, we can quickly show that ABC is an equilateral triangle. The measure of all three angles must equal 60 degrees.

---

[18] As a consequence of this, when two vectors are parallel their dot product is the same as the product of their magnitudes, and when they are orthogonal (a word more or less meaning perpendicular) their dot product is 0. This is closely related to the earlier footnote on the area of parallelograms.

and *CA* must all have the same length, therefore triangle *ABC* is equilateral.

Angle *ABC* must measure 60 degrees.

If you think of the plane formed by *ABC* as a saw cutting through the cube, you might wonder how many cuts of this type are possible, and what kind of shape multiple cuts would produce. What do you suppose can be said about that shape?

The size of a tetrahedron

If each side of a cube has length 1, what is the volume of the tetrahedron in Figure 5.94, created by cutting the cube four times through its corners?

There are many potential approaches to this problem, depending on how you see this figure, and what you remember about the volume of pyramids.

Method 1. *Direct calculation.*

The volume of a pyramid is one-third of the product of its height and the area of a base.

$$V = \frac{1}{3} A_b h$$

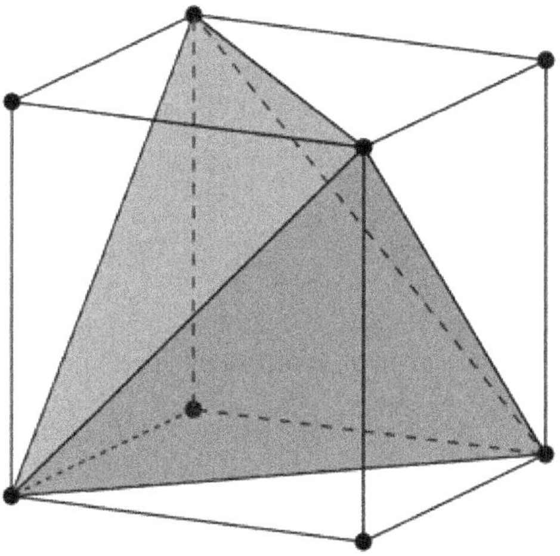

Figure 5.94. A tetrahedron is formed by cutting a cube through its vertices.

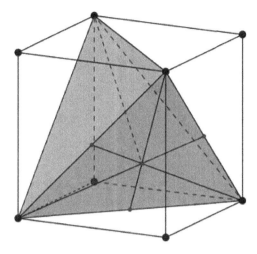

Figure 5.95.   The three altitudes of one face meet at that face's orthocenter. We can connect this point to the opposite vertex to find the height of the tetrahedron (dashed line).

The tetrahedron is a triangular pyramid, so to calculate its volume we need to first find the area of a base, and then its height.

To find the area, we need to find the altitude of one of the triangular faces. Next, we can find the height by calculating the distance between the center of a face and the opposite vertex. (We already know that the length of one edge is $\sqrt{2}$ from the previous problem.) In Figure 5.95 you can see that we have drawn three altitudes across one face, and the height from the center of the face to the appropriate vertex.

We can use the Pythagorean Theorem to calculate the altitude of a face (Figure 5.96). Here, half of one edge of the tetrahedron forms one side of the right triangle, and another edge is the hypotenuse.

You can find the relationship between the location of the center of an equilateral triangle and its altitude using a variety of methods. We will use, without proof, the fact that the center of an equilateral triangle divides its altitudes into parts in the ratio 1:2. Then, we can find the height of the tetrahedron using one of the two right triangles shown in Figure 5.97. The first has for hypotenuse the altitude of one face, and a base one-third the length of an altitude; the hypotenuse of the other is one edge of the tetrahedron, and its base is two-thirds the length of an altitude.

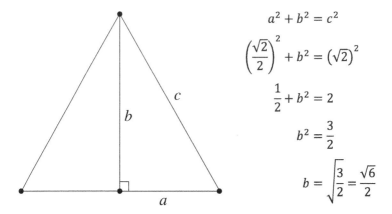

$$a^2 + b^2 = c^2$$

$$\left(\frac{\sqrt{2}}{2}\right)^2 + b^2 = \left(\sqrt{2}\right)^2$$

$$\frac{1}{2} + b^2 = 2$$

$$b^2 = \frac{3}{2}$$

$$b = \sqrt{\frac{3}{2}} = \frac{\sqrt{6}}{2}$$

Figure 5.96.   Finding the altitude (b) of a single face of the tetrahedron.

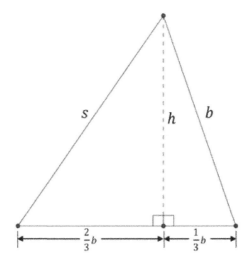

Figure 5.97.   A slice of the tetrahedron running through one edge (s), the altitude of a face (b), the height (h), and the orthocenter. The orthocenter lies two-thirds of the way across the altitude of a different face.

Both calculations should produce the same result.

$$\left(\frac{2}{3}b\right)^2 + h^2 = s^2 \qquad\qquad \left(\frac{1}{3}b\right)^2 + h^2 = b^2$$

$$\left(\frac{2}{3}\cdot\frac{\sqrt{6}}{2}\right)^2 + h^2 = \left(\sqrt{2}\right)^2 \qquad\qquad h^2 = b^2 - \left(\frac{1}{3}b\right)^2 = \frac{8}{9}b^2$$

$$\frac{2}{3} + h^2 = 2 \qquad\qquad h^2 = \frac{8}{9}\left(\frac{\sqrt{6}}{2}\right)^2 = \frac{8}{9}\cdot\frac{6}{4} = \frac{4}{3}$$

$$h^2 = 2 - \frac{2}{3} = \frac{4}{3} \qquad\qquad h = \frac{2\sqrt{3}}{3}$$

$$h = \frac{2\sqrt{3}}{3}$$

Putting this all together, we can compute the volume of the tetrahedron.

$$V = \frac{1}{3}A_b h = \frac{1}{3}\cdot\left(\frac{1}{2}\cdot s\cdot b\right)\cdot h$$

$$= \frac{1}{3}\left(\frac{1}{2}\cdot\sqrt{2}\cdot\frac{\sqrt{6}}{2}\right)\cdot\frac{2\sqrt{3}}{3} = \frac{1}{3}\cdot\frac{\sqrt{12}}{4}\cdot\frac{2\sqrt{3}}{3}$$

$$= \frac{1}{3}\cdot\frac{2\sqrt{36}}{12} = \frac{1}{3}\cdot\frac{2\cdot 6}{12}$$

$$= \frac{1}{3}$$

Method 2. *Indirect calculation.*
If we take the formula for the volume of a pyramid as given, then there are four identical pyramids that have been subtracted from the original cube. Each one is composed of a right triangular base with sides of length 1, and height of 1. (see Figure 5.94.)

Calculating the volume of a single pyramid is a straightforward process.

$$V_{\text{pyramid}} = \frac{1}{3}A_b h$$

$$= \frac{1}{3}\left(\frac{1}{2}\cdot 1\cdot 1\right)\cdot 1$$

$$= \frac{1}{6}$$

Given that there are four of these pyramids, their sum of their volumes is 4 times the previous result. We can subtract this number from the volume of the cube to find our answer.

$$V_{\text{tetrahedron}} = V_{\text{cube}} - 4 \cdot V_{\text{pyramid}}$$
$$= 1 - 4 \cdot \frac{1}{6}$$
$$= 1 - \frac{2}{3}$$
$$= \frac{1}{3}$$

This method requires you to already know the formula for the volume of a pyramid. There is a way to calculate this volume without knowing that formula, though!

Method 3. *The Euclidean method.*
View the cube from the side, as shown in Figure 5.98. The plane running through the center of the diagram (toward the viewer) divides the tetrahedron in half. Each of these halves form shorter tetrahedrons whose volumes are equal.

The plane cutting through the other four vertices contains the altitudes of both the smaller tetrahedron, and one of the pyramids that has been cut away. Using parallel lines (the long dashed line in

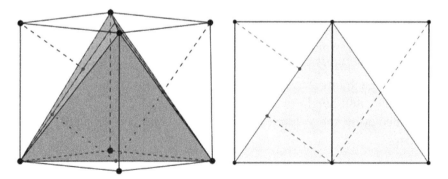

Figure 5.98. Looking at the cube from a particular angle allows us to identify parallel lines and congruent triangles, demonstrating that half the tetrahedron has the same altitude as one cut-away volume.

Figure 5.98), or congruent triangles, we can show that these altitudes are of equal length.

These two solids have a common base, and their altitudes are equal, so they have the same volume. Also, the half pyramids in front of and behind the half tetrahedron have equal volumes, and the sum of their volumes is equal to the volume of one full pyramid.

Each half of the cube has been divided into three equal parts, one of which originated from the tetrahedron. So the tetrahedron has one third the volume of the original cube.

This method is a modified version of Proposition 12.7 from Euclid's *Elements*, where he first demonstrated the formula for the volume of a pyramid. So, in a way, the previous methods used a kind of circular logic — we used a formula in a situation, which was used to derive the formula in the first place!

In actuality, all three of these methods are perfectly valid, since Euclid's proof is sufficient to show that we can rely on the formula in any situation. In a situation like this, the approach you take comes down to the way you view the problem, and the tools you have available.

Many magnets

Nine magnets of equal strength are placed in a cubic test chamber. One is suspended exactly in the exact center of the chamber, and the other eight are embedded in the chamber's eight corners.

A piece of metal floating at a random point in the chamber will be drawn to the magnet that it is closest to. If you select a random point inside the chamber, what are the odds that it is closer to the central magnet (Figure 5.99) than to any one of the corners?

Calculating the exact volume of this region would require some tricky geometry, indeed. As always, there is a better way to find the solution.

First, consider what would happen in analogous situations in lower dimensions. If two magnets were placed on a line (that is, one dimensional space), that space would be divided into two halves (Figure 5.100, left). Everything to the right of the midpoint would be

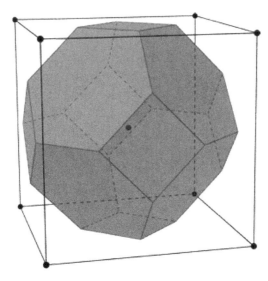

Figure 5.99. Nine magnets in a cubic chamber. Eight are embedded in the cube's vertices, and one is suspended in the center of the chamber. The volume shaded gray is the region of the chamber that is closest to the central magnet.

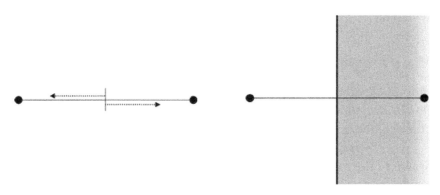

Figure 5.100. (Left) In one dimension — on a line — two equal strength magnets divide the space in half at a single point. (Right) In two dimensions, the same magnets divide the plane in half at a line.

pulled right, and everything to the left of the midpoint would be drawn left.

If we move up to two dimensions while keeping the number of magnets the same, the plane would be divided in half by a line

(Figure 5.100, right). Continuing the analogy, it seems reasonable to believe that two magnets arranged in a three-dimensional space would divide that space in half at a plane.

Our next step will be to construct one- and two-dimensional versions of Figure 5.99.

The one-dimensional analogue of the problem produces a diagram very similar to the one just depicted in Figure 5.100. A reasonable conclusion to draw from Figure 5.101 is that half the length of the line is closer to the center magnet than the ones on the outside.

Similarly, the two-dimensional analogue — using a square, as seen on the right half of Figure 5.101 — seems to cover about half the plotted region. Dotted lines drawn through the center magnet and perpendicular to the sides show that each of these sub-regions are symmetrical, with half of each smaller square closer to the center than the outside, demonstrating that the outside square has been divided in half.

If you have any expectation at all regarding the original, three-dimensional, version of this problem, it is probably a belief that the test chamber is divided in half by the magnets. How can we prove it? Well, just as we can use analogies to carry an idea from lower to higher dimensions, we can sometimes use the same strategies in both lower and higher dimensions!

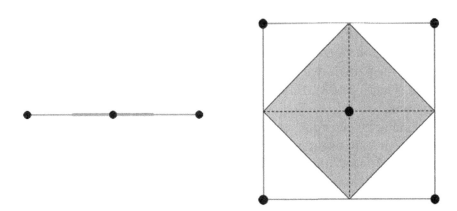

Figure 5.101. Magnets on a line (left) and on a square (right) split these spaces exactly in half.

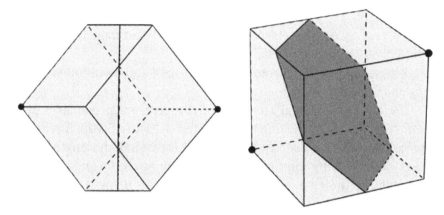

Figure 5.102. Two magnets at opposing vertices of a cube split the cube in half.

If we divide the chamber into eight cubic sub-chambers, each one having the central magnet at one vertex and one of the corner magnets at the opposing vertex, we can see that each of these sub-regions has been divided exactly in half by the two magnets (Figure 5.102). Since half of each cube is closer to the center than the corner, the volume of the original shape must be half of that of the original cube!

Therefore, the odds that a randomly selected point is closer to the central magnet than the corners are 1 in 2.

We have our answer, but before we close this investigation, we will make two quick observations from the previous illustration. First, by symmetry, each edge of the original shape must have the same length, whether it's a side of a square and a side of a hexagon or of two hexagons. You can see this by "unrolling" a single cube like a die and tracing the path drawn by the intersection of the cube and the cutting plane (Figure 5.103). This may not have been clear from the original diagram!

Finally, by extending each hexagon past the walls of the bounding cube, you can begin to imagine the basis of the original shape (Figure 5.104). Without the chamber in the way, the magnets would describe an icosahedron!

It was no accident when we phrased this problem as a question of probabilities. Sometimes, a shape is just a shape. Other times, a shape represents something deeper.

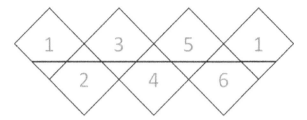

Figure 5.103. The six faces of a cube, unrolled. The line created by the intersection of the cutting plane and each of the cube's faces is illustrated with a bold line.

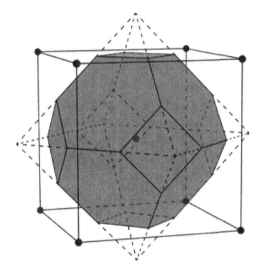

Figure 5.104. Extending the surfaces of the original shape through the faces of the cube to demonstrate that the magnets create an icosahedron.

The collaborating colleagues

At a particular company, there are two employees, A and B, who work in the Research and Development department. A and B perform their best work when they collaborate. They each spend their mornings checking email and attending meetings, and meet up at the cafeteria at lunch.

Neither one is good at planning ahead, so they rely on serendipity to arrange their meetings. A and B each go to the cafeteria somewhere between noon and 1 pm, and spend exactly 15 minutes eating. If their colleague doesn't arrive in that time, each of them assumes that the other is not coming, and returns to their office.

On any given day, what is the probability that they will meet up?

It's not clear whether there is a good analytical solution. Almost all the probability problems we have addressed so far started by counting something and finding a ratio. But what, if anything, could we count here?

Eventually, you might decide to randomly test pairs of times. For instance, if A arrives at 12:17 and B arrives at 12:46, they won't meet. Choose two times, check to see if they are within 15 minutes of each other, and repeat. This approach, known as the Monte Carlo method,[19] is perfectly valid, but requires a truly large number of tests to approach the true value of a solution. Because of this, Monte Carlo methods work best when performed by computers. We are mere humans, unable to perform the millions of calculations this approach would require. A systematic approach will work best here.

If A and B both show up at noon, they will meet. If they both show up at 12:30, they will meet. In fact, if they both have lunch at the same time, they will always meet.

If A has lunch at noon and B shows up at 12:30, they won't meet. By symmetry, the reverse is also true: if B eats at noon and A shows up at 12:30, they won't meet.

We can quickly test all the combinations and multiples of a half hour, but that doesn't reveal much: the data tells us they meet one third of the times tested (i.e., at 3 results out of 9). So, we can zoom in, and look next at all the multiples of a quarter hour.

Something interesting happens when you look at intervals of a quarter hour: you begin to ask yourself, if A shows up at noon and B shows up at 12:15, do they meet? That is, if A eats for 15 minutes, wouldn't she see B as he's showing up? How punctual are they, really?

Eventually, you should decide to take the problem at its word: A is wrapped up in her research, or leaves through a different door, so she doesn't see B as he arrives, *if* he shows up even a fraction of a second

---

[19] Named after the famed Monte Carlo Casino in Monaco.

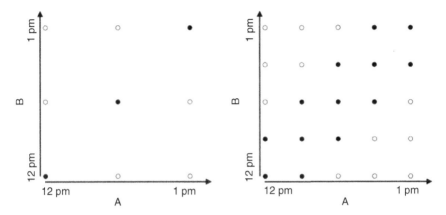

Figure 5.105. Plotting the colleagues' arrival times at different levels of granularity. (Left) If checked at half-hour intervals, the employees meet 3 out of 9 times. (Right) At quarter-hour intervals, the employees meet in 13 of 25 tested cases.

late. A difference of 15 minutes draws a sharp divide between the possibility of A meeting B, or them missing one another.

Using this determination, we can count results to find that A and B have a 13/25, or 52%, chance of meeting up.

Somewhere during this process, you may have realized that A and B represent independent variables, and used that fact to plot results. The first two iterations of this process — at 30, then 15-minute intervals — are shown in Figure 5.105.

There were 9 different combinations of times at the half-hour level, and 25 at the quarter-hour level. The next interval we might try after this is five minutes. But there are 169 combinations possible at that level of granularity.[20] This amount of testing demanded by this approach increases quickly!

You could, nevertheless, begin to fill in intermediate points. The diagonal lines that began to form at 15 minutes granularity represent the boundary between A and B meeting, or not. At 5 minutes of

---

[20] This calculation is similar to the one for calculating the number of corners in a chessboard. There are 13 different times A could arrive at the lunchroom: 12:00, 12:05, 12:10... 13:00, and 13 different times B could arrive. Multiplying together yields 169 different times.

granularity, the diagonal lines would need two more black points added between each pair, and every other grid point would be colored in accordance with which region it falls in.

There are ways to compute the number of points of each kind at increasing levels of detail without explicitly testing each one.[21] But again, all the problems in probability we have seen so far involved counting, and it's not clear when there will be enough points to produce an accurate estimate.

What if we greatly increase the number of points under testing? What if we test all the points at intervals of a microsecond, or a nanosecond, or the smallest possible unit of time?[22] It would be impossible to count these billions and trillions of points by hand! But, we might be able to "count" them using a different method than we're used to. What if we measure the *area* of each region to determine the probability of each occurrence?

Figure 5.106 illustrates the end result of this train of thought. One way to look at this is that there are *countless* points under consideration. Another way to see it is to say that if you choose a point randomly, with each *x*-coordinate corresponding to A's arrival time and each *y*-coordinate corresponding to B's arrival time, you can see if they meet based only on which *region* the point falls in.

Either way, calculating the relative areas now is a breeze. The probability that the two colleagues meet is 7/16, or 43.75%.

You can see how this problem relates to the previous one. Both measured the sizes of particular spaces, but while it was clear that was what was happening in the magnet problem, here, the mental leap may not have been so obvious.

---

[21] For instance, you could find the appropriate triangular numbers on the top and bottom and subtract them from the square. If we test every point in a one-minute grid, there are $61 \cdot 61 = 3{,}721$ total grid points, and $\frac{44 \cdot 45}{2} \cdot 2 = 1{,}980$ points where A and B don't meet. This results in a $\frac{1{,}741}{3{,}721} = 0.4679\ldots = 46.79\%$ chance that they meet.

[22] Readers familiar with calculus will probably think that we could take the limit of a function here — and indeed we could!

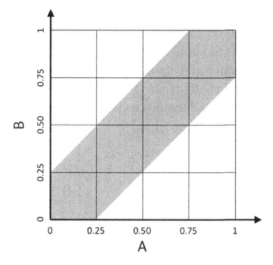

Figure 5.106.   Turning discrete intervals into continuous regions. The hour between noon and 1 pm has been translated into a region of the plane between 0 and 1; the three regions here can be called "B arrives later than A," "A and B meet," and "A arrives later than B." The probability that they meet is related to the area of the shaded region.

From maps to graphs to plots, this chapter has demonstrated many of the ways problems can be seen, and solved, visually. In the process, we have seen that many seemingly unrelated problems can be solved using the same abstraction, despite their differences.

Visual problem solving is our bridge between real-world problems and the abstract symbol manipulations many students think of as "real math." Visualizations connect the real and the abstract, and act as a check on our intuition. You can't easily map a mathematics problem to smell, or sound, or taste, or touch, but it's often possible to draw a little picture of a mathematics problem to tie your thoughts and senses together.

Sadly, there are limits to visual problem solving. The visual proof of the Pythagorean Theorem from this chapter is a good example of this. The image we provided — while useful, memorable, and enlightening — cannot easily stand on its own, without explanation or justification. And, outside certain branches of mathematics, the most broadly useful methods of proof and problem solving would be impossible without the use of abstract symbols.

We will continue to use visuals for all reasons mentioned in this chapter: they allow us to cut a problem down to its essential elements; they help us see connections between problems; and they help deepen understanding and memory. But only looking at problems that can be solved visually is an artificial limitation, and in the next chapter we will go beyond the mere visual to experience the power of symbolic problem solving.

Chapter 6

Problem Solving for Mathematics in General Usage

Each of the previous chapters tackled a specific aspect of problem solving. Whether entertaining, or counterintuitive, or important, or visual, those problems all had at least one thing in common with each other.

By contrast, everyday problem solving can feel like it has none of these things. "Mathematics in general usage" probably makes you think about dreary calculations of budgets and sale prices. But these need not be the only types of problems you solve in your day-to-day life!

If the previous chapter was about finding a way to translate problems into pictures, this chapter extends the idea of translation, but from written and spoken words and sentences into mathematical "sentences." As you will see, the inspiration for these sentences can come from anywhere: from your life, the news, or simply from the world around you.

All About Numbers

A necessary step toward developing effective mathematical thinking is to learn how to use the most basic mathematical "words" — numbers.

You may think you know how to use numbers, but there are always ways to enhance our knowledge!

What is a number? We explored the history of the simplest numbers in Chapter 4. As you may recall from that chapter, in ancient times, mathematics was most commonly used to account for physical goods. But the number systems we discussed then were not the earliest form of numbers, and to this day in some remote corners of the earth, we can find systems that are even more basic.

Correspondences

Isolated tribes exist to this day that do not have words for large numbers. We don't mean this in the sense that Archimedes struggled to talk about numbers larger than a myriad (10,000) in *The Sand Reckoner* — typically, these tribes do not have words for any number larger than two! You may have heard of these systems by reference to the saying "one, two, many."

This is not to say that ancient peoples and remote tribes could not, or cannot, solve problems involving larger numbers. Imagine if someone approached you with a basket of fruit, and told you that you could take one piece for each person in your family. Without using any number larger than two, how could you make sure you have enough for everyone?

The simplest method, which has been observed by anthropologists, is to enumerate each person. Therefore, you might take one piece for your mother, and one for your father; one piece for your sister, and one for your brother. Counting yourself, you would take five pieces of fruit. (But you're not allowed to say "five!")

In small, close-knit communities where everyone knows each other's name and no one accumulates property, this system can suffice. For better or worse, no one reading this book lives in such a society!

Using a slightly more advanced strategy, it is possible to work with larger quantities. Imagine that you are an illiterate shepherd living in ancient Mesopotamia. A shepherd's first duty is to ensure that every sheep in his flock is safe and accounted for. Without using numbers, how could you ensure that all your sheep have returned at the end of the day?

One method you could use would be to correlate each sheep to an object, such as a pebble.[1] Each morning, as you let the sheep out of their pen, you could place a pebble into an empty bucket. At the end of the day's grazing, you could remove a pebble from the bucket as each sheep returned to the pen. If the bucket is empty as the last sheep returns, you can be confident that you have the same number of sheep in the evening as you did in the morning!

Natural numbers, that is, positive integers, are the first type of number learned by every schoolchild. They are *natural* because they reflect a one-to-one correspondence between objects. If you have five pieces of fruit and five family members, you can give one piece of fruit to each person, even though fruit and people aren't similar in any meaningful way.

What do numbers themselves gain us? Unlike pebbles, numbers weigh nothing. If a royal tax collector approached the illiterate shepherd from the previous story and asked how many sheep he had, it would be far easier for both of them to share sounds or symbols ("one-hundred," or 100) than for the tax collector to carry away the shepherd's pebbles. In this regard, using numbers is like attaching a "handle" to the *idea* of 100 pebbles, without needing to carry the pebbles themselves.

Numbers are not the things they represent, but they are so closely intertwined that for many people, it's difficult to think of numbers of objects without thinking of the words we use to describe those numbers. Sometimes, we can get so caught up on the words of a problem that we forget to think about what the problem represents. Our next two examples demonstrate how helpful it can be to "think like a shepherd."

---

[1] A version of this method was used well into the 19th century by shepherds in the British Isles, although in these instances pebbles were used to count scores of sheep (that is, groups of 20 sheep).

The dishonest bellhop

This classic puzzle demonstrates how some tricky wording can obscure a simple solution.

Three mathematicians, Alice, Roberta, and Eve, head to a distant city for a mathematics conference. A single night at the conference hotel is listed as costing $30. To save money, they agree to share a room, each putting in $10 for the first night's lodging.

That evening, the hotel manager is looking over the books when he realizes that the conference rate is actually $25, $5 less than what the guests paid. He hands the difference to the bellhop, and instructs him to carry it up to the guests' room.

On the way, the bellhop realizes that there's no easy way to split $5 three ways. He pockets $2, and hands the remaining $3 to the mathematicians. They are happy to get a lower rate, and split the money evenly.

After the refund, each mathematician has paid $9 for the room, for a total of $27; the bellhop has $2. These numbers add up to $29, but we started with $30. What happened to the missing dollar?

This problem is both counterintuitive and entertaining, and could easily fit in either of those chapters. It is the kind of puzzle that can engage and frustrate a room full of people for hours, but the approach we are choosing to present here makes it especially suited for inclusion in this chapter.

First, using the habits you learned in Chapter 3, you should be inclined to think that there is no "missing" dollar. So the question now is, what about the *phrasing* of this problem makes it seem like a dollar went missing?

In a party setting, often someone will strike upon the idea of using real money to follow each step of the problem. And sure enough, *thinking like a shepherd* here makes the solution obvious!

Let's trace the flow of the money. (You can follow along in Figure 6.1.) In the first step, Alice, Roberta, and Eve each give $10 to the hotel clerk. They have $0, and the hotel has $30, as expected.

Next, the manager hands $5 to the bellhop. Now, the hotel has $25 and the bellhop has $5, once again adding up to $30.

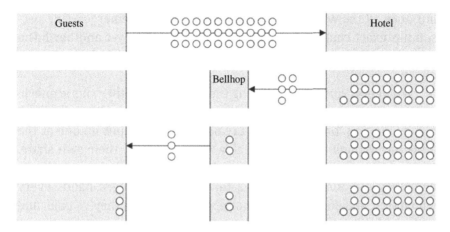

Figure 6.1.   From top to bottom: The guests pay $30 to the hotel; the manager gives $5 to the bellhop; the bellhop gives $3 to the guests. At the end, the guests have $3, the bellhop has $2, and the hotel has $25 — and there is no "missing" dollar.

Now, the bellhop pockets $2, and hands $1 to each of the mathematicians. The hotel has $25, the bellhop has $2, and the mathematicians have $3 combined — naturally adding up to the original sum!

Where is the missing dollar? It doesn't exist, as we have just demonstrated. So what happened? The answer should jump out at you when we examine the previous diagram in detail. In the final configuration, we can see that while the guests "paid" $27 (that is, $25 + $2) to the hotel — they actually paid that sum to the hotel *and the bellhop*. After performing some arithmetic, you can find that each guest paid $$\frac{25}{3}$$ (or about $8.33) to the hotel, and $$\frac{2}{3}$$ (about $0.67) to the bellhop. Adding the bellhop's $2 to the guests' $27 is non-sensical, since his $2 is *already* part of that $27. It's easy to overlook this detail when hearing the problem out loud!

This problem may feel contrived, but it resembles situations you may have found yourself facing in the past. Imagine you're at a restaurant with a group of friends, and the check arrives. After taxes and tip, the total works out to an odd number — for instance, $13 per person. One of your friends brought only a credit card, and offers to take cash from everyone and put the entire bill on their card. Everyone agrees to this, but once all the wallets are open, a new challenge emerges. One person has exact change; another only has large bills; yet another has a mix of bills (say, a $10 bill, some $5 bills and several $1 bills), and so on.

When everyone starts pooling their money together, this scenario can quickly descend into chaos. Your friend with a wallet full of $20 bills is looking for someone to break one of them, but no one at the table has enough change to break a $20 and to pay their own share. Meanwhile, someone else traded a $5 bill for four $1 bills, and says now they only need to pay $12 total. Money changes hands every which way, and soon no one is quite sure how much they've paid, and how much they owe.

In a situation like this, it's possible to track the movement of money in a way that *is* fair, but more importantly *feels* fair to everyone. Instead of turning the table's negotiations into one big story problem, treat each transaction individually. "I put $20 in," says one of your friends, "so I should get $7 back." When he gets his change, his involvement in this check problem ends. One by one, everyone at the table puts down *at least* enough money to pay their share (preferably in smaller denominations), and take the change they need from the accumulated stack of bills.

At the end, you will be faced with one of two cases. Either you were able to make exact change for everyone, and can close the bill; or, some subset of the original group is still owed change, at which point you can politely ask the waiter to make change for just a few large bills.

The individual components of these transactions — breaking a bill with friends, and the bellhop problem — come straight out of grade school arithmetic, demanding nothing more than a little addition and subtraction. But in both of these cases, it's easy to get tripped up by words, and to lose track of the things those words represent!

Number sense

We can use all kinds of words and symbols to represent numbers. Natural numbers — 1, 2, 3, and so forth — are just one type of

Figure 6.2. The number line. Every real number (0, 2, 4.3 trillion, π, −24.17) exists somewhere on this line.

"handle" we can use to come to grips with numerical problems. Different numerical expressions have distinct strengths, weaknesses, and behaviors despite the fact that they (in some sense) all talk about the same thing.

Once again we ask, what is a number? In the vast majority of situations,[2] the problems you solve will have solutions that can be found on the number line, as shown in Figure 6.2.

A fundamental property of the number line is that every point represents a quantity that is either greater, smaller, or equal to every other. If we select two points (not necessarily different) from the number line, they will be *comparable* (that is, greater than, lesser than, or equal to one another) even if we can't say precisely what their values are. We can state that every number, even those that cannot be written down,[3] exist somewhere on this line.

You may recall this as similar to the strategy used by the Greeks, as demonstrated in Chapter 4. Many of the problems the Greeks solved started by comparing the sizes of lines without ever considering their numeric lengths. They took this approach to avoid thinking about numbers, such as the square root of two, whose mere existence was controversial. This had the unfortunate side effect of limiting the practical applications of Greek mathematics.

---

[2] Ignoring other types of numbers, such as complex numbers, quaternions, and so forth.

[3] Most of the numbers you have ever dealt with have finite decimal expression (like 2.314), or repeating decimal expressions (like 0.66666...). If you choose a point at random on the number line, however, it will almost certainly have an infinite, non-repeating decimal expression!

Consider the number pi. As far as we know, the digits of pi never repeat: 3.14159265358979... Almost every number on the number line has an infinite number of digits!

We do not want to be limited in that way! Our goal is to be able to understand, and solve, as many different types of problems as possible. To do so, we must cultivate three distinct skillsets. First, we need to be able to make sense of numbers, no matter their form — in a sense, to humanize both questions and answers. Second, we need to be able to perform mechanical operations on the symbols we use to represent numbers. This skill is the one most frequently developed in mathematics classes, with or without the others. Finally, we must develop intuitions about which types of symbols (whole numbers, fractions, algebraic expressions, etc.) will yield the best results for a given problem.

The final skill here will be the most difficult to adequately communicate. As always, practice and play are your allies when trying to develop this skill. For the moment, we will point to a single concrete example of this, which you can meditate on in your own time.

In Chapter 5, we investigated the relative areas of a rectangle and a parallelogram that "kissed" at two common points (Figure 6.3). The first time we attempted to solve that problem, we used decimal numbers rounded to two places in the intermediate steps. At the end we discovered that the parallelogram's area was 39.98, which is remarkably close to 40, a whole number. When we took a different approach, we discovered that this was no coincidence — the two shapes had the exact same area, 40 square units! Treating the problem symbolically instead of numerically brought us to a solution that was "more true."

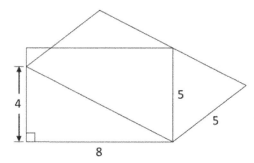

Figure 6.3.   The "kissing" quadrilaterals from Chapter 5 each had the same area — 40 square units.

Choosing the "best" representation of a problem or number often produces a "nice" solution, which can deepen our understanding of the problem itself.

The three skills we have identified — humanizing numbers, performing mechanical operations, and developing intuitions about numerical representations — are not easily extricated from one another. Instead of trying to take them on directly, we will turn the question around, examining these themes in the context of different number systems. We have addressed many of these representations in isolation, but now it's time to bring them all together.

A numerical menagerie

How many different ways can you think of to use a word like "four" to describe different numbers?

If we restrict ourselves to natural numbers, four has a single, clear meaning. You can visualize "four apples" quite easily. What about "four bushels of apples"? If you have never been to a farm, you might struggle to visualize the size of a bushel. Nevertheless, you might recognize that a bushel is a measure of size, like a liter or gallon. So it's not much of a stretch to imagine four identical containers, each one filled with "a bushel" of apples.

These examples are superficially similar, as in each case the number used correlates to some visualizable collection of objects. The first case uses the unit "apples" (or an implied unit "objects"); the second uses the unit "bushels."

We can also talk about more complex units of measure. You might see apples in a store priced at four dollars per pound. There's a good chance you don't know how many apples there are in a pound, in which case you could use the store's scale to weigh your produce and estimate how much it will cost. We can combine these different measures to solve problems, such as "if an apple tree produces 5 bushels of apples on average, and there are 48 pounds of apples per bushel, how much money will a farmer make by selling the apples for $4 per pound?" As we will soon see, handling units of measure is a lot like performing algebra, and can ease the transition from arithmetic to algebra.

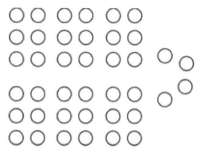

Figure 6.4. Six 6-packs, plus four sodas: One way to imagine the number forty using cans of soda.

How else can "four" be used? We can talk about larger numbers, like "forty" (four tens). Using a little imagination, you should be able to visualize forty objects. Cans of soda can be purchased in six-packs, and for most readers it shouldn't be much of a stretch to imagine six of these lined up, with four more cans sitting next to them (Figure 6.4).

What about larger numbers, such as "four hundred," "four thousand," or even larger — the planet Neptune is over four *billion* miles away, and the United States Government had a budget of four *trillion* dollars in 2017. Are you able to visualize these numbers in your head?

These are all natural numbers, at least by the mathematical definition. If you count upwards, you will eventually reach each one of these numbers, but it might take a very, very long time! In a human sense, we're not *naturally* equipped to conceptualize or handle such large numbers — "natural" or not.

What about smaller numbers? The first number that comes to mind might be "point four," that is, 0.4. This also goes by another name: "four tenths," written 4/10. If we were instead speaking of percentages, you might call it "forty percent," or 40%. From the perspective of the number line, these three names all describe the same point, but they are best suited to very different problems. For instance, it's easier to compare the sizes of two numbers in decimal notation than two fractions, especially if the fractions don't share a common denominator.

Moreover, there are nuances within a single representation. "Four percent" and "four percentage *points*" are very different things.

Much to the chagrin of statisticians, laypeople frequently conflate the two! If a political candidate, who was polling at 50% last week, is up four percentage points (or "four points") in the polls, then this week she should be polling at 54%. But if she is up four *percent* in the polls, then this week her numbers should have risen by $0.50 \cdot 0.04 = 0.02$, or two *points*, to 52%.

If you think of some of these numbers as mathematical operations (4/10 is a division by ten; four billion is the result of multiplying four by one billion), then what can we say about other operations? If we tell you that a particular cube has sides of length 1.587 and ask you to compute its volume by hand, it might take you quite a while. If, on the other hand, we told you that its sides are of length "cube root of four," or $\sqrt[3]{4}$, that same calculation becomes almost trivial. The cube must have a volume of 4.

We can also extend the idea of the same number having different names. What we call "four" in the decimal system is written as 100 in the binary system, or can be written as 11 in the ternary (base-3) system, and 10 in the base 4 system. How you choose to say each of these numbers depends on what it most useful at the time. In base 4, is the number that is written "10" called "ten," or is it still called "four"? More importantly, do the properties of a number change when we write it differently?

These are just some of the ways we can talk about numbers. In every case we've identified, the "handle" metaphor holds true. The words we use allow us to "pick up" specific numbers from the number line, and drop them into mathematical problems.

But how can we make sense of all of these systems? Happily, each of the preceding examples can be grouped into one of three major categories. There are numbers that are attached to *units*, such as "bushels," "dollars per pound," "billions," and "trillions." There are numbers that are associated with *operations*, such as division, multiplication, and roots. And there are numbers, and problems, that are closely related to the *number system* used. Since we have already dealt with many problems involving number systems, it's only natural that we start with them.

Applied number theory

Leonardo of Pisa, also known as Fibonacci, was one of the foremost mathematicians of his time. The first sentence of his book *Liber Abaci* (1202) described the Hindu–Arabic decimal numbers for the first time in the Western World, and he used these numbers to introduce European readers to a formidable system of computation, which we know as arithmetic.

His popularization of the decimal numbers revolutionized arithmetic. The rules for working with positional decimal numbers simplified accounting and trade across Europe. These rules have been disseminated from generation to generation for over 800 years; you no doubt learned them in grade school.

It's possible to perform basic arithmetic just like the shepherd from the earlier story, by adding and removing pebbles from a bucket. First grade mathematics classes begin with one-digit addition and subtraction because these operations are easy: children can check their work by counting on their fingers. At this age, a mathematics teacher's greatest challenge can be convincing students to learn the methods, because eventually they'll run out of fingers!

By the middle of grade school, students have generally learned the four basic operations: addition, subtraction, multiplication, and division. In theory, at this age children should be quite capable of applying arithmetic operations to numbers of any size, provided they have enough time. But arithmetic "by the rules" can be quite tedious, and many students become receptive to learning methods that allow them to "short circuit" the normal rules.

What arithmetic shortcuts do you use? Oddly, you can view memorization as the one of the first major shortcuts. We are not advocates of rote learning for its own sake, but — hearkening back to an earlier analogy — if you have to look up the same word in a dictionary five times, in the long run you can save time by memorizing it! Students who memorize the multiplication tables early save countless hours throughout their mathematical careers.

The next most common shortcuts developed by (and increasingly, taught to) children involve borrowing, or grouping. The general principle here is surprisingly simple: powers of ten are easier to work with than other numbers, so if we take "pebbles" out of one number and add them to the other, we can build powers of ten to find solutions with greater confidence.

A brief demonstration follows. How would you add the numbers 83 and 39 in your head?

We can take 7 from 39 (leaving 32) and add the 7 to 83, making 90. Next, we can take 10 from 32, reducing it to 22, and add the 10 to 90, making 100. Adding 100 to 22 is trivial; the answer is 122!

$$83+39=83+7+32=83+7+32$$
$$=90+32=90+10+22=90+10+22$$
$$=100+22=\mathbf{122}$$

If you can keep track of three numbers at a time, grouping allows you to solve problems in your head that would normally require pencil and paper.[4]

The next level of shortcuts generally involve multiplication and division. Every child learns that you can multiply a number by 10 by adding a 0 to the right of that number. Children also typically learn a few rules for divisibility. A number is even (that is, divisible by two) if its last digit is even, and a number is divisible by five if its last digit is 5 or 0. These rules can be composed: you can multiply a number by 5 by first dividing by 2 (a less time-consuming operation), and then multiplying the result by 10!

$$278\times5=278\times\left(\frac{1}{2}\times2\right)\times5=\frac{278}{2}\times10=139\times10=\mathbf{1{,}390}$$

You can also reverse this. Say you want to leave a tip of 20% (that is, 1/5) on a restaurant bill of $45.70. How could you calculate this quickly?

---

[4] This is similar to the algorithm cashiers use to make change. If the total displayed on the cash register is $33.17 and I hand you $40, you could grab (in order) three pennies, one nickel, and three quarters to make $34; then, a $1 bill to make $35; and finally, a $5 bill to make $40 total.

With practice, you can make change at a cash register without ever needing to look at the display!

In base-10, it's easy to calculate 10% by shifting decimal places. And 20% is just double this value!

$$\$45.70 \times 20\% = 45.7 \times \frac{1}{5} = 45.7 \times \frac{1}{5} \times \left(\frac{1}{2} \times 2\right) = 45.7 \times \frac{1}{10} \times 2$$
$$= \frac{45.7}{10} \times 2 = 4.57 \times 2 = \$9.14$$

What do all these types of shortcuts have in common? The times tables we memorize, and the shortcuts we develop are dependent on the number system used — in this case, the decimal system.

The choice to use decimal numbers is not, strictly speaking, arbitrary. We have ten fingers and ten toes, and as we have already seen, using a *consistent* number system presents huge advantages. Performing arithmetic using a hybrid system, such as the Babylonian numbers (which blended base-10 and base-60) or the United States customary system (which uses feet, inches, pounds, and so forth), can be a cumbersome affair. With time and practice, you can develop shortcuts to work with any number system, but it is far more efficient — in terms of both time and effort — to use a single system at every scale. The wide adoption of the metric system throughout the world[5] has proven that most people prefer to work in a single base, so that the arithmetic they know from everyday counting can be applied

---

[5] The metric system of weights and measures is the *de facto* standard for international exchange. Curiously, the same committees of the French Revolution that developed the metric system also advocated for the decimalization of time, breaking each day into 10 hours and each hour into 100 minutes. French Revolutionary Time (and the related French Republican Calendar) never gained the same traction as the metric system, and were mostly abandoned in the early part of the 19th century.

A possible explanation for this difference is that, prior to the late 18th century, systems of weights and measures varied not only by country, but also by industry and across time. Notably, the ruler of a nation could increase revenues without raising taxes simply by redefining units of measure. So a pound of flour in 1733 might not be the same quantity as a pound of flour in 1734!

The metric system was appealing first and foremost because it was a uniform, international system of weights and measures, which happened to be based on decimal numbers. Timekeeping was already standardized prior to this date, and decimalization itself did not provide sufficient justification for adopting a new system.

uniformly to problems of any kind, from daily life, to science, to finance,[6] and trade.

Since most of the mathematics you will ever perform is likely to use decimal numbers, it's worth getting to know them well! Many mathematical puzzles depend on the base you choose, and these problems provide excellent opportunities to develop computational fluency.

Counting zeroes

How many zeroes are there at the end of the decimal expansion of the number (30!)?

First, recall the definition of the factorial (!) function. The number 30! is equal to the product of the first 30 whole numbers. Thus,

$$30! = 1 \cdot 2 \cdot 3 \cdot \ldots \cdot 28 \cdot 29 \cdot 30$$

You could try to find the exact value of this number using a calculator. The result has 33 digits, while most calculators can display at most 16 digits.

What kinds of numbers will add a 0 to our final result? You might guess that there will be three 0s, one contributed by each of the multiples of 10 found between 1 and 30: 10, 20, and 30. Let's test that instinct by computing the first few factorials.

$$1! = 1$$
$$2! = 1! \cdot 2 = 2$$
$$3! = 2! \cdot 3 = 6$$
$$4! = 3! \cdot 4 = 24$$
$$5! = 4! \cdot 5 = 120$$

The first 0 appeared earlier than we predicted! We thought that we would have to multiply by a multiple of 10 to have a zero appear on the left, but it appeared after multiplying by 5! Can you see why?

---

[6] Even these changes are somewhat modern. Until the early 1970s, the United Kingdom used a system of pounds, shillings and pence, wherein there were 12 pence per shilling and 20 shillings per pound. And, until 2001, stock exchanges in the US quoted stock prices in units of 1/16 of a dollar.

When a number ends in one or more 0s, that indicates that it is divisible by 10 one or more times. The number 128,000 ends in three 0s, so we know that we can divide it by 10 three times. The number 10 is in turn divisible by 2 and 5. Any time we multiply by 2 and 5, or *any numbers divisible by* 2 and 5, another 0 will appear on our product.

Numbers divisible by 2 — even numbers — are much more common than those divisible by 5, so the number of 0s produced will depend on the latter rather than the former. (That is, we will always have a surplus of 2s.) There are six multiples of 5 between 1 and 30 (inclusive), so by this logic there should be at least six 0s at the end of our final result.

However, consider the number 25, which is equal to 5^2. This number provides us with *two* 5s, which will add two 0s to the right side of the result. In particular, $4 \cdot 25 = 100$. None of the other numbers contains more than one 5, so we can confidently state that the final total will have seven 0s on the end:

$$30! = \ldots\ldots\ldots 0,000,000$$

You can extend this logic even further: the first number that contains three 5s is 125, so (for example) 130! is the product of 26 multiples of 5, five multiples of 25, and one multiple of 125. Therefore, the decimal expansion of 130! Has $26 + 5 + 1 = 32$ 0s at the end:

$$130! = \ldots\ldots\ldots 00,000,000,000,000,000,000,000,000,000,000$$

The poor college student

A young college student realizes that he needs an infusion of cash from his parents to survive until the end of the semester. Trying to save some of his pride, he sends his parents a puzzle by email. The message reads:

```
      S  E  N  D
  +   M  O  R  E
  ---------------
   M  O  N  E  Y
```

If you substitute each letter for a unique digit, it is possible to turn this into a valid decimal sum. How much money is he requesting from his parents?

You may have encountered this puzzle before. It is the most famous example of this type of mathematics puzzle, called "alphametics," "verbal arithmetic" or "cryptarithmetic," first published in 1924 by Henry Dudeney. Verbal arithmetic puzzles don't have much to teach us about deep mathematical insights, but they can help you exercise your logical faculties and test your understanding of how arithmetic works.

Observe that, while there are 13 total letters in this puzzle, there are only eight distinct letters: D, E, M, N, O, R, S, and Y. If each letter corresponds to a distinct digit, then it should certainly be possible to find a base-10 solution. Naturally, two of the ten decimal digits will go unused.

We typically perform addition by adding the digits in each column, moving from right-to-left and carrying as necessary. If you look at column one, at the far right, it's not clear how to begin finding the values for the numbers short of guessing. Since we want to approach this puzzle logically, it seems reasonable to instead examine the opposite side of the equation, starting with the fifth column.

The sum has more digits than either of the addends, which tells us something about the value of that number. Consider how carrying works. The largest four-digit number is 9,999. Even if both the addends in this puzzle were this large (which is clearly not possible!) the largest sum they could produce would be 19,998. So the letter M must be equal to 1.

There is also an M at the left side of the second addend. In order for there to be a 1 carried into column five, S must have a value of either 8 or 9. In either case, the letter has a value of O or 1. Since we've already used 1, O must be equal to 0.

If S had a value of 8, then there would have to be a carry from the sum in column three. In column three, E is added to O (that is, the digit 0), to produce a value different than E — namely, N. For this to happen, there must be a carry from column two to column three. But if there were a carry from column three to column four, E would have to have a value of 9, and *N* would have to have a value of 0, which has already been used! Therefore, there is no carry between columns three and four, and *S* must be 9.

We will pause for a moment to take stock of the problem so far, as the next steps will be a bit more involved.

```
    1       1   ?
        9   E   N   D
    +       1   0   R   E
        1   0   N   E   Y
```

We know that $N = E + 1$. Depending on whether there is a carry into column two, one of the following is true:

$$1 + N + R = E + 10$$
$$N + R = E + 10$$

We can replace *N* with $E + 1$ in both of these to test for contradictions.

$$1 + (E + 1) + R = E + 10 \qquad\qquad (E + 1) + R = E + 10$$
$$2 + R = 10 \qquad\qquad\qquad\qquad 1 + R = 10$$
$$R = 8 \qquad\qquad\qquad\qquad\qquad R = 9$$

We know that *R* can't be 9, since we already used that value for *S*. So *R* equals 8, and there must be a carry from column one.

We can't get any more information out of columns two and three for the moment since they reduce down to the same equation ($1 + N + 8 = E + 10$ and $1 + E = N$), so we'll examine instead column one.

Y can't be 0 or 1, so $D + E$ is at least 12. *D* or *E* can be at most 7, since we've already identified the letters corresponding to 8 and 9. Since all letters have distinct values, these two can be either 5 and 7, or 6 and 7. Either way, one of *D* or *E* has to represent the number 7.

If *E* were 7 then, in column three, *N* would be 8, which is impossible. So *D* is 7. Because of this, if *E* were 6 then *N* would be 7, which became impossible the moment we found the value of *D*. So *E* is 5, *N* is 6, and *Y* is 2!

The final equation is shown below, revealing that the college student asked his parents for $10,652.

$$
\begin{array}{ccccc}
1 & & 1 & 1 & \\
& 9 & 5 & 6 & 7 \\
+ & & 1 & 0 & 8 & 5 \\
\hline
1 & 0 & 6 & 5 & 2 \\
\end{array}
$$

Now that you've had a chance to warm up your arithmetical muscles, you should be in a good position to handle the following problems.

Decimal exercises

Problem 1

A particular five-digit number has the following property: if you put the digit 1 at the end (that is, on the right) of it, it is three times larger than if you put a 1 at the beginning. What is the original five-digit number?

First, ask yourself what this question is saying. "Putting a 1 at the end" is equivalent to multiplying by 10, then adding 1, whereas "putting a 1 at the beginning" is equivalent to adding the first power of 10 that is larger than the number. In this case, that means adding 100,000 to the original number.

If we call the number n, then the two expressions can be written as shown.

$$10n+1$$
$$100,000+n$$

One of these is three times larger than the other. Students often struggle to convert word problems into algebraic expressions, so see if you can write the equation described in the prompt above on your own.

The complete equation corresponding to this problem is

$$3(100,000+n)=10n+1$$

We can solve this algebraically.

$$300,000+3n=10n+1$$
$$299,999=7n$$
$$\mathbf{42,857}=n$$

We can use this result to check our work by verifying that both sides of the original equation are equal:

$$3 \cdot (100{,}000 + 42{,}857) = 3 \cdot (142{,}857)$$
$$= \mathbf{428{,}571}$$

$$10 \cdot 42{,}857 + 1 = 428{,}570 + 1$$
$$= \mathbf{428{,}571}$$

Problem 2

If you reverse the digits of a certain five-digit number, the resulting number is four times larger. What is the original number?

This problem is not nearly so straightforward as the previous one. We can, however, use some of the ideas exposed in the previous three problems to work toward a solution.

First, write the original number as *ABCDE*. We can turn the problem into a sort of equation:

$$ABCDE \cdot 4 = EDCBA$$

Is this really an equation, though? We will need to do one more transformation to turn this into a proper equation.

Recall that a number, written in a particular base, is the sum of the digits multiplied by successive powers of that base. To figure out how many "pebbles" a decimal representation indicates, you can count out the number of "ones" (10^0) equal to the leftmost digit, the number of "tens" (10^1) equal to the second digit, and so forth. Applying this principle here, we can write the previous equation more mathematically.

$$4 \cdot (10^4 A + 10^3 B + 10^2 C + 10^1 D + 10^0 E)$$
$$= 10^4 E + 10^3 D + 10^2 C + 10^1 B + 10^0 A$$

If you remember your algebra, there might be a little voice in the back of your head right now, screaming something about "one equation and five unknowns." That voice is correct! There isn't enough information contained in this equation to find a solution using "typical" algebraic methods.

What other approaches can we try? Once again, we can begin by looking at the left and right sides of these numbers. Each of them has five digits, so *EDCBA* must be less than 99,999, and since this number is 4 times larger than *ABCDE*, dividing by 4 we find that *ABCDE* must be *strictly* less than 25,000. *ABCDE* must be somewhere between 10,000 and 24,999, so the letter *A* can be either 1 or 2. Working in reverse, we can see that *EDCBA* must be greater than 40,000.

We can find the value of each digit by applying logic to the written multiplication shown in Table 6.1.

If we knew each of the digits of *ABCDE*, we would begin performing this multiplication by multiplying 4 by *E*, writing the ones digit below the line (in the space occupied by the letter *A*), and carry some number of tens to the space above the letter *D*. So, we can say that *E* times 4 is equal to *A* plus some number of tens.

If you multiply 4 by the numbers 1 through 9, you get 4, 8, 12, 16, 20, 24, 28, 32, and 36. We already know that any number multiplied by an even number must be even, which is evidenced by our list of multiples. So, *A* must be even. And we have already seen that *A* must be either 1 or 2, so *A* must be 2.

EDCBA is greater than 40,000, and the only values of *E* that result in a value of 2 for *A* are 3 and 8. *E* cannot be 3, so it must be 8.

On the right side of Table 6.1, we have inserted all the values we know so far.

Looking at the fifth column of this multiplication, $4 \times 2 = 8$. It's logically impossible for there to be a digit carried from the fourth column to the fifth, since if that digit were greater than 2, the sum of 8 and the carry would produce a *six*-digit number with a leading 1. And, if that digit were 1, then the value of *E* would be 9.

Table 6.1.

Reformulating the problem into a more traditional form we can see that ABCDE is multiplied by 4 to get EDCBA	The right state of the equation after solving for A and E
A B C D E × 4 ——————— E D C B A	? ? 3 2 B C D 8 × 4 ——————— 8 D C B 2

Since nothing is carried to the fifth column from the product $4 \times B$ in the fourth column, B must be less than 3. (Otherwise, $4 \times B \geq 12$.)

Looking now at column two, $4 \times D$ must be even, since 4 is even. So when we add the 3 carried from column one to the (even) result of $4 \times D$ we can see that B must be odd, since an even number plus an odd number is odd. B can't be 0 or 2; so it must have the value 1.

The only way to have a 1 as the result of the arithmetic in column two is if the result of multiplying D by 4 ends in the digit 8, since $8 + 3 = 11$. The only two one-digit multiples of 4 that end in 8 are $4 \times 2 = 8$ and $4 \times 7 = 28$. Therefore, D can be either 2 or 7.

Looking again at the fourth column, recall that $4 \times B$ plus whatever is carried over from column three is less than 10. B is 1, so $4 \times B$ must equal 4; D can't be smaller than 4, so it can't be 2. So D must be 7.

If we represent the carry from column three to column four using the letter x, we can swap the values we have just found into the ensuing equation to find the carried value.

$$(4 \times B) + x = D$$
$$(4 \times 1) + x = 7$$
$$x = 3$$

So we know a 3 is being carried from column three to column four.

Finally, we can express C in terms of the carries in each column. C multiplied by 4, plus the 3 carried from column two, is equal to C plus the 30 carried into column four. It may be easier to see what this sentence means when reading the following algebra:

$$4C + 3 = C + 30$$
$$3C = 27$$
$$C = 9$$

We have identified the values of all five digits, which you can be verified in the following. The original number $ABCDE$ must have been 21,978.

```
        3  3  3
     2  1  9  7  8
  ×              4
  _____
     8  7  9  1  2
```

Problem 3

Two numbers are constructed using each of the digits 1 through 9 exactly once such that, when multiplied together, they produce the largest possible product. What are the two numbers?

(More concretely: 147 and 298,365 together use each of the digits 1 through 9 exactly once. Their product is 43,859,655, which is not the largest possible, since we can find at least one other pair of numbers that produces a larger product, such as 147 × 928,365 = 136,469,655.)

We should use the digits in reverse order, so that the largest digit sits furthest to the left. The two numbers must start with 9 and 8. The next digits we should use are 7 and 6, but it's not clear which number we should attach each of these digits to!

Here, you can play around with a calculator to try to identify trends. You'll notice that 87 · 96 is larger than 86 · 97, but can you tell why?

Choose a number at random. Multiply it by itself, then compare that result to the product of one more than your number, and one less than your number. Add one to the larger number, and subtract one from the smaller, and multiply again. Keep repeating this process until you see a pattern emerge. For instance:

$$100 \cdot 100 = 10,000$$
$$101 \cdot 99 = 9,999$$
$$102 \cdot 98 = 9,996$$
$$103 \cdot 97 = 9,991$$

Given pairs of numbers whose sum is constant, the closer together they are, the larger their product will be! We can also demonstrate this algebraically:

$$(x+0)(x-0) = x \cdot x = x^2$$
$$(x+n)(x-n) = x^2 + nx - nx - n^2 = x^2 - n^2$$

Therefore, our goal at each step should be to make sure the distance between the two numbers is as small as possible. Continuing to add digits in the manner identified previously, we end up with 9,642 and 8,753, with the digit 1 left over. Which number should we add this to?

The same principle applies here: 87,531 is closer to 9,642 than the reverse.

$$87,531 - 9,642 = 77,889$$
$$96,421 - 8,753 = 87,668$$

So, the two numbers 87,531 and 9,642, when multiplied together, produce the largest possible product.

$$87,531 \times 9,642 = 843,973,902$$

Fluency in arithmetic is its own reward. The more natural and instinctive arithmetic manipulations become to you, the easier it is to learn higher math, because less of your brain is engaged re-learning "old" material, or figuring it out from scratch. (This principle applies equally at every stage of mathematical learning. When students perform poorly in calculus class, it is more often than not a product of weak algebra skills.)

Curiously though, these "toy" problems have some practical applications. When combined with some ideas from modular arithmetic (last seen in Chapter 3), we can come up with rules to determine when one number is divisible by another, which will help make your arithmetic calculations more fluent.

Build your own divisibility rules

When we want to know whether a number is evenly divisible by 2, why do we only examine the rightmost digit?

We can write any number and you can instantly determine that it is odd. Take the number 38,742,187: since it ends in 7, it must be odd. You don't even need to hear or see any of the initial digits. For all you're concerned, we could say "blah blah blah, blah blah blah and seven," and you would only need to hear the final syllables to make your determination. If a number's last digit is 0, 2, 4, 6, or 8, then it is divisible by two.

In the problem "Counting Zeroes," we noted that each multiple of 10 is composed of both a 2 and a 5. We can turn that observation around, decomposing a whole number like *ABCDE* into products of successive powers of ten:

$$ABCDE = 10^4 A + 10^3 B + 10^2 C + 10^1 D + 10^0 E$$

If we replace the final digit E with a 0, the 10^0 term disappears.

$$
\begin{aligned}
ABCD0 &= 10^4 A + 10^3 B + 10^2 C + 10^1 D \\
&= 10 \cdot (10^3 A + 10^2 B + 10^1 C + 10^0 D) \\
&= 5 \cdot 2 \cdot (10^3 A + 10^2 B + 10^1 C + 10^0 D)
\end{aligned}
$$

Any number whatsoever that ends with a 0 must be divisible by 10 — but it must also be divisible by 5, and by 2! If *E* is divisible by 10 (i.e., if it has the value 0), then the number *ABCDE* is also divisible by 10. If *E* is divisible by 5 (if it has the value 0 or 5), then again, *ABCDE* is divisible by 5. And if *E* is 0, 2, 4, 6, or 8, then *ABCDE* is divisible by 2. What may not be obvious is *why* this occurs.

To say that one number is divisible by another is equivalent to saying that the former can be expressed as a whole number multiple of the latter.[7] In the previous example, you can observe that the entire left half of the equation can be replaced by 5 times some other number, which we will call *k*.

$$n = 5 \cdot 2 \cdot (10^3 A + 10^2 B + 10^1 C + 10^0 D) = 5 \cdot k$$

We can ignore the specific value of *k*, noting only that so long as it is a whole number, *n*/5 will also be a whole number. So when we say *ABCDE* is divisible by 5, we are in effect telling you that it can be written in the form of some number multiplied by 5.

Using this observation, can you find equivalent rules for other common numbers? What determines divisibility by 4 or 8? What about 3, 6, and 9?

---

[7] This section can also be understood in the language of *modular arithmetic*, which we covered briefly in Chapter 3.

Dividing by 4 and 8

According to the *Fundamental Theorem of Arithmetic*, every integer can be factored into a unique collection of prime numbers.[8] Since $10 = 2 \cdot 5$, the number 10 can only ever "contain" a single 2, and a single 5. There is no way to evenly divide 10 by any other prime numbers, so it's not possible to find some magical power of 10 that will be evenly divisible by other primes, such as 3 or 7.

However, we can use *two* tens to create a 4. That is, $10^2 = 10 \cdot 10 = (2 \cdot 5) \cdot (2 \cdot 5) = 4 \cdot 5 \cdot 5$. Again using $ABCDE$ to demonstrate:

$$ABCDE = ABC \cdot 100 + DE$$
$$= 4 \cdot 25 \cdot ABC + DE$$
$$= 4 \cdot k + DE$$

Translating this into words: to test for divisibility by four, we need only look at the two rightmost digits of a given decimal number. This is because the left digits will *always* be divisible by 100, which is divisible by 4. The number 13,4**36** is divisible by 4 because it ends in 36, and $4 \cdot 9 = 36$; the number 714,5**14** is not, since 14 is not divisible by 4.

The rule for divisibility by 8 is a natural extension of this demonstration. If the last *three* digits of a number are divisible by 8, then so is the entire number. So 63,**832** is divisible by 8, because 832 $(800 + 32)$ is.

Do you think you can demonstrate the rules for 3 or 9? It helps if you already know those rules, since that knowledge will direct your efforts toward a specific target. We will start with the number 9.

Dividing by 3 and 9

A decimal number is divisible by 9 if the sum of its digits is divisible by 9. According to this rule, each of the following numbers are divisible by 9.

72	$7 + 2 = 9$	$9 = \mathbf{9} \cdot 1$
378	$3 + 7 + 8 = 18$	$18 = \mathbf{9} \cdot 2$
123,456,789	$1 + 2 + 3 + 4 + 5 + 6 + 7 + 8 + 9 = 45$	$45 = \mathbf{9} \cdot 5$

---

[8] Recall that 1 is neither a prime number, nor a product of prime numbers.

Why should this rule be true? After all, the process we use seems rather arbitrary. Let's look again at our example number.

$$ABCDE = 10^4 A + 10^3 B + 10^2 C + 10^1 D + 10^0 E$$

We can expand on the previous approach. Instead of splitting this number at a single point, we can examine each digit individually. After a little experimentation, you should notice something curious. No matter the number, each of its digits can be expressed in terms of a multiple of 9, plus 1. When the expression is simplified, what remains is the digit itself, plus a multiple of 9.

$$\begin{aligned} A \cdot 10,000 &= A \cdot (1 + 9,999) \\ &= A \cdot (1 + 9 \cdot (1,111)) \\ &= 1 \cdot A + 9 \cdot (A \cdot 1,111) \\ &= A + 9 \cdot k \end{aligned}$$

When we consider all of a number's digits together, the resulting terms fall into two categories: single digits and multiples of 9.

$$\begin{aligned} ABCDE &= A \cdot 10,000 + B \cdot 1,000 + C \cdot 100 + D \cdot 10 + E \cdot 1 \\ &= A \cdot (1 + 9,999) + B \cdot (1 + 999) + C \cdot (1 + 99) + D \cdot (1 + 9) + E \cdot (1 + 0) \\ &= (A + 9 \cdot k_1) + (B + 9 \cdot k_2) + (C + 9 \cdot k_3) + (D + 9 \cdot k_4) + (E + 9 \cdot 0) \\ &= (A + B + C + D + E) + 9 \cdot (k_1 + k_2 + k_3 + k_4) \\ &= (A + B + C + D + E) + (9 \cdot k) \end{aligned}$$

Here, k does not have the same value it did earlier. Rather, in both cases this letter acts as a handle for a box containing a singular idea: "k is some whole number, whose value doesn't affect the portion of the expression we care about." The expression in the left parentheses stands for *some* multiple of 9, and that is all we need to know about it.

What about the contents of the right parentheses? Well, *if and only if* $(A + B + C + D + E) = 9 \cdot m$ for some whole number m, we can further reduce the previous equation:

$$\begin{aligned} ABCDE &= (A + B + C + D + E) + (9 \cdot k) \\ &= 9 \cdot m + 9 \cdot k \\ &= 9 \cdot (m + k) \end{aligned}$$

As you can now recognize, the last line here says that *ABCDE* is divisible by 9, so long as its digits add up to a multiple of 9. This is exactly what we set out to demonstrate! If a number's *digital sum* is a multiple of 9, the original number is too — and vice versa!

The divisibility rule for 3 is structurally similar. Remember, to determine whether a number is divisible by 3, you check to see if the digital sum of its digits produces a multiple of 3. It should come as no surprise that you can prove the divisibility rule for 3 by applying the same steps that we used for 9, but replacing every occurrence of 9 with 3!

$$\begin{aligned} A \cdot 10{,}000 &= A \cdot (1 + 9{,}999) \\ &= A \cdot (1 + 3 \cdot 3{,}333) \\ &= 1 \cdot A + 3 \cdot (A \cdot 3{,}333) \\ &= A + 3 \cdot k \end{aligned}$$

Combining rules: Dividing by 6

We said we would address one other divisibility rule, for the number 6. Can you see how to put all these pieces together?

The number 6 is divisible by two primes, 2 and 3. So a number is divisible by 6 if its digital sum adds up to a multiple of 3 (the divisibility rule for 3) *and* if its last digit is even (the divisibility rule for 2).

For example, you can observe that the last digit of the number 324 is even, so it is divisible by 2. Its digital sum produces a multiple of 3:

$$n = 32\mathbf{4}$$
$$3 + 2 + 4 = 9 = \mathbf{3} \cdot 3$$

Therefore, 324 is divisible by 6!

As a final comment, these principles are applicable in any number system. We've focused here on the decimal system because they are the numbers in "General Usage." In the future, if humanity encounters an alien race with 12 fingers, we will be able to predict what their arithmetic looks like! It might be worth your time to explore the rules for different numerical bases on your own. Like learning a foreign language, doing arithmetic in a different base can teach you a lot about your "native mathematical language:" base-10.

The principles in this section will help you when faced with manageable decimal numbers. But you will need a different set of techniques to handle the second class of numbers, those that are very large (or very small)!

The size of a number

How do we make sense of very small and very large numbers?

As it turns out, there is strong evidence that we are all born with a natural quantitative sense that deepens over time.

According to child psychologists, children who are old enough to understand numbers, but too young to attend school can put numbers in the proper order, but they will generally "compress" the number line, placing smaller numbers far apart and larger numbers closer together. A five-year-old might place the number "5" halfway between 0 and 100, and lump the numbers "70" and "90" together on the far right edge (Figure 6.5). He knows 5 is bigger than 0 and less than 100, but he doesn't know exactly how *much* bigger.

As students progress through the grades, they become adept at drawing the number line more and more linearly. By the fourth grade, an average student can place numbers on a number line that runs from 0 to 1,000 according to their sizes. A ten-year-old will usually place 500 halfway between 0 and 1,000, and 10 very close to 0, for instance (Figure 6.6).

Figure 6.5.   How a five-year-old might draw the number line.

Figure 6.6.   By ten years old, most children can comfortably place numbers in correct positions along the number line.

In elementary school, progressive understanding of the sizes of numbers has a strong correlation to mathematical performance. You don't have to perform the addition $273 + 140$ to guess that the answer is close to 400. Having intuition about the sizes of numbers in a problem makes estimating the size of the correct answer nearly instantaneous.

We expect most readers of this book have passed this stage, and can make reasonable guesses about magnitudes of sums, differences, and products involving numbers of only a few digits.

Inevitably, though, our intuition for numbers begins to break down. For you, this might happen when the numbers get very large, or very close to zero, or become negative. We have words and prefixes that allow us to talk about all these numbers (e.g. *tenth*, *milli-*, *million*, *giga-*), but there is a vast difference between speaking words and understanding them.

Your comprehension will also vary based on context. You can envision four pebbles, or four cars, or four elephants with similar ease. Now, think about 10,000 pebbles. How tall would a pile of 10,000 pebbles be? How much should 10,000 pebbles weigh?

If we change the context slightly, that same quantity suddenly becomes tractable. If you live in the United States, $10,000 is a reasonable unit of measure. An apartment that costs $10,000 per month must be fancy; a car that costs $10,000 is cheap; and $10,000 per year is a very low salary. Adults can generally handle numbers that are relevant to their lived experiences, but when doing so, there's actually something very subtle going on in their heads.

Here, the amount of *information* available once again becomes pertinent. Consider the difference between the following two statements:

"I'll be there in five minutes."
"I'll be there in seven minutes."

If you were waiting for your friend and she sent you the first message, how would you interpret it? Most people would read that as "I am nearby, and will arrive anywhere from three to ten minutes from

Figure 6.7.   How most people actually think about large numbers. $10,000 and $35,880 are trimmed to "$10k" and "$36k."

now." However, the second statement suggests a high degree of accuracy. Maybe your friend knows the route to your location so well that she can predict exactly how long it will take. Or, perhaps she is using a GPS system and reading the expected travel time directly off the screen. Either way, there is a large *semantic* difference between these two statements, even though the *syntax* is similar.

A fresh college graduate just getting started in New York City might earn a wage of $17.25 per hour. If he works 40 hours per week and 52 weeks per year, his expected salary would be $35,880 (Figure 6.7). If you asked him what his annual salary was, which of the following two answers would you expect to hear?

"I make thirty-five thousand, eight hundred and eighty dollars per year."

"I make thirty-six thousand per year."

Unless you're an accountant or a scientist, most of the numbers you use will be rounded in this way. And there is absolutely nothing wrong with this! We round numbers for social and linguistic reasons ("I'll be there in five"), because we don't have more accurate data ("the temperature is going to be in the mid 70s today"), because our audience doesn't care about the exact value ("I make $36 grand"), or to communicate a general sense of scale ("the US population is 330 million.") Indeed, using overly precise numbers can come across as pretentious at best and ignorant at worst. Take the following joke, for example.

> *While visiting the Museum of Natural History, a family of tourists comes upon the enormous skeleton of a Tyrannosaurus Rex. After several minutes spent gazing in awe at the beast's bones, one of the children asks his mother when the dinosaur lived.*

"I don't know, honey. Maybe one of the workers knows." Looking *around briefly, she spots a janitor.*

"Excuse me, sir," she says, *"but do you know how old this skeleton is?"*

"Sure thing! It's sixty-five million and four years, five months, and thirteen days old."

"That's remarkable! How do you know its age so accurately?"

"Well," says the janitor, *"on my first day of work, I asked the very same question of the museum director, and she told me that it was sixty-five million years old. And I've been working at this museum for four years, five months, and thirteen days..."*

Round numbers can be *incredibly* useful, as long as we are able to make sense of them, and don't ask them to do too much!

You'll note that most of these examples use two (or sometimes three) significant digits, followed by 0s. If you think about examples from your own life, or if you were to page through a newspaper, you will find that this is all the detail most people ever want!

For the moment, we will focus on how to calculate and understand the general scale of answers by combining two different skills. You know how to multiply and divide by powers of 10 by adding and removing 0s; we can combine this skill with your ability to make sense of familiar numbers to bring almost any number down to a human scale.

On scaling

Coming to grips with large (and small) numbers requires that you be able to climb up and down through orders of magnitude with ease. What is easy will be determined in part by your own "natural" numbers, as we will demonstrate with a quick exercise.

What is the result of the following division?

$$\frac{1000000000000}{10000}$$

The most straightforward way to solve this problem is to cancel out pairs of 0s.

$$\frac{1000000000000}{10000} = \frac{100000000000}{1000}$$
$$= \frac{10000000000}{100} = \frac{1000000000}{10} = 100000000$$

This is similar to how a child learning to perform addition can count on her fingers. It's not fast, nor is it pretty, but it gets the job done!

The previous division has another weakness. It's almost impossible to determine what the numbers are; larger numbers seem to be a blur of 0s. It is customary to insert commas[9] between groups of digits. This can simplify cancellation, since you can "chunk" groups of three 0s at a glance.

$$\frac{1,000,000,000,000}{10,000} = \frac{1,000,000,000}{10} = 100,000,000$$

As we have seen, many of the numbers you will encounter in daily life will be expressed in words! What is the best way to handle numbers like these?

$$\frac{\text{one trillion}}{\text{ten thousand}} = ?$$

You could convert every number to the appropriate decimal representation (including commas), and cancel groups of 0s as before. The more efficient method is to learn the scale of numbers, and use *linguistic* properties to avoid having to do mathematics in the first place!

To begin, one thousand thousands makes one million; one thousand millions makes one billion; and one thousand billions makes one trillion. So, this problem can be re-written, and we can cancel the common *words*.

---

[9] Or periods, as in Europe.

$$\frac{\text{one thousand billion}}{\text{ten thousand}} = \frac{\text{one billion}}{\text{ten}}$$

This process might seem overblown for this example. But consider: it's much easier to do arithmetic using smaller numbers, and a problem you're interested in might involve many steps. So if you can "set aside" the scale of the problem, perform all the intermediate steps with smaller numbers (only numbers between 0 and 1000, for instance), and bring the scale back only at the *end*, then you don't have to worry about all the "big" numbers during the intermediate steps!

To demonstrate that this differs from the "typical" approach, we can use real numbers that are nonetheless foreign to your experience.

If you travel to India, you might hear people using the number words *lakh* and *crore*. These words behave just like the number words you're familiar with. So, where an American might say "I make 36 *thousand* dollars per year," an Indian might say "I make 4 *lakh* rupees per year."

Suppose you're going to give a talk on economics to an Indian audience, and you look up some statistics on the state of the economy. According to government sources, the country's Gross Domestic Product in 2018 was Rs[10] 142 lakh crore, while the population was 135 crore people. What is the average GDP per capita in India?

One way to approach this problem would be to look up the definition of each number word, convert to decimal numbers, cancel 0s, perform the relevant arithmetic, and finally *convert back* to local units for your presentation. But that's a lot of unnecessary work!

Instead, we can re-arrange terms to cancel number *words*, and "set aside" the scale of the problem until the end — all without knowing what the words *lakh* or *crore* mean![11]

---

[10] "Rs" is one possible abbreviation for Indian Rupees.

[11] To save you from having to look these words up on your own:
  Lakh, abbreviated L, is equal to 100,000, but written in India as 1,00,000.
  Crore, abbreviated cr, is equal to 100L or 10,000,000, but written as 1,00,00,000.
  You can verify on your own that "1 lakh crore" is equal to 1 trillion.

$$\left(\frac{\text{Rs } 142 \text{ lakh crore}}{135 \text{ crore people}}\right) = \left(\frac{142 \text{ crore}}{135 \text{ crore}}\right)\left(\frac{\text{lakh Rs}}{\text{person}}\right)$$

$$= \left(\frac{142}{135}\right) = \text{Rs } 1.05 \text{ lakh per capita}$$

The practice of separating the scale of numbers from their significant figures can greatly simplify many types of calculation. Until recently, this was a skill every student had to learn.

A brief history of estimation

Before smartphones and pocket calculators, the calculating machine of choice was a device called a *slide rule*. At the time of writing, it has been over 40 years since slide rules fell out of favor. It's no wonder: calculators (and later, computers) are faster to operate, less error-prone, and more accurate than slide rules could ever be.

A slide rule, which superficially resembles a ruler, offered one advantage that calculators could never match. When using an electronic calculator, there are no easy checks on user carelessness or input errors.[12] If you miss a digit when punching in a number, and then read the (erroneous) result off the display, you won't have the slightest clue that a problem occurred! When using a slide rule, the user is *forced* to estimate the size of the answer before performing calculations, so even before he picks up his "calculator" he must already have an estimate of the size of the result in his head.

In order to compute the answer to a problem, the *operands*[13] first had to be reduced to the scale marked on the slide rule, which generally ran from 1 to 10 (for reasons we will explain later). You would

---

[12] This concern is not merely academic. For three months in 2017, a popular mobile phone operating system had a bug where, if a user hit one button quickly after another, the second button press would not register. One wonders how many times nimble-fingered users read an incorrect result from their calculator's screen before this bug was squashed!

[13] An operand is a term (generally a number) to which a mathematical operation is applied. When performing addition, the two operands are called *addends*, which,

then manipulate the rule to find the result of the operation, and then multiply or divide, as appropriate, to scale the final result.

We can safely assume that you've never used a slide rule, so a more concrete example is required. Suppose you wish to divide one number by another:

$$\frac{456,000,000}{12,300} = ?$$

Remove enough 10s from each number to bring that number in the range between 1.0 and 10.

$$456,000,000 = 4.56 \times 100,000,000$$
$$12,300 = 1.23 \times 10,000$$

Now, perform a division *using only these powers of 10*.

$$\frac{100,000,000}{10,000} = \frac{100,000}{10} = 10,000$$

(Here, we first cancel a group of three 0s from the numerator and the denominator to make the source of the final result more apparent.)

Whatever the result of this division, we can estimate that it will lie somewhere between 10,000 and 100,000. With an educated guess, we can narrow the estimate to an even smaller range, somewhere between about 30,000 and 40,000.

---

when added, produce a sum. In multiplication, two *multiplicands* are multiplied, resulting in a product.

Both addition and multiplication are *commutative*, which is to say that the order of the operands doesn't matter.

Non-commutative operations, such as subtraction and division, use different names for the operands. So, when subtracting, we subtract the *subtrahend* from the *minuend*, producing a difference; while in division, we divide the *numerator* by the *denominator*, producing a quotient (and possibly a remainder).

Manipulating the slide rule,[14] we discover that 4.56 divided by 1.23 is 3.707. We can multiply this result by our estimate to produce the full answer, which we compare to the output generated by a calculator.

$$3.707 \times 10,000 = 37,070 \approx 37,073.17$$

Not bad! The result of using "outdated" technology produced a value within 1/10,000 of the true answer.

Practical estimation

Nowadays you don't have to learn to use a slide rule to benefit from ballpark estimates. You do, however, have to deliberately *choose* to perform estimates.

A fun way to practice estimation is to answer hypothetical questions. When slide rules were still the most widely-used way to perform calculations, physicist Enrico Fermi popularized a type of estimation that we now call *Fermi Estimates* or *Fermi Problems*.

Fermi would pose a question to his colleagues, such as "how many piano tuners are there in Chicago?" The goal was to use reference materials, educated estimates, and rough calculations to get a sense of the scale of a problem.

From our point of view, the canonical Fermi Problem suffers three key flaws: it has been covered extensively in other sources, it is not relevant enough to "General Usage" for our tastes, and the popularity of pianos has declined significantly since Fermi's time. Fortunately for us, this method can be applied to all kinds of questions!

We can choose a problem in the same spirit as Fermi's. One question we might ask is, how many people work in the restaurant industry in the US?

If you don't have a good idea where to start in a situation like this, it's best to break the question down into smaller questions.

---

[14] Ironically lacking access to a slide rule, here we had to rely on a book of logarithms from 1962.

One approach would be to ask "how much revenue does the restaurant industry bring in per year?"

There are approximately 300 million people in the US. Let's say that each person eats out twice per week, and spends $20 per meal. Remember, these are average, ballpark figures. Some people, such as infants and college students, might not eat out even once per *month*, while others might buy lunch everyday. And, for every person buying a fast-food value meal for under $10, there is another person somewhere having a fancy dinner and drinks for $30 or $40. These figures don't have to be *exact*, just educated.

If everyone in the US eats out twice per week, and there are 50 weeks per year (remember, we're using estimates), then the typical person eats out 100 times per year. Right now, you might be saying "aren't there 52 weeks per year?" You would be correct, but we already *know* that the rest of our values are rounded estimates, so there's no real harm in rounding off two weeks per year (which amounts to less than 4% of the total). By streamlining all our calculations in this way, we can speed up the process of estimation.

We can now calculate how much money the restaurant industry makes per year. At each step, we will try to break up numbers to maximize the number of 10's we have to work with.

$$300 \text{ million people} \times \left(20\frac{\text{dollars}}{\text{meal}} \times 2\frac{\text{meals}}{\text{week}} \times 50\frac{\text{weeks}}{\text{year}}\right) \text{per person}$$
$$= (300 \times 20 \times 2 \times 50)\text{million dollars/year}$$
$$= (300 \times (2 \times 10) \times (2 \times 50)) \times \$1,000,000$$
$$= (300 \times 2) \times 10 \times 100) \times \$1,000,000$$
$$= (600 \times 1000) \times \$1,000,000$$
$$= \$600,000,000,000$$
$$= \textbf{\$600 billion/year}$$

There's a lot going on here! First, we've dropped in each of our estimated values, but we have also included some strange units. These act as a check on whether we're performing the calculation we think we are.

We know that the answer we're looking for is expressed in terms of dollars per year. Then, we can associate each term in our formula with a valueless unit, cancelling at each step to verify that we have accounted for each factor.

$$\text{people} \times \frac{\frac{\text{dollars}}{\text{meal}} \times \frac{\text{meals}}{\text{week}} \times \frac{\text{weeks}}{\text{year}}}{\text{person}}$$

$$= \text{person} \times \frac{\left(\frac{\text{dollar} \times \text{meal} \times \text{week}}{\text{meal} \times \text{week} \times \text{year}}\right)}{\text{person}}$$

$$= \text{person} \times \frac{\left(\frac{\text{dollar}}{\text{year}}\right)}{\text{person}} = \frac{\text{dollars}}{\text{year}}$$

The four factors we identified — number of people, dollars spent on average per meal, number of visits to restaurants per week, and number of weeks per year — when combined, produce an answer with the correct unit.

We also continue grouping multiplications in a way that favors powers of 10. For completeness' sake, we included the value "1,000,000" (one million) at each step. As we have shown, once you've practiced with this method, it is easier to recall that you're working in units of millions or billions or trillions at the end than to keep track of them at each step. With experience, then, the previous process "feels" more like this:

$$300 \times 20 \times 2 \times 50 = 6,000 \times 100$$
$$= 600 \times 1,000$$
$$= 600 \times 1,000 \, \text{million}$$
$$= \textbf{\$600 billion/year}$$

Now that we have a sense of how large the industry is, we need to figure out how many people it employs. Suppose a typical restaurant spends 1/3 of its revenue on labor. Then altogether, American restaurant workers are paid $200 billion per year.

Federal minimum wage is around $7.50 per hour, but some workers make more than that because of seniority. Let's say that the average worker earns $10 per hour. There are 40 hours per work week, and 50 weeks per year; that means there are 2,000 working hours per year for full-time employees. So the average restaurant worker might gross $20,000 per year. Dividing the annual cost of labor by the average wage, we can estimate the number of food service workers.

$$\frac{200\ \text{billion}}{20{,}000} = \frac{200\ \text{million}}{20} = \frac{100\ \text{million}}{10} = 10\ \text{million}$$

To put this in more human terms, 1 out of every 30 people you've ever met works in a restaurant. Average class size in American schools is 30 students, so if you can visualize a class that you have attended, the ratio 1:30 is like saying that one of your former classmates works in the restaurant industry.

How accurate are our estimates, really? Using the internet, we can find official sources for some of them. According to the National Restaurant Association, the entire industry was projected to make $799 billion in 2017, while according to the Bureau of Labor Statistics, in 2016 there were 13,206,100 (about 13 million) people nationwide working in "food preparation and serving related occupations." So our estimates weren't far from the truth![15]

Apples to apples

Fermi Estimation is an effective tool to have in any problem solving toolkit. But to make the efficient use of it, you need to know how to turn problem statements into mathematical equations.

In the previous example, we arranged several valueless units as a "sanity check" on our math. You can use this skill on its own to build formulas, even in the absence of numbers! This is a process known as *dimensional analysis*. As long as two units are *commensurable* — that is, expressed in the same dimension, such as length or time — there are transformations you can apply to convert or otherwise manipulate between them.

By far the most common application of this principle is in every-day conversions, like the one in the following example.

---

[15] Believe it or not, we did these calculations *first*, and only then did we look up the actual numbers online!

Problem 1

You are training to run a 5k (i.e., 5 kilometer) race, but your pedometer only displays distances measured in miles. How many miles do you have to run to complete a 5k?

Assume you vaguely remember that the conversion factor between miles and kilometers is 1.6, but you can't recall which direction the conversion goes. Should you multiply or divide the number of kilometers by 1.6 to find the number of miles?

Even if you can't remember the "formula," you might be able to reconstruct it from experience. If your car was made for the US market, its speedometer probably lists both miles and kilometers. Which scale has greater values? Or, if you have vacationed outside the US, you may have noticed that posted speed limits had different values than what you are used to.[16] Are highway speeds outside the US "higher" or "lower?"

Regardless of the method you use to recall the relationship, the result is the same: one kilometer is smaller than one mile, and so there is more than one kilometer per mile. The unit associated with the value 1.6 must be **kilometers per mile**. That is, when converting, you must apply *some* operation to the ratio:

$$\frac{1.6 \text{ km}}{1 \text{ mile}}$$

This question concerns converting kilometers to miles, so you can multiply and divide the respective units (without values) to see which result makes more sense.

$$\text{km} \times \frac{\text{km}}{\text{mile}} = \frac{\textbf{km}^2}{\textbf{mile}} \qquad \frac{\text{km}}{\frac{\text{km}}{\text{mile}}} = \text{km} \times \left(\frac{\text{mile}}{\text{km}}\right) = \textbf{miles}$$

The multiplication on the left produces a strange unit, but the two units of "kilometers" on the right cancel each other out, leaving only

---

[16] Or the reverse, if you live outside the US.

miles! Having identified the correct operation, we can now perform the correct conversion:[17]

$$\frac{5\,\text{km}}{\dfrac{1.6\,\text{km}}{1\,\text{mile}}} = 5\,\text{km} \times \frac{1\,\text{mile}}{1.6\text{km}} = \frac{5\,\text{miles}}{1.6} \approx \mathbf{3.1\ miles}$$

If you already understand this principle, this example might seem ridiculous, trivial, or both. Even so, if you perform computations like this on a calculator, that device will gladly allow you to input terms in any order whatsoever, even if the operation you plug in makes no sense! First and foremost, learning to cancel units protects against careless mistakes.

While useful, this principle would not merit inclusion in this book if it had no other use. Through repeated application, we can build up equations of arbitrary complexity, as we will now demonstrate!

Problem 2

Steve is considering replacing the lightbulbs in his home office with energy-efficient bulbs. At the moment, he has 4 lights, each fitted with a 75W bulb. On average, he has the lights on for 4 hours per day. According to his power bill, he pays $0.20 per kilowatt-hour of electricity.

LED lightbulbs consume 12.5W, and cost $5 apiece. Assuming he doesn't change his habits, how long will it take for him to make up the purchase price of the new bulbs through savings on his power bill?

---

[17] The value here is not quite accurate, because we used the approximate value 1.6 instead of the actual value of 1.609344 km/mile.

We reached that value by performing the following conversion. First note that, by definition, 1 inch = 2.54 cm.

$$\frac{5280\,\text{feet}}{1\,\text{mile}} \times \frac{12\,\text{inches}}{1\,\text{foot}} \times \frac{2.54\,\text{cm}}{1\,\text{inch}} \times \frac{1\,\text{meter}}{100\,\text{cm}} \times \frac{1\,\text{km}}{1000\,\text{meters}} = 1.609344\,\text{km/mile}$$

You can confirm that every unit, except "km" and "mile," exists in both a numerator and a denominator, thus canceling out.

That's a lot to take in! You can see how this is representative of "general" problem solving, though: this is definitely the kind of problem that a responsible homeowner might endeavor to solve! Although daunting, we can address the steps one-by-one, using what we've learned so far, to find a solution.

You might be tempted to start by punching numbers into a calculator, but resist that urge! Finding an answer that we can be confident in demands that we approach this problem deliberately.

Most of the figures in this problem should be familiar to you, with one exception. You may not recognize what a "kilowatt-hour" is (or maybe even what a "watt" is). If you think back to the example of Indian numbers (lakh, crore), you might recall that *this doesn't matter*. If we are careful, we can perform mathematical cancellations on these units without *needing to* understand what they mean.

Our goal is to produce some value with a unit of time ("how long will it take..."). Every other unit should therefore be canceled by a conversion, or some other provided value. Let's take this step-by-step.

Steve has 4 lights, each fitted with a 75W (read "75 watt") bulb.

$$4 \text{ bulbs} \times \frac{75 \text{ watt}}{1 \text{ bulb}} = 300 \text{ watt}$$

He uses the lights for 4 hours per day.

$$300 \text{ watt} \times \frac{4 \text{ hours}}{1 \text{ day}} = \frac{1200 \text{ watt hours}}{1 \text{ day}}$$

He pays $0.20 per kilowatt-hour of electricity.

$$\frac{1200 \text{ watt hours}}{1 \text{ day}} \times \frac{1 \text{ kilowatt hour}}{1000 \text{ watt hours}} \times \frac{0.20 \text{ dollars}}{1 \text{ kilowatt hour}} = \frac{0.24 \text{ dollars}}{1 \text{ day}}$$

The 4 new LED bulbs consume 12.5W each.

$$4 \text{ bulbs} \times \frac{12.5 \text{ watt}}{1 \text{ bulb}} = 50 \text{ watt}$$

Assuming that Steve doesn't change his habits, the new bulbs will be powered for 4 hours per day.

Here, we can either repeat the same computations we applied previously:

$$50 \text{ watt} \times \frac{4 \text{ hours}}{1 \text{ day}} \times \frac{1 \text{ kilowatt hour}}{1000 \text{ watt hours}} \times \frac{0.20 \text{ dollars}}{1 \text{ kilowatt hour}} = \frac{0.04 \text{ dollars}}{1 \text{ day}}$$

Or, we can observe that the new bulbs use 1/6 the energy of the old bulbs, and set up a proportion that divides the previous result by 6.

$$\frac{50 \text{ watt}}{300 \text{ watt}} = \frac{x}{\dfrac{0.24 \text{ dollars}}{\text{day}}}$$

$$16 \times \frac{0.24 \text{ dollars}}{\text{day}} = x$$

$$x = \frac{0.04 \text{ dollars}}{\text{day}}$$

These results are equivalent.

Once he replaces the bulbs, his savings will be the difference between these two figures. Where he would have spent $0.24 per day on lights before, now he will spend $0.04 per day, for a savings of $0.20 per day.[18]

$$\frac{0.24 \text{ dollars}}{1 \text{ day}} - \frac{0.04 \text{ dollars}}{1 \text{ day}} = \frac{0.20 \text{ dollars}}{1 \text{ day}}$$

The LED bulbs cost $5 each.

---

[18] It's worth emphasizing that the only reason we can subtract these numbers is that they both have common dimensions (that is, they are commensurable). When multiplying and dividing, we can cancel out units as necessary; when adding or subtracting, it doesn't make sense to add apples to oranges, or watts to dollars.

Intuitively this makes sense: the question "what is longer, 10 minutes, or 38 degrees Celsius?" is absurd when written out. However, when we perform calculations without paying attention to units, we sometimes miss these absurdities.

$$4 \text{ bulbs} \times \frac{5 \text{ dollars}}{1 \text{ bulb}} = 20 \text{ dollars}$$

How long will it take for Steve to make up the purchase price of the new bulbs?

$$\frac{20 \text{ dollars}}{\dfrac{0.20 \text{ dollars}}{1 \text{ day}}} = 20 \text{ dollars} \times \frac{1 \text{ day}}{0.20 \text{ dollars}} = \mathbf{100\, days}$$

Note that this process was almost entirely mechanical. We needed to use a little common sense to make sure that each operation produced the result we needed, but ultimately most of the "mental effort" here was used to ensure that each fraction was arranged in a way that units canceled out.

The final way this method can be applied is to "reverse engineer" formulas you find in real life to discover what they're describing. The following example demonstrates an application of this process taken from real-life experiences.

Problem 3

Sarah is taking a physics exam when she encounters a problem about the period of a pendulum's swing. Unfortunately, she can't recall the proper formula to calculate this value. Fortunately, her professor has provided the students with an unlabeled formula sheet.

One of the formulas on the sheet looks familiar to her:

$$T = 2\pi \sqrt{\frac{\ell}{g}}$$

Sarah thinks that it's reasonable to assume that ℓ corresponds to length (in meters); looking elsewhere on the sheet, she sees that the constant g has a value $g = 9.8 \ m/s^2$.

If period is measured in units of seconds, how can she verify that this formula returns a result that has the correct unit?

Here, we can once again insert just the units corresponding to each value, and simplify accordingly. Note, here, that both 2 and π are dimensionless constants — "2" is just 2 here, not 2 feet or seconds or liters — and since neither one modifies our analysis, they can be removed.

$$T = \sqrt{\frac{m}{\frac{m}{s^2}}}$$

$$T = \sqrt{m \cdot \left(\frac{s^2}{m}\right)}$$

$$T = \sqrt{s^2}$$

$$T = s$$

Sure enough, this formula produces values with the correct dimension!

A more realistic scenario might entail differentiating between twenty different formulas on a single formula sheet. With some educated guesswork and some applications of dimensional analysis, a test-taker can intuit the purpose of almost any formula!

So far, we have been addressing the "words" of mathematical problems — numbers and units. But you may have noticed that the techniques we have been using have become increasingly involved.

If numbers are the "words" we use to describe mathematical problems, then the equations we have been constructing form the basic "sentences" of problem solving — algebra.

The Art of Algebra

What does algebra mean to you?

You might struggle to answer this question! Assuming you followed a typical curriculum, it's quite possible that algebra was spread out across six (or more) school years.[19] In light of this, it might be

---

[19] In the United States, mathematics curricula cover algebraic concepts of increasing complexity every year from the 6th grade (age 11–12) until 11th grade (age 16–17), with one year set aside for geometry.

challenging to distill everything you learned into a few simple words. Regardless, think back on your schooling and try to find a few words to summarize your experiences.

Maybe you remember particular skills you learned, like "graphing" or "functions." Or perhaps what you remember most is "memorization" — of terms, operations, and formulas. If you struggled with math, you might remember experiencing feelings of "frustration" or "pointlessness."

No matter your experiences — good, bad, or in between — you're not alone in feeling that way! The truth is, mathematics, as it is commonly taught, often fails to connect with students' goals or experiences. Contrast mathematics with the humanities: despite the sometimes rarified presentation of other subjects, such as English, Foreign Languages, or History, students generally understand that constructing an argument, speaking to people from other countries, or learning about the past are useful skills to develop. By contrast, many students come away from high school not understanding the reason why mathematics is useful!

Ignoring all the definitions, terminology, and confusing notation, try to recall some of the common threads running through the past five chapters. What have you learned about *how* and *why* we do math?

We have offered a few answers already. Sometimes, we solve puzzles for fun. At other times, we just want to find an answer to a problem. Most importantly, though, sometimes we yearn to understand *why* a problem has a particular solution, and how that process relates to real life.

In light of this, algebra is no more than an intermediate form, a medium (or process) via which we can come to understand the world. As it turns out, almost every "general" problem — every problem that relates to the real world — passes through five stages. All of these problems:

- are based on observed data;
- that can be converted into symbols;
- which can be manipulated using general rules;

- to find an unknown quantity;
- that corresponds to some real phenomenon.

Your memories of algebra almost certainly center around stages 3 and 4. If mathematics is a language, then stages 3 and 4 are akin to that language's grammar. But language is only interesting inasmuch as it allows us to talk about real things!

On its own, algebra is both so *powerful* and *general-purpose* that it is easy for teachers (and those who already understand the "point," like book authors) to present that power in isolation, forgetting to supply *justification* to students as to why we might *want* to do these particular things. Good notation makes mathematics easier, but focusing on it to the exclusion of the other stages misses the point. When students ask their teachers "when will we ever have to know this?" their frustrations bring the flaws of common teaching styles into sharp relief.

If you have ever been that student, once again you are in good company! When Gottfried Leibniz first published his calculus, Christiaan Huygens (who you may remember from Chapter 4) followed the new mathematics with interest. In 1693 Huygens, then in his 60s, wrote several letters to Leibniz asking for elaboration on his methods. Huygens made it clear that he understood *how* to perform the calculations Leibniz invented, but pressed the latter repeatedly to justify *why* one might want to use certain methods. Leibniz contritely provided several specific examples of how his methods had been applied to real-world problems, and when further pressed by the elder mathematician conceded that there is little value in studying contrived problems for their own sake.

A new notation that makes certain problems *possible* does not suddenly make those problems worth solving. Indeed, throughout history, the motivation for solving problems generally preceded the development of new methods and notation.

The inventor of algebra, Muhammad Ibn Musa al-Khwarizmi (ca. 780–850) was motivated to invent algebra to solve problems involving Islamic inheritance laws. He expressed problems using literal words and sentences, and referred to the unknown using the word "thing" instead of the more modern "x." His mathematics was algebra *by definition*, despite the absence of modern notation.

Later mathematicians realized that his rules could apply to problems from other domains, including astronomy, physics, and probability. As they refined his methods, successive *generations* of mathematicians abbreviated commonly used operations into more concise symbols. The plus sign that is so familiar to readers today was first written in the mid-14th century (a full 150 years *after* the publication of Fibonacci's *Liber Abaci*) as an abbreviation of the Latin word *et*. The invention of the equals sign can be traced to the year 1557, when a mathematician named Robert Recorde, tired of writing out "is equal to" over and over, decided to abbreviate the phrase using two lines of equal length. Even then, his notation wasn't broadly accepted for over 100 hundred years. So as you can see, every piece of mathematics that we see as "obvious" today — including mathematical notation — had to be invented!

The algebra you know is the distillation of thousands of years of experimentation, *ad hoc* notation, custom, and habit, so it's no wonder that what we think of as the "essence" of algebra can seem almost impenetrable at first glance.

Lest it be said that we are being too hard on educators, we admit that they are faced with a nearly impossible task. If teachers could reliably determine at 11 years of age that *this* child will go on to be an engineer, while *that* child will spend their entire adult life working as an accountant, motivating mathematics problems during class would be much easier. We don't live in that world. Instructors teach general algebraic methods so that when each student studies the next course in his or her area of focus, the most relevant portions of what they learned will "click" into place.

In the same spirit, we (as authors of this text) can't possibly predict all the mathematical problems you will want to solve in your own life. What we can do, then, is to present you with an alternative way to view algebraic problem solving, so that you can make sense of the problems you will discover in your own life.

Variations on a theme

Quite possibly the simplest algebraic "sentence" we can explore[20] is the distance formula. You might recall it off-hand; it expresses the relationship between distance, rate, and time:

$$d = r \cdot t$$

You should already be thinking about how you can apply dimensional analysis to problems of this form. Distance and time are expressed in simple units, like "kilometers" and "hours" respectively. In this example, *rate* would be expressed in units of *kilometers per hour*.

$$km = \frac{km}{hour} \times hours$$

Naturally, you can choose to express the terms in any units (converting as appropriate), without affecting the structure of this "sentence." Students take this approach far too often: they memorize this as a "formula," dropping values plucked straight out of a problem statement into the appropriate slots. Treating the problem this way misses the richness of what is being said.

We can use an analogy to relate this idea to something you are no doubt familiar with. Every language has general rules for generating syntactic sentences — that is, sentences that "follow the rules" of grammar and word order.

In English, a simple declarative sentence can be composed of a subject (that is, a noun) and a verb.

The boy walked.

If you were restricted to only speaking sentences of this form, your communications would be extremely limited! Happily, there are all kinds of ways to expand on this sentence. We can elaborate by adding an object to the verb:

---

[20] Or, at least, the simplest *non-trivial* one. There exist simpler examples, such as $x = 2$, but such examples cannot provide us with nearly enough food for thought!

*The boy walked **to school**.*

Or we can add detail by adding adjectives and adverbs.

*The **young** boy walked to school **quickly**.*

We can also replace or supplement individual words with phrases.

*The young boy **and his neighbor** walked to the school **they would be attending next year**.*

Written mathematics has rules of syntax just like English. And as in English, those rules allow us to perform certain kinds of substitutions to tailor formulas to fit our problems. We must be attentive, though! It can be all too easy to form a "sentence" that has no meaning. In English, the canonical example of this type of sentence was created by the linguist Noam Chomsky:

Colorless green ideas sleep furiously.

Although all the words are in "correct" positions in relation to one another (adjectives modify a noun, the verb follows the subject, etc.), this sentence doesn't relate to anything real! In technical terms, sentences like this have correct *syntax*, but no *semantic* meaning. Many common mathematical errors are the result of constructing expressions that are structurally correct, but that have no meaning.

In mathematics, we can prevent these errors by using a little common sense, verifying that each step corresponds to some *real* phenomenon (i.e., that each step has meaning), and by applying the principle of dimensional analysis.

Using these principles, let's look at some of the ways we can construct "sentences" using mathematics!

There and back again

Chris enjoys biking. One day, he decided to test his new GPS app by riding his bike out to a nearby hill, up the hill the same distance, then

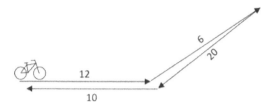

Figure 6.8.   Riding up a hill and back down. All four segments have the same length.

back down via the same route before returning home (Figure 6.8). When he checked his app upon his return, it told him that he rode four segments of equal length at the following speeds:

- 12 mph on flat ground on the way to the hill
- 6 mph climbing
- 20 mph downhill
- 10 mph on the return trip (since he was tired)

What was Chris' average speed over the course of his ride?

Note that we haven't told you the length of the segments, nor have we said how long he spent riding! Nevertheless, this problem is solvable.

The distance formula can be applied directly to each segment. Here, we use a subscript i to represent the indices of each of the four rates and times. Thus, $r_1 = 12$ mph, $r_2 = 6$ mph, and so forth.

$$d = r_i \cdot t_i$$

We can find the values for each time t_i by dividing by the rates. Thus:

$$\frac{d}{r_i} = t_i$$

$$\frac{d}{12} = t_1 \qquad \frac{d}{6} = t_2 \qquad \frac{d}{20} = t_3 \qquad \frac{d}{10} = t_4$$

These four times can be added together to find the total duration of Chris' ride in terms of the unknown d.

$$t_{total} = \frac{d}{12} + \frac{d}{6} + \frac{d}{20} + \frac{d}{10}$$
$$= \frac{5d}{60} + \frac{10d}{60} + \frac{3d}{60} + \frac{6d}{60}$$
$$= \frac{24d}{60}$$
$$t_{total} = \frac{2d}{5}$$

We can easily find the total distance he rode his bike: it is 4 times the length of a single segment ($d_{total} = 4d$). We can substitute these two values into the distance formula, and solve for the average speed, or rate:

$$d_{total} = r_{avg} \cdot t_{total}$$
$$4d = r_{avg} \cdot \frac{2d}{5}$$
$$\frac{4d}{\frac{2d}{5}} = r_{avg}$$
$$r_{avg} = 10\,\text{mph}$$

A second trip

After training on the first hill, Chris decided to take on a different hill. His plan this time was to ride to the hill, climb it until he got tired, then turn around and descend the hill along the same route before riding home.

He again checked his app at the end of his ride, and found the speeds on the four segments were:

- 12 mph on the level segments (both out and back)
- 10 mph climbing the hill
- 15 mph on the descent

His roundtrip time was 2 hours. However, this time the distance he rode up (and down) the hill was not equal to the distance from his house to the hill (Figure 6.9).

Figure 6.9.   Riding to a hill and back. The level segments have the same length as each other, but not necessarily the same as the distance rode up the hill.

How far did Chris ride?

Here, we have rates and a time, but no sense of the distances he rode. Since the distance he rode up the hill is the same as the distance he rode down, we can certainly say *something* about the relationship between these two values.

$$d_{\text{hill}} = 10 \cdot t_{\text{up}}$$
$$d_{\text{hill}} = 15 \cdot t_{\text{down}}$$

It's tempting to try to take the arithmetic mean of these two speeds to find the average speed on the hill. Be careful, though! That approach might not be valid.

When you suspect you can apply a mathematical operation, but you're not certain, it's often helpful to test a simpler example.

Pick two speeds, like 1 mph and 3 mph. The arithmetic mean of these two is $(1 + 3)/2 = 2$ mph. Now, pick a convenient distance that both divide, like 3 miles. Would walking 3 miles at 1 mph and then 3 miles at 3 mph take the same amount of time as walking 6 miles at 2 mph?

$$3\,\text{miles} = 1\,\text{mph} \cdot t_1 \qquad\qquad 6\,\text{miles} = 2\,\text{mph} \cdot t_2$$
$$\frac{3}{1} = 3\,\text{hours} = t_1$$
$$3\,\text{miles} = 3\,\text{mph} \cdot t_3$$
$$\frac{3}{3} = 1\,\text{hour} = t_3 \qquad\qquad \frac{6}{2} = 3\,\text{hours} = t_2$$
$$t_1 + t_3 = 4\,\text{hours} \qquad\qquad \frac{t}{2} = 3\,\text{hours}$$

The two times differ, so we've shown that the arithmetic mean of two speeds isn't the same as traveling at two different speeds! How can we find the true value of the average speed?

We will need to find an expression for the average speed over this distance using facts we *do* know. First, let's write down a version of the distance formula that sums up where we're trying to go.

If Chris travels up the hill a certain distance (d_{hill}) at a rate (r_{up}) for a certain length of time (t_{up}), and down the same distance at a different rate (r_{down}) for a different length of time (t_{down}), that is equivalent to him traveling the total distance ($2 \cdot d_{hill}$) at an average speed (r_{avg}), the value we wish to find), for the same amount of time ($t_{up} + t_{down}$).

$$(2 \cdot d_{hill}) = r_{avg} \cdot (t_{up} + t_{down})$$
$$r_{avg} = \frac{2 \cdot d_{hill}}{t_{up} + t_{down}}$$

The two time terms are troublesome, so let's try to express one in terms of the other. Going back to the original pair of equations, we can see that both distances are the same, so the right terms are equal to each other. We can solve for either one in terms of the other.

$$10 \cdot t_{up} = d_{hill} = 15 \cdot t_{down}$$
$$10 \cdot t_{up} = 15 \cdot t_{down}$$
$$t_{up} = \frac{3}{2} \cdot t_{down}$$

Now, we can insert this value in the equation for r_{avg}.

$$r_{avg} = \frac{2 \cdot d_{hill}}{\frac{3}{2} \cdot t_{down} + t_{down}} = \frac{2 \cdot d_{hill}}{\frac{5}{2} \cdot t_{down}} = \frac{4 \cdot d_{hill}}{5 \cdot t_{down}}$$

The justification for the next step will be more clear if we re-write the previous equation another way.

$$r_{avg} = \frac{4}{5} \cdot \left(\frac{d_{hill}}{t_{down}} \right)$$

If we can find an expression for the term in parentheses, we can solve for r_{avg}. Luckily, we have just such an equation available from a previous step!

$$d_{hill} = 15 \cdot t_{down}$$
$$\left(\frac{d_{hill}}{t_{down}} \right) = 15$$

Finally, we can find a value for the average rate.

$$r_{avg} = \frac{4}{5} \cdot 15 = \mathbf{12\,mph}$$

How fortunate! Chris' average speed on the hill is exactly the same as his speed on flat ground. This means that we can effectively ignore the distance he traveled up and down the hill — no matter how tall it was, his average speed for the entire duration of the ride was 12 mph!

Now, we can solve for d:

$$d = 12\,mph \cdot 2\,hours = \mathbf{24\,miles}$$

In case you wonder if there's a more efficient way to find the average of several speeds — there is! We can apply the distance formula to general speeds r_i, as shown.

$$d = r_i \cdot t_i$$
$$t_i = \frac{d}{r_i}$$

The sum of all the distance terms is the same as the average rate times the sum of all the times, by definition.

$$d + d + \cdots + d = (r_1 \cdot t_1) + (r_2 \cdot t_2) + \cdots + (r_n \cdot t_n)$$
$$n \cdot d = r_{avg} \cdot t_1 + t_2 + \cdots + t_n$$

We can replace each time term t_i with its corresponding expression in terms of d and r_i, and then find an expression for r_{avg} by dividing.

$$n \cdot d = r_{avg} \cdot \left(\frac{d}{r_1} + \frac{d}{r_2} + \cdots + \frac{d}{r_n} \right)$$

$$r_{avg} = \frac{n \cdot d}{\left(\frac{d}{r_1} + \frac{d}{r_2} + \cdots + \frac{d}{r_n} \right)}$$

It's possible to clean this up a bit by recognizing that every single term on the right side contains a multiplication by the distance d, meaning that we can cancel them all.

$$r_{avg} = \frac{d \cdot (n)}{d \cdot \frac{1}{r_1} + \frac{1}{r_2} + \cdots + \frac{1}{r_n}}$$

$$= \frac{n}{\frac{1}{r_1} + \frac{1}{r_2} + \cdots + \frac{1}{r_n}}$$

Unsurprisingly, it turns out this result has a name. It is called the *harmonic mean*, and is one of the three *Pythagorean means*. As you can guess from their name, they have been studied for thousands of years. We are already very familiar with the arithmetic mean from our discussion of expected values. The third mean, which we haven't explicitly discussed, is known as the *geometric mean*.

The harmonic mean finds use in electronics for calculating the resistance of resistors wired in parallel or capacitors wired in series, in certain financial and statistical calculations — and, of course, when finding the average of several rates.

We can demonstrate this by using this to re-compute the solution to the earlier problem "there and back again":

$$r_{avg} = \frac{4}{\frac{1}{12} + \frac{1}{6} + \frac{1}{20} + \frac{1}{10}}$$

$$= \frac{4}{\frac{5}{60} + \frac{10}{60} + \frac{3}{60} + \frac{6}{60}}$$

$$= \frac{4}{\frac{24}{60}}$$

$$= 10\,\text{mph}$$

Using the harmonic mean produces a much cleaner solution!

Head to head

Chris and his friend Evan live on opposite sides of a park. One day, Evan challenged Chris to a race: at an agreed-upon time, they both set out from their respective houses and headed across the park at top speed. When each of them reached the other's house, they turned around and headed back home.

Chris and Evan passed each other in the park twice, headed in opposite directions. The first time they passed, Chris glanced at his app and saw that he was 800 m from where he started (Figure 6.10). After turning around, Chris met Evan going the other direction only 450 m from Evan's house.

Assuming that both Chris and Evan maintained a constant speed the entire time, and that turning around took them no time, how wide is the park?

This problem asks us to find a distance given neither rates nor times! However, it is possible to solve this problem with only the information provided. In fact, there are multiple routes to the solution; we will demonstrate a slow, easy method and a clever method that might be a bit more difficult to spot at first.

Method 1

We can first set up some equations summarizing what we know so far. We will use r_E to refer to Evan's speed, and r_C to refer to Chris'. Additionally, we can call the time they first meet t_1 and the time of their second meeting t_2. The park's width is, naturally, d.

The following four equations re-state what we learned from the problem statement: at some time t_1, Evan was 800 m from the edge of the park. At the same moment in time, Chris had traveled 800 m.

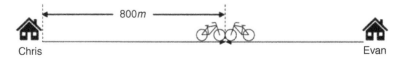

Chris   800*m*   Evan

Figure 6.10.   Crossing the park, Chris meets Evan 800 m from his house.

Later, at time t_2, Evan had traveled 450 m less than twice the width of the park, while Chris was only 450 m into his return trip.

$$d-800=r_E \cdot t_1$$
$$800=r_C \cdot t_1$$
$$2d-450=r_E \cdot t_2$$
$$d+450=r_C \cdot t_2$$

The next step we want to take should reduce the number of variables somehow. In truth, there are quite a few possible paths forward here; some of them tell us more than others. You might be tempted to keep the only numeric information we have for as long as possible, but remember — our goal is to rephrase one of the *variables* in terms of the others. So, let's add the first and second equations, and the third and fourth.

$$(d-800)+(800)=(r_E \cdot t_1)+(r_C \cdot t_1)$$
$$d=r_E +r_C \cdot t_1$$

$$(2d-450)+(d+450)=(r_E \cdot t_2)+(r_C \cdot t_2)$$
$$3d=(r_E +r_C)\cdot t_2$$

Next, we can divide the second of these equations by the first.

$$\frac{3d}{d}=\frac{(r_E +r_C)\cdot t_2}{(r_E +r_C)\cdot t_1}$$
$$3=\frac{t_2}{t_1}$$
$$3t_1 =t_2$$

We can substitute this value in any of the four earlier equations. Which one should we choose? Again, our goal is to reduce the number of "free" variables by elimination (either division or subtraction). Chris' equations have fewer uses of the d term, so these present a tempting target. Let's see what happens when we substitute this value of t_2 into Chris' equations, and divide.

$$\frac{d+450}{800} = \frac{r_C \cdot 3t_1}{r_C \cdot t_1}$$

$$= 3 \cdot \frac{(r_C \cdot t_1)}{r_C \cdot t_1} = 3$$

$$\frac{d+450}{800} = 3$$

$$d+450 = 2400$$

$$d = \textbf{1950 m}$$

Method 2

For our second solution, it may be helpful to look at the diagram in Figure 6.11. Note that when Chris and Evan first meet, *together* they have covered a distance of d meters. When they next meet, they have together covered three times that distance. Since their individual rates are constant, their combined rate is also constant, so we can be certain that $t_2 = 3t_1$.

Each of the distances in this figure can be expressed in terms of four values. For instance, between the times t_1 and $3t_1$, Chris rode

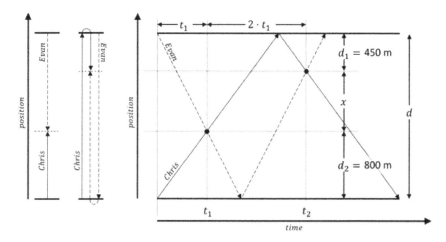

Figure 6.11. (Left) Chris and Evan pass each other two times. First, 800 m from Chris's house; next, 450 m from Evan's. (Right) Chris and Evan's paths, graphed as a value of distance from Chris's house versus time.

$x + 2 \cdot 450$ m, while Evan rode $2 \cdot 800 + x$ m. We can choose any pair of comparable distances and divide one expression of the distance formula by another to cancel unknowns. Here, we use the expression of the distance Chris covers in time t_1, and the distance that he covers between t_1 and t_2.

$$800 = r_C \cdot t_1$$

$$\begin{aligned} x + 2 \cdot 450 &= r_C \cdot (t_2 - t_1) \\ &= r_C \cdot (3t_1 - t_1) \\ &= 2 \cdot r_C \cdot t_1 \end{aligned}$$

Note what happens when we divide these two results and simplify: we are left with only x, and we can find its value using a little bit of algebra.

$$\frac{x + 2 \cdot 450}{800} = \frac{2 \cdot (r_C \cdot t_1)}{(r_C \cdot t_1)}$$

$$\frac{x + 2 \cdot 450}{800} = 2$$

$$x + 2 \cdot 450 = 2 \cdot 800$$

$$x = 2 \cdot 800 - 2 \cdot 450$$

$$\mathbf{x = 2 \times (800 - 450)}$$

We have found the value of x in terms of the other two distances. But, more importantly, we could easily replace both 800 and 450 at each step with other variables, say d_1 and d_2, and describe a general solution to all problems of this type!

$$d = d_1 + 2 \cdot (d_1 - d_2) + d_2$$

Of course, you can (and should!) plug in the values 800 and 450 here to verify that this produces the same solution as before: 1950 m.

Students commonly tell their instructors that they are able to follow the demonstrations in class, but when it comes to the test, they're completely lost. We have made a special effort to ensure that the standard form of the distance formula ($d = r \cdot t$) appeared as frequently as possible, but make no mistake — these demonstrations have been tailored to be easy to follow!

To offer you a little peek behind the curtain, we tested dozens of approaches to the previous problem, before settling on one that was sufficiently illustrative and clean. The others produced the same result, but most didn't do a good job of demonstrating *why* that answer is correct.

Your own work will have its own share of false starts and dead ends. Solving the same problem by multiple paths will show you that your answer is correct (and not the result of a spurious error!), but, more importantly, it will give you a chance to understand the mathematics in great detail. Your learning need not end once you have found a single route to a solution. There is as much, or more, to be learned in solving the same problem a second time!

We have solved three problems related to speeds. The next problem will be the last of this type. If you haven't worked through the previous problems on your own, this will be your chance to do so, and put your learning to the test!

The old man and the river

Elmer is paddling upstream on a river. As he is passing under a bridge, the wind picks up and blows his hat off his head. He proceeds up the river for another 15 minutes before noticing his hat is gone. He turns around (with no loss of time) and paddles back the way he came, toward his hat. When he finally catches up to his hat, it is 1 mile downstream of the bridge.

What is the speed of the river's current, in miles per hour?

Method 1

Let's set up some variables. The speed of the river's current is r_C. The boat's speed (relative to the river) is r_B. We will choose t to represent the time it took Elmer to catch up to his hat (after turning around), while the distance he traveled in that time is d_D (with "D" representing "downstream").

Elmer's speed when paddling upstream is the difference $r_B - r_C$, and he paddled for 1/4 of an hour.

$$d_U = r_B - r_C \cdot \frac{1}{4}$$

He traveled downstream at a speed of $r_B + r_C$ for a time of t hours.

$$d_D = r_B + r_C \cdot t$$

We also know that d_D is 1 mile greater than d_U; that is, Elmer had to paddle 1 mile more downstream than he did upstream.

$$d_D = d_U + 1$$
$$d_U + 1 = r_B + r_C \cdot t$$

We can replace d_U here with an earlier expression.

$$\left((r_B - r_C) \cdot \frac{1}{4} \right) + 1 = (r_B + r_C) \cdot t$$

We don't have an expression for t, so we'll need to express t in terms of some other variables in order to cancel it out. Luckily, we know that Elmer's hat traveled 1 mile downstream for $\left(t + \frac{1}{4} \right)$ hours at a rate of r_C. We can solve this equation for t:

$$1 = r_C \cdot \left(t + \frac{1}{4} \right)$$
$$\frac{1}{r_C} = t + \frac{1}{4}$$
$$\frac{1}{r_C} - \frac{1}{4} = t$$

Finally, we can substitute this value of t in the previous equation and simplify to find the speed of the current.

$$\left((r_B - r_C) \cdot \frac{1}{4} \right) + 1 = (r_B + r_C) \cdot \left(\frac{1}{r_C} - \frac{1}{4} \right)$$

$$\left(\frac{r_B}{4}-\frac{r_C}{4}\right)+1=\frac{r_B}{r_C}-\frac{r_B}{4}+\frac{r_C}{r_C}-\frac{r_C}{4}$$

$$\frac{r_B}{2}+1=\frac{r_B}{r_C}+1$$

$$\frac{r_B}{2}=\frac{r_B}{r_C}$$

$$r_C=2\,\text{mph}$$

Method 2

Imagine that instead of paddling on a river, Elmer was walking through a train. "Paddling upstream" would be equivalent to moving from the front of the train toward the back (against the movement of the train). When he dropped his hat on the floor, it moved in the same direction as the train.

No matter how fast the train was moving, Elmer spent equal amounts of time walking away from his hat and returning to it. Since he "walked toward the back of the train" (paddled upstream) for 15 minutes, he must have also spent 15 minutes walking in the opposite direction. His round trip must have taken 30 minutes total.

In that period, the hat was carried 1 mile by the train. Moving 1 mile in 30 minutes is the same as 2 miles per hour, which is the speed of the train — and, by way of analogy, of the river's current.

The distance formula isn't only about distances, or at least not in the sense you might think. We can use it to solve problems involving rates of any kind, as the next three problems demonstrate.

Two painters

Janet paints new apartments for a living. She discovers that she can cover the walls of a single apartment with two coats of paint per day — the first coat before lunch, and the second after.

One morning she is joined in her work by another painter, Richard. He starts helping her two hours before lunch, and leaves as soon as the first coat is complete.

Janet immediately begins putting on a second coat of paint over her first, then takes lunch at her normal time. That afternoon, an hour before her normal quitting time, she observes that she has repainted the entire apartment with the exception of the section Richard painted that morning.

Richard paints at a rate of 800 square feet per day. If both Janet and Richard paint at a constant rate, and have the same length work-day, what is the area of the classroom's walls, in square feet?

If your natural inclination is to only think of applying the distance formula to distances — kilometers, miles, feet, and so forth — you might not spot the parallel here. But if you consider the units here, the similarity should leap out!

$$\text{miles} = \frac{\text{miles}}{\text{hour}} \cdot \text{hours} \qquad \text{ft}^2 = \frac{\text{ft}^2}{\text{day}} \cdot \text{days}$$

We can treat this problem just like any of those we've already been dealing with. In fact, let's frame it in terms more like those we have been dealing with so far.

Janet is a postal worker responsible for moving mail from the regional hub to the branch office staffed by Richard. On a typical day, she travels all morning from her office to Richard's, has lunch, and spends the afternoon on the return journey.

Two hours before lunch on a particularly nice day, Richard sets out to meet Janet to have lunch outdoors. He rides his bike at 8 miles per hour, and meets her along the road slightly before Janet's normal lunch time.

After lunch, Janet heads back to her office, arriving an hour earlier than she usually does.

How far apart are the two post offices?

It may be easier for you to make sense of the latter version of this problem. Hopefully, though, you can see that these two statements describe problems that are *very* similar in structure, even though their units are different!

In any case, we can transform these problem statements into diagrams, similar to what we did in the previous problem, as seen in

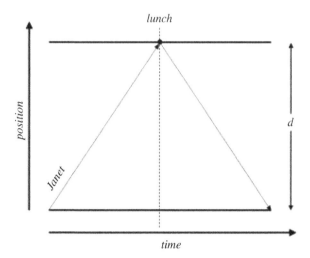

Figure 6.12. Janet's normal day. She travels to Richard's office, has lunch, and then travels back in the afternoon.

Figures 6.12 and 6.13. You can think of each representing Janet's position over time, or what portion of the apartment she has painted, as you prefer. In either case, on a typical day, Janet finishes the first part of her journey by lunch (which has been edited out, similar to how we were able to remove the river from the bridge problem in Chapter 5).

On the day in question, however, Richard "meets her partway," and she finishes all but his portion of the work an hour earlier than normal (Figure 6.13). We can look at this question from a different point of view (Figure 6.14). From Janet's perspective, Richard met her a half hour from her destination, if traveling at her normal speed. This saved her a round-trip time of one hour.

Since Richard set out two hours before lunch, and met up with Janet a half hour before lunch, he traveled for 1.5 hours total. Since Janet could have covered that same distance in 0.5 hours, her speed must be 3 times Richard's.

$$8 \text{ mph} \cdot 1.5 \text{ hours} = r_J \cdot 0.5 \text{ hours}$$
$$8 \text{ mph} \cdot \frac{1.5 \text{ hours}}{0.5 \text{ hours}} = r_J$$
$$8 \text{ mph} \cdot 3 = r_J$$
$$\mathbf{24 \text{ mph}} = r_J$$

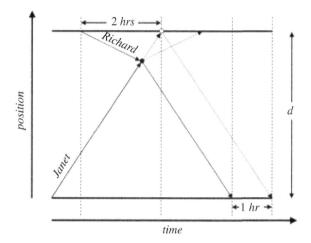

Figure 6.13. Richard leaves his office 2 hours before lunch, meeting Janet part way. On this day, Janet gets back to the office one hour early, since her return journey is shorter than normal.

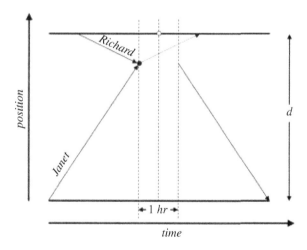

Figure 6.14. We can "move" the hour Janet saved from the end of her day to the middle of the day. By symmetry, Richard saved her a half hour in the morning and a half hour in the afternoon.

We can translate the problem back to the original terms, using the same steps to find Janet's painting speed.

$$800\,\text{ft}^2/\text{day} \cdot 1.5\,\text{hours} = r_J \cdot 0.5\,\text{hours}$$

$$800\,\text{ft}^2/\text{day} \cdot \frac{1.5\,\text{hours}}{0.5\,\text{hours}} = r_J$$
$$800\,\text{ft}^2/\text{day} \cdot 3 = r_J$$
$$\mathbf{2400\,ft^2/day} = r_J$$

While it's not necessary to translate problems in this fashion, this can be a useful tool to have in your toolbox. A powerful side-effect of mathematical abstraction is that problems that can be expressed in the same form can usually be treated using the same processes. When you encounter an unfamiliar problem, try to translate it into mathematical terms, and see if it resembles another problem you *do* understand. With practice, you should learn how to spot familiar structures that trigger your intuition!

Chickens and eggs

One and a half chickens can lay one and a half eggs in one and a half days. How long would it take for ten chickens to lay ten eggs?

This is a classical counterintuitive problem that would be right at home in Chapter 3. As we will demonstrate, it fits in here just as well!

You might initially assume that ten chickens can lay ten eggs in ten days. If we have succeeded in teaching you how to detect a counterintuitive problem, you should know that the easy answer is almost never the correct one! So, what is the correct way to approach this?

Let's pretend we're back in Chapter 3. One way to alter this problem would be to put it in more familiar terms, and very few things are more natural than people. Try to fill in the gap in the following statement.

"One and a half *women* can have one and a half *babies* in___*months.*"

Naturally, the answer here is nine months! One woman can have one baby in nine months, and ten women can have ten babies in...

nine months. You can find the answer to the original problem analogously.

This change in perspective will allow you to solve the original problem. However, this section is about rates, or more specifically, rate equations. Equipped with an intuitive solution, how can we translate this approach into mathematical language?

We possess three distinct pieces of information: we know that a certain number of chickens can lay a certain number of eggs in a certain number of days. There is a problem, though. Our normal approach is to relate a rate and a time to produce a distance (or another analogous measure). What happens when we try to put together the facts we have available?

The natural unit suggested by dimensional analysis is "eggs per day."

$$\text{distance} = \text{rate} \times \text{time}$$
$$\text{eggs} = \frac{\text{eggs}}{\text{day}} \times \text{days}$$

But this doesn't account for the number of chickens! Perhaps we can adjust the equation a bit, this time considering the number of chickens.

$$\text{eggs} = \frac{\text{eggs}}{\text{chicken}} \times \text{chickens}$$

Now, we are not accounting for the length of time! Is there some way to combine these two to produce a unified solution?

Once again, it can help to think about how languages behave. In English, we can turn an entire sentence into a *clause* (a part of a sentence with its own subject and verb), and use that clause as part of a larger sentence.

The boy used to take the bus to school.

The boy walked to school.

The *boy who used to take the bus* walked to school.

In this example, we have embedded the first sentence into the second (in italics) as a separate clause. The mathematical equivalent to this is to embed one equation into the other. But which?

Due to the mathematical properties of multiplication and division, it turns out that it doesn't particularly matter! However, for the sake of maintaining parallels to the distance formula, we will use the equation involving time on the outside, and the equation that only concerns chickens and eggs on the inside. Doing so results in a nested equation. You can verify that all the units behave in accordance with the principles of dimensional analysis. Then, we can insert all the values we know into this equation.

$$\text{eggs} = \frac{\left(\frac{\text{eggs}}{\text{chicken}} \times \text{chicken}\right)}{\text{days}} \times \text{days}$$

$$e\,\text{eggs} = \frac{\frac{1.5\,\text{eggs}}{1.5\,\text{chickens}} \times c\,\text{chickens}}{1.5\,\text{days}} \times d\,\text{days}$$

We can simplify some of the terms, and more importantly, we can simplify the equation itself by extracting the "chickens" term c from the fraction.

$$e = \frac{\frac{1\,\text{egg}}{1\,\text{chicken}} \times c}{1.5\,\text{days}} \times d$$

$$e = \frac{1\,\text{egg/chicken}}{1.5\,\text{days}} \times c \times d$$

To make sense of the rate term, we can express it in words: 1 egg per chicken, every 1.5 days; or 1 egg per chicken per 1.5 days.

We can re-insert the values from the problem statement to verify that this equation produces the correct result.

$$e = \frac{1\,\text{egg/chicken}}{1.5\,\text{days}} \times (1.5\,\text{chickens}) \times (1.5\,\text{days})$$

$$e = \frac{1}{1.5} \times 1.5 \times 1.5$$

$$e = 1.5\,\text{eggs}$$

This equation seems correct, so we can go ahead and use it to find the solution to the original problem!

$$10\,\text{eggs} = \frac{1\,\text{egg/chicken}}{1.5\,\text{days}} \times (10\,\text{chickens}) \times (d\,\text{days})$$

$$(10\,\text{eggs}) \times \frac{1}{10\,\text{chickens}} \times \frac{1.5\,\text{days}}{1\,\text{egg/chicken}} = d\,\text{days}$$

$$1.5\,\text{days} = d$$

In this chapter, we shared with you a variety of points-of-view that should make your own everyday interactions with numbers and mathematical problems much more natural.

When students struggle to understand the purpose of math, it's often because they do not see how the formulas, symbols, numbers, and units of mathematical expressions relate to the answers they wish to find.

We hope that these problems have made you more fluent in your interactions with numbers, with arithmetic, and with translating ideas into words and expressions, expressions that now have much more meaning to you than ever before!

Printed in the United States
By Bookmasters